D0138272

Geometry

with Geometry Explorer™

Michael Hvidsten

Gustavus Adolphus College

Higher Education

Boston Burr Ridge, IL Dubuque, IA Madison, WI New York
San Francisco St. Louis Bangkok Bogotá Caracas Kuala Lumpur
Lisbon London Madrid Mexico City Milan Montreal New Delhi
Santiago Seoul Singapore Sydney Taipei Toronto

Higher Education

GEOMETRY WITH GEOMETRY EXPLORER™

Published by McGraw-Hill, a business unit of The McGraw-Hill Companies, Inc., 1221 Avenue of the Americas, New York, NY 10020.

This book is printed on acid-free paper.

1 2 3 4 5 6 7 8 9 0 DOC/DOC 0 9 8 7 6 5 4

ISBN 0-07-294863-9

Publisher, Mathematics and Statistics: *William K. Barter*
Executive Editor: *Robert E. Ross*
Editorial Assistant: *Dan Seibert*
Senior Marketing Manager: *Nancy Anselment*
Project Coordinator: *April R. Southwood*
Senior Production Supervisor: *Sherry L. Kane*
Senior Media Project Manager: *Sandra M. Schnee*
Lead Media Technology Producer: *Jeff Huettman*
Senior Coordinator of Freelance Design: *Michelle D. Whitaker*
Cover/Interior Designer: *Rokusek Design*
(USE) Cover Image: *Sierpinski's Gasket, visualized in the Poincare model of Hyperbolic geometry,* created by the author
Supplement Producer: *Brenda A. Ernzen*
Compositor: *Lachina Publishing Services*
Typeface: *11/13 Computer Modern*
Printer: *R. R. Donnelley Crawfordsville, IN*

Library of Congress Cataloging-in-Publication Data

Hvidsten, Michael.
Geometry with geometry explorer™ / Michael Hvidsten.—1st ed.
p. cm.
Includes index.
ISBN 0–07–294863–9
1. Geometry. 2. Geometry Explorer™ (Computer file). I. Title.

QA445.H84 2006
516—dc22 2004058155
CIP

www.mhhe.com

Contents

Preface

> It may well be doubted whether, in all the range of science, there is any field so fascinating to the explorer, so rich in hidden treasures, so fruitful in delightful surprises, as Pure Mathematics.
> —Lewis Carroll (Charles Dodgson), 1832–1898

An *explorer* is one who seeks out new worlds and ideas. As Lewis Carroll would probably agree, exploration is not always easy—the explorer can at times find the going tough. But, the treasures and surprises that active exploration of ideas brings is worth the effort.

Geometry is one of the richest areas for mathematical exploration. The visual aspects of the subject make exploration and experimentation natural and intuitive. At the same time, the abstractions developed to explain geometric patterns and connections make the subject extremely powerful and applicable to a wide variety of physical situations. In this book we give equal weight to intuitive and imaginative exploration of geometry as well as to abstract reasoning and proofs.

As any good school teacher knows, intuition is developed through *play*, the sometimes whimsical following of ideas and notions without clear goals in mind. To encourage a playful appreciation of geometric ideas, we have incorporated many computer explorations in the text. The software used in these explorations is *Geometry Explorer*, a virtual geometry laboratory where one can create geometric objects (like points, circles, polygons, areas, etc.), carry out transformations on these objects (dilations, reflections, rotations, and translations), and measure aspects of these objects (like length, area, radius, etc.). As such, it is much like doing geometry on paper (or sand) with a ruler and compass. However, on paper such constructions are static—points placed on the paper can never be moved again. In *Geometry Explorer*, all constructions are *dynamic*. One can draw a segment and then grab one of the endpoints and move it around the canvas, with the segment moving accordingly. Thus, one can construct a geometric figure and test out hypotheses by experimentation with the construction.

The development of intuitive notions of geometric concepts is a critical first step in understanding such concepts. However, intuition alone cannot provide the basis for precise calculation and analysis of geometric quantities. For example, one may know experimentally that the sides of a right triangle follow the Pythagorean Theorem, but data alone do not show *why* this result is true. Only a logical proof of a result will give us confidence in using it in any given situation.

Throughout this text there is a dual focus on intuition/experimentation on the one hand and on explanation/proofs on the other. This integration of exploration and explanation can be seen most clearly in the use of major projects to tie together concepts in each chapter. For example, the first project explores the golden ratio and its amazing and ubiquitous properties. Students not only experimentally discover the properties of the golden ratio, but are asked to dig deeper and analyze why these properties are true.

The goal of the projects is to have students *actively* explore geometry through a three-fold approach. Students will first see a topic introduced in the text. Then, they will explore that topic using *Geometry Explorer* or by means of in-class group projects. Finally, they will review and report on their exploration, discussing what was discovered, conjectured, and proved during the course of the project.

The beginning of each project is designated by a special heading—the project title set between two horizontal lines. The conclusion of each project is designated by an ending horizontal line. Projects are illustrated with screen shots from the *Geometry Explorer* program, which comes bundled with the text.

Using *Geometry Explorer*

Each project includes a series of specific geometric activities using *Geometry Explorer*. The following conventions will be used for directing computer explorations:

- **Menu References** All menu references will be in bold face type and will reference the menu option to click on. Parent menus will be listed in parentheses to assist in navigating to the correct menu. For example, the phrase "Click on **Hide** (**View** menu)" means to go to the **Hide** menu under the **View** menu and select that menu.

- **Selection** When asked to select an object on the screen, first check that the Select button (the one with the left arrow in the Create panel of buttons) is pressed, and then click on the object to select it.

- **Multi-selection** To select more than one object, hold down the Shift key when selecting.

- **Creating Objects** When asked to *create* an object on the screen, use one of the buttons in the Create panel. To create a point, for example, first click on the Point button in the Create panel and then click on the screen. To create a circle, click on the Circle button and click and drag to create a circle.

- **Constructing Objects** When asked to *construct* an object, use one of the buttons in the Construct panel. These buttons will work only if the correct objects for the construction have already been selected. For example, to construct the intersection of two circles, first multi-select the circles and then click on the Intersect button (first button in first row of Construct panel of buttons).

- **Dragging or Moving Objects** To move an object, use the Select tool. Click on the object and drag the mouse to move the object.

- **Attaching Points to Objects** When asked to attach a point to an object such as a circle, create a point on top of a portion of the circle. To test whether a point is attached to an object like a circle, drag the point with the mouse. The point should move *only* along the circle; that is, it is *attached* to the circle.

Keeping these few conventions in mind will solve many, if not most, of the user interaction issues that come up when doing the projects of the text. A more complete reference guide to *Geometry Explorer* can be found in Appendix B.

Audience

This text is designed for use by mathematics students at the junior or senior collegiate level. The background in geometry required is that of elementary high school Euclidean geometry. Prior experience with proving mathematical results is highly recommended. Some experience with matrix algebra and the notion of *group* from abstract algebra is also highly desirable.

The arrangement of topics in the text was designed to give as much flexibility as possible. While Chapters 1 and 2 are fundamental, many of the other chapters can be covered independently from one another. Chapter 3 covers basic analytic geometry of vectors and angles, as well as complex numbers and analytic functions. Unless review of such matters is necessary, Chapter 3 can be viewed as optional foundational material.

Chapter 4 covers Euclidean constructions and depends only on the material in Chapters 1 and 2. Chapter 5 is devoted to transformational geometry and requires only a basic understanding of vectors and angles from Chapter 3, beyond the material covered in Chapters 1 and 2. Chapter 5 is a pre-requisite for all subsequent chapters. Chapters 6, 7, and 9 can be covered in any order. Chapter 7 is a pre-requisite for Chapter 8.

A suggested syllabus for a one-semester course for prospective high-school geometry teachers would include Chapters 1, 2, 4, sections 5.1–5.6, sections 7.1–7.6, and as much of the first four sections of Chapter 9 as time permits.

A suggested syllabus for a one-semester course for math majors of better than average ability would include Chapters 1, 2, 5, 6, 7, and 9.

A suggested syllabus for a one-semester course focusing on non-Euclidean geometry would include Chapters 1, 2, and section 3.5, as well as Chapters 5, 7, and as much of Chapter 8 as time permits.

Technical Requirements

The software that accompanies this book, *Geometry Explorer*, runs on Macintosh, Windows, and Linux computers, and also on any other computer that has a Java Virtual Machine (Java 1.2 or above). At least 128 MB of RAM is needed for the program to effectively function. On Macintosh computers the operating system must be 9.0 or above. On PCs the operating system must be at least at the level of Windows 98/NT. For an acceptable level of performance, *Geometry Explorer* should be installed on computers that have clock speeds of at least 200 MHz. To install the software, follow the instructions on the *Installation Guide* that comes with the software CD. For software updates and bug fixes, check the *Geometry Explorer* web site http://www.gac.edu/~hvidsten/gex.

Acknowledgments

This text and the accompanying *Geometry Explorer* software have evolved from the many geometry courses I have taught at Gustavus Adolphus College. I am deeply grateful to my students, who have graciously allowed me to experiment with different strategies in discovery-based learning and the integration of technology into the classroom. Their encouragement and excitement over these new approaches have been the primary motivating factors for writing an integrated learning environment for the *active* exploration of geometry.

I am especially grateful to those who helped out in the early phases of this project. Alicia Sutphen, a former student, provided valuable assistance in the early stages of designing the software. Special thanks go to those faculty who field-tested early drafts of the text: Steve Benzel, Berry College; Jason Douma, University of Sioux Falls; George Francis, University of Illinois at Urbana-Champaign; Dan Kemp, South Dakota State University; Bill Stegemoller, University of Southern Indiana; Mary Wiest, Minnesota State University at Mankato; and Stephen Walk, St. Cloud State University.

I would also like to thank the following reviewers: Nick Anghel, University of North Texas; Brian Beaudrie, Northern Arizona University; David Boyd, Valdosta State University; Anita Burris, Youngstown State University; Victor Cifarelli, University of North Carolina at Charlotte; Michael Dorff, Brigham Young University; Gina Foletta, Northern Kentucky University; Matthew Jones, California State University–Dominguez Hills; Tabitha Mingus, Western Michigan University; Chris Monico, Texas Tech University; F. Alexander Norman, University of Texas at San Antonio; Ferdinand Rivera, San Jose State University; Craig Roberts, Southeast Missouri State University; Philippe Rukimbira, Florida International University; Don Ryoti, Eastern Kentucky University; Sherrie Serros, Western Kentucky University; Wendy Hageman Smith, Radford University;

The material in this text is based upon work supported by the National Science Foundation under Grant No. 0230788.

Finally, I am grateful to my wife, Rebekah Richards, for her encouragement, her help in editing, and her understanding of the many hours needed to complete this project.

Chapter 1

Geometry and the Axiomatic Method

We owe geometry to the tax collector.

—J. L. Heilbron, *Geometry Civilized* [20]

Let no one ignorant of geometry enter here.

—Inscription over the doors to Plato's Academy

1.1 Early Origins of Geometry

In a fundamental sense, geometry is a natural outgrowth of our exposure to the physical universe and in particular to the natural world. In our interactions with our environment, we encounter physical shapes, such as rocks and mountains, that we then organize by patterns into groups and classes. Rocks get put into the "round" category and mountains into a separate category. As our powers of perception become more refined, we notice other patterns of objects, such as the symmetries found in nature. An example of a natural symmetry is that of the rotational symmetry found in the California poppy (Fig. 1.1).

Fig. 1.1 California Poppy, Mimi Kamp, Southwest School of Botanical Medicine

It is not surprising that human beings, being embedded in the natural world, should be inspired by and curious about geometrical shapes. For example, when constructing shelters our ancestors invariably chose to use precise geometric figures—most often circles or rectangles. There were very practical reasons to use these shapes; rectangular structures are easily laid out and circular huts provide a maximum of living space for the area they enclose.

While ancient peoples used geometric shapes for quite utilitarian purposes, they also surrounded themselves with patterns and designs that did not have any functional purpose.

In this ancient Navajo rug, there is no *practical* need to decorate the fabric with such an intricate design. The decoration met a different need for the individual who created it, the need for beauty and abstraction.

From the earliest times geometric figures and patterns have been used to represent abstract concepts, concepts that are expressed through the construction of objects having specific geometric shapes. A good example of this connection between the abstract and the concrete is that of the pyramids of ancient Egypt (Fig. 1.2).

Fig. 1.2 The Giza Plateau Complex, Copyright 1997 Oriental Institute, University of Chicago

The pyramids were built primarily as tombs for the pharaohs. However, a tomb for a pharaoh could not be just an ordinary box. The pharaoh was considered a god and as such his tomb was designed as a passageway connecting this life to the afterlife. The base of each pyramid represented the earth. It was laid out precisely with four sides oriented to face true north, south, east, and west. From the base the sides reached a peak that symbolized the connection with the Egyptian sun god.

While the design and construction of the pyramids required very specific geometric knowledge—basic triangle geometry and formulas for the volume of four-sided pyramids—the Egyptians also developed simple geometric rules for handling a quite different task, that of surveying. The arable land of ancient Egypt lay close to the Nile and was divided into plots leased to local Egyptians to farm. Each year, after the Nile had flooded and wiped out portions of the land, tax collectors were forced to re-calculate how much land was left in order to levy the appropriate amount of rent. The Egyptians' study of land measurement was passed on to the Greeks and is evidenced by the word *geometry* itself, which in Greek means "earth measure."

A special class of Egyptian priests arose to handle these two types of geometrical calculations—practical surveying and the more abstract and spiritual design of monuments and tombs. The ancient Greek philosopher Aristotle believed that the existence of this priestly class motivated a more abstract understanding of geometry by the Egyptians than by any of their predecessors. However, this higher study of geometry was still quite primitive by our standards. A truly abstract and logical understanding of geometry would come with the Greeks' absorption of Egyptian ideas through the schools of Thales and Pythagoras.

1.2 Thales and Pythagoras

The Egyptians had remarkably good formulas for the volume of a truncated pyramid (not surprisingly) and had a good approximation (about $3\frac{1}{6}$) for the constant π. However, there was a serious problem with Egyptian geometry. They did not delineate between values that were approximations and those that were exact. Indeed, nowhere in Egyptian geometry is there a concern for what the "actual" value of a computation is. Their method of solving a problem was to take the numbers involved and follow a recipe of adding, subtracting, and so on, until they ended up with a final number. For example, they knew that certain triples of integers, say 3, 4, and 5, would form the lengths of a right triangle, but they had no notion of what the relationship between the sides of a right triangle were in general.

In the period between 900 and 600 BC, while the Egyptian empire was waning, a new seafaring and trading culture arose in Greece. As the Greeks traveled throughout the Mediterranean, they interacted with a diverse set of cultures, including the Egyptians. The entrepreneurial spirit of the Greeks was reflected in the creation of independent schools of learning led by master teachers such as Thales and Pythagoras. This was in stark contrast to the centralized monopoly of the priestly class in Egypt. The Greeks created a marketplace of ideas where theories were created and debated vigorously.

In this society of ideas, two notable schools arose—the Ionian school founded by Thales of Miletus and the Pythagorean school founded by Pythagoras of Samos. Much about both men is lost to history and comes down to us through myths and legends. However, the impact that their schools had on mathematics, science, music, and philosophy was revolutionary.

To understand the profound change that occurred in how we think of mathematics now versus before the Greeks, let us consider a simple problem—that of computing the area of a triangle. The Egyptians were aware that if the triangle were a right triangle, then one could "flip" the triangle across the longest side and get a rectangle. The triangle's area is then half the product of the two shortest sides.

They also knew that in some special triangles, with certain fixed lengths of sides, the area was half the length of one side times the height of the triangle. However, this is all that the Egyptians knew, or wanted to know, about the area of a triangle in general. Geometry to them was an *empirical* subject, needed only to analyze those triangles found in real objects.

The Greeks, on the other hand, viewed the abstraction of things as the ideal and perfect form of reality. Sides and edges in this world were imperfect. By abstracting edges into segments having no width and having *exact* length, the Greeks could talk about the exact answer to questions such as the area of a triangle. And not only that, they could discuss in general what the area should be for *any* triangle that is built of three abstract segments.

This abstraction, freed from the constraints of empirical foundation, allowed for the study of *classes* of objects, rather than the objects themselves. For example, if one supposed the earth was a sphere, then one could deduce many properties of the earth just by using what was known about abstract spheres.

But this way of thinking required a new way of determining the veracity of statements. For the Egyptians, the statement "a triangle with sides of length 3, 4, and 5 has a right angle" would be accepted as true if one drew a 3-4-5 triangle in the sand and measured the angle to be approximately a right angle. However, a statement about abstract triangles cannot be proved this way—one cannot draw a *perfect* 3-4-5 triangle, no matter how hard one tries. The greatest achievement of the Greeks was the development of a precise and logical way of reasoning called the *deductive method*, which provided sound rules of argument to be used for abstract systems of objects. The deductive method has formed the basis for scientific reasoning from the time of the Greeks until the modern age.

The Greeks' focus on abstraction and their creation of the deductive method can be traced to the two great schools of Thales and Pythagoras.

1.2.1 Thales

Thales (ca. 624–548 BC) lived in the coastal city of Miletus in the ancient region of Ionia (present day Turkey). He is reported to have visited Egypt in the first half of the sixth century BC. While there, he studied Egyptian mathematics and measured the height of the Egyptian pyramids by the use of shadows and similar triangles.

Thales was, in many respects, a bridge between the empirical and mythical world of the Egyptians and the abstract and rational world of Greek civilization. To the Egyptians, reality was infused with the actions of spiritual and mystical beings, and thus, while one could discover some basic facts about how things worked, the true nature of reality was not important—the gods would do as they wanted to with the world. Thales is known as the first scientist because he believed in a rational world where one could discover universal truths about natural phenomena by abstracting properties of the world into ideal terms. This was a fundamental shift in how nature was perceived. No longer were natural processes simply the whims of mythical beings; they were products of processes that could be described in terms that could be debated and proved true or false. Thales was perhaps the first thinker to seriously consider the question, What is matter composed of? His answer was *water*, which we know today to be incorrect, but it was an answer that Thales could back up with a logical argument and could, in principle, be shown to be true or false.

Thales is known for being the first mathematician, the first to use deductive reasoning to prove mathematical results. As best we can tell, his methods of proof actually involved a combination of deductive reasoning and inductive reasoning from examples. The full development of deductive reasoning from first principles (axioms) would come later, with its most complete expression found in the work of Euclid.

The five geometric theorems attributed to Thales are

1. A circle is bisected by a diameter.

2. The base angles of an isosceles triangle are equal.

3. The pairs of vertical angles formed by two intersecting lines are equal.

4. Two triangles are congruent if they have two angles and the included side equal.

5. An angle inscribed in a semicircle is a right angle.

The last result has become known as "The Theorem of Thales." It can be proved in several ways. In Fig. 1.3 we suggest one method. Try proving this theorem. That is, show that $\beta + \gamma = 90$, using Thales' theorem on isosceles triangles and the fact that the sum of the angles in a triangle is 180 degrees.

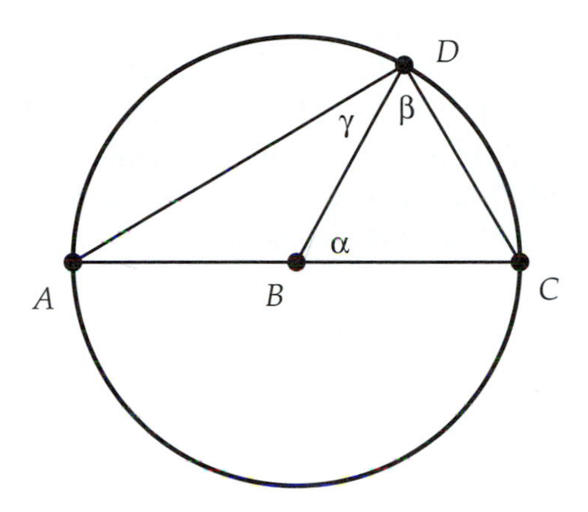

Fig. 1.3

1.2.2 Pythagoras

According to legend, Pythagoras (ca. 580–500 BC) was a pupil of Thales. Like Thales, Pythagoras traveled widely, gathering not only mathematical ideas, but also religious and mystical teachings. To the Pythagoreans, philosophy and mathematics formed the moral basis of life. The words *philosophy* (love of wisdom) and *mathematics* (that which is learned) were coined by the Pythagoreans.

In the Pythagorean worldview, mathematics was the way to study the ideal, that which was truly harmonious and perfect. "Numbers rule the universe" was their motto. They believed that all of nature could be explained by properties of the *natural* numbers $1, 2, 3, \ldots$. The search for harmony and perfection were of utmost importance and translated directly to a focus on ratios and proportion—the harmonious balance of numbers.

The theorem historically attributed to Pythagoras, that *in a right triangle the square on the hypotenuse is equal to the sum of the squares on the two sides*, was known as an empirical fact to the Egyptians and others in the ancient Orient. However, it was the Pythagoreans who first provided a logical proof of this result.

The Pythagoreans' development of geometry and their focus on the theory of numbers in relation to music, astronomy, and the natural world had a profound effect on Greek culture. As Boyer and Merzbach state in their history of mathematics [7, page 57], "Never before or since has mathematics played so large a role in life and religion as it did among the Pythagoreans."

One of the most important features of the schools of both Pythagoras and Thales was their insistence that mathematical results be justified or proved true. This focus on proofs required a method of reasoning and argument that was precise and logical. This method had its origins with Pythagoras and Thales and culminated with the publication of *The Elements* by Euclid in about 300 BC. The method has become known as the "Axiomatic Method." Howard Eves in [14] records the invention of the axiomatic method as one of the very greatest moments in the history of mathematics. Before we investigate the axiomatic method, let's take a little side trip into one of the most elegant constructions used by the Pythagoreans—the construction of the golden section.

1.3 Project 1 - The Ratio Made of Gold

Leonardo Da Vinci called it the "Sectio aurea." Luca Paccioli, an Italian mathematician of the 1500s, wrote a book, which Da Vinci illustrated, called *De Divine Proportione*. Johannes Kepler, in the late 1500s, said:

> Geometry has two great treasures: one is the theorem of Pythagoras; the other, the division of a line into extreme and mean ratio. The first we may compare to a measure of gold; the second we may name a precious jewel.

What these great thinkers were referring to is one of the simplest, yet most aesthetically pleasing, geometric constructions—that of the golden ratio.

The first study of the golden ratio is attributed to the Pythagoreans. In their search for harmony of number, they sought the ideal figure, one in which the dimensions were in perfect harmony of proportion. Proportion has to do with ratios of measurements. We say that two ratios are in *proportion* if they are equal to one another. The simplest geometric ratio is that of two segment lengths. The simplest way to create a proportion of ratios is to start with a single segment and cut it into two parts. We create one ratio by looking at the ratio of the total segment to the longer of the two sub-pieces, and we can create another ratio by that of the longer sub-piece to the smaller. The splitting of the segment will be most harmonious if these two ratios are equal, that is, if they are in proportion.

Our first project will be to construct geometric figures that have this perfect harmony of proportion. We will do so by using the software environment *Geometry Explorer* that came bundled with this text. Before you start this project, review the brief notes on using *Geometry Explorer* found in the preface. In particular, review the notes on *creating* versus *constructing* geometric objects and the information about *attaching* points to objects. As you progress through the lab, refer to the software reference guide found in Appendix B for any other user interface issues that arise.

You should work through all the constructions and exercises until you come to the end of the project, where you will find instructions on how to write up a report of work done for this project.

1.3.1 Golden Section

Start *Geometry Explorer*. You should see a window like the one in Fig. 1.4. Our first task will be to create a segment $\overline{AB}$ on the screen. Using the Segment tool in the Create panel of buttons, click and drag on the screen to create a segment. (You can use the Label tool—the one with the "A" on it in the Create panel—to click on the endpoints of this segment and make labels visible.) Using the Point tool, *attach* a point C to segment $\overline{AB}$. Note that C is always "stuck" on $\overline{AB}$; if we drag C around, it will always stay on $\overline{AB}$. Holding down the Shift key, multi-select points A, B, and C (in that order). Now choose **Ratio** (**Measure** menu). We have measured the ratio of the length of segment $\overline{AB}$ to the length of segment $\overline{BC}$, and this ratio measurement will appear on the screen, as shown in Fig. 1.4.

Now, multi-select points B, C, and A (in that order) and choose **Ratio** (**Measure** menu). The length ratio of segment $\overline{BC}$ to $\overline{CA}$ will now appear on the screen below the previous ratio. Drag point C around and see if you can get these two ratios to match up; that is, see if you can create a *proportion* of ratios.

Interesting!! The ratios seem to match at a magic ratio of about 1.6. Let's see why this is. For the sake of argument, let's set the length of $\overline{BC}$

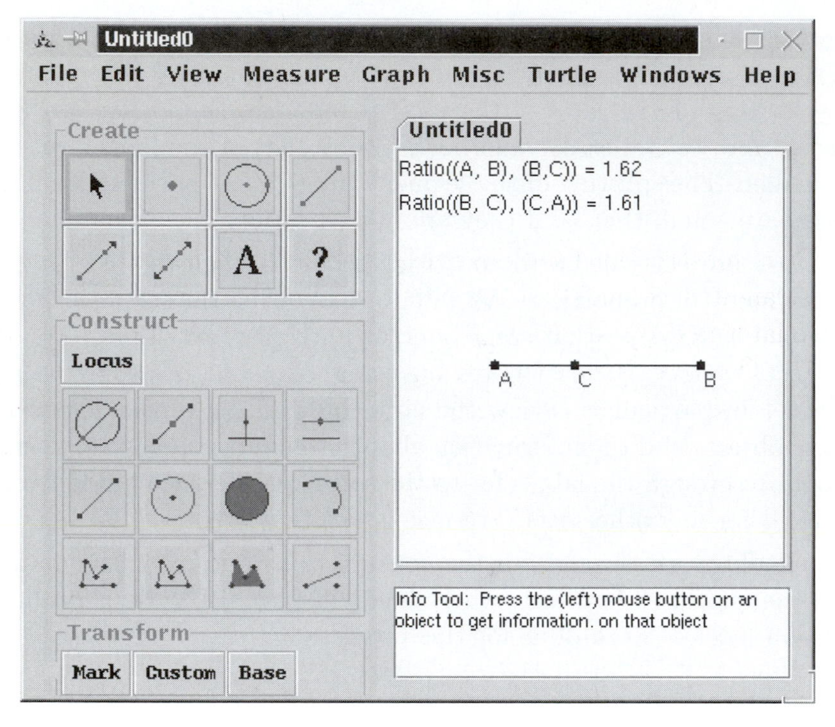

Fig. 1.4 *Geometry Explorer* Main Window

equal to 1. Let x be the length of $\overline{AB}$. Then, what we are looking for is a value of x that satisfies the equation

$$\frac{x}{1} = \frac{1}{x-1}$$

If we solve this equation, we get that x must satisfy $x^2 - x - 1 = 0$. This has two roots $\frac{1\pm\sqrt{5}}{2}$. The positive solution is $\frac{1+\sqrt{5}}{2}$, which is about 1.62. This ratio is the perfection of balance for which the Pythagoreans were searching. The segment subdivision having this ratio is what Da Vinci called the *Golden Section*.

Construction of the Golden Ratio

We can see from the previous discussion that it is not too hard to approximate the golden ratio by moving point C. However, in the true Pythagorean spirit, is it possible to construct two segments whose ratio of lengths is *exactly* the golden ratio?

Traditionally, the question of geometric construction of numerical values or geometric figures has played a key role in the development of geometry. (Euclid's notion of *construction* encompasses *both* of the ideas of creating and constructing used in *Geometry Explorer*.) Euclidean constructions are carried out by drawing lines (or segments) and circles and by finding the intersections of lines and circles. Such constructions are called *straightedge and compass* constructions as they represent pencil and paper constructions using a straightedge (line) and compass (circle). A review of constructions will come later in the text. For now, we point out that all of the tools available in the Create and Construct panels of buttons in *Geometry Explorer* are valid Euclidean constructions.

To construct the golden ratio of a segment, we will need to split a segment into two sub-segments such that the ratio of the larger to the smaller sub-segment is exactly $\frac{1+\sqrt{5}}{2}$. How can we do this? Since the fraction $\frac{1}{2}$ appears in the expression for the golden ratio, it might be useful to construct the midpoint of a segment. Also, an easy way to construct $\sqrt{5}$ would be to construct a right triangle with base lengths of 1 and 2. We'll keep these ideas in mind as we explore how to construct the golden ratio.

Golden Ratio Construction Step 1

To get started, we need a segment. Clear the screen (**Clear** (**Edit** menu)) and create segment $\overline{AB}$.

Since we already discussed the need for midpoints, let's construct the midpoint C of $\overline{AB}$ by selecting the segment and clicking on the Midpoint button (second button in first row of the Construct panel). To make life easier for ourselves, let's assume the length of $\overline{AB}$ is 2. Then, we have one base of the triangle we discussed above.

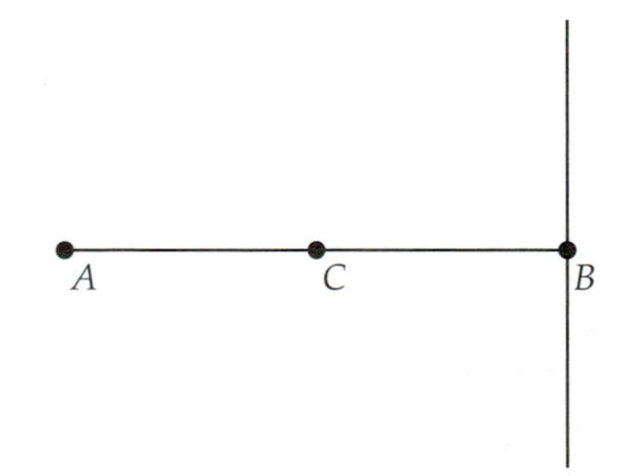

To get the other base, we need a right angle at B. Multi-select segment $\overline{AB}$ and point B (remember to hold the Shift key down while selecting) and construct the perpendicular (third button in first row of Construct panel) to $\overline{AB}$ at B (refer to the preceding figure).

Golden Ratio Construction Step 2

Now, we need to create a segment up from B along the vertical line that has length 1. This can be done using a circle centered at B of radius 1. To create this circle, select the Circle tool (in the Create panel) and click on point B. Keeping the mouse button down, drag the cursor until it is directly over point C. When the cursor is over C, that point will become highlighted (a small circle will pop up around C). The cursor represents the radius point of the circle we want to construct, and we want this radius point to be *equal* to C so that the circle has radius exactly equal to 1. Release the mouse button and drag points A and B around the screen. Notice how the circle radius is always of constant length equal to the length of $\overline{BC}$.

The technique of dragging a point of a new circle or line until it matches an existing point will be a common technique used throughout the labs in this text. It is an easy way to create objects that are in synchronization with existing objects.

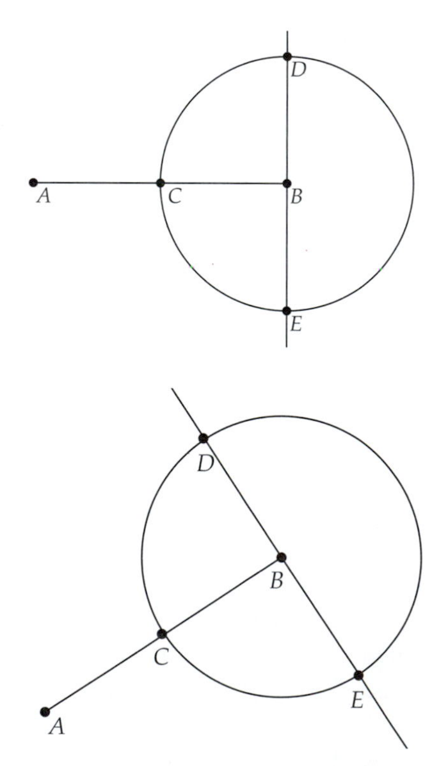

To construct the vertical leg of the desired right triangle, return A and B to the position shown and multi-select the circle and the line $\overleftrightarrow{BD}$. Construct the intersection points D and E by using the Intersect tool in the Construct panel of buttons.

At this point, you should check that you have constructed all of the figures correctly. Drag points A and B around the screen and check that your circle moves accordingly, with center B and radius to C, and that your intersection points D and E move with the circle and the perpendicular.

A great advantage in using *Geometry Explorer* over paper and pencil construction is that you can dynamically move figures, exploring how their properties change (or stay constant) as you do so.

Golden Ratio Construction Step 3

Point D is the point we are after, so hide the circle (click on the circle and choose **Hide Object** (**View** menu)). Hide point E in a similar fashion. Move $\overline{AB}$ back to a horizontal position and create a segment connecting A to D. (Use the Segment tool. Click on A and drag the cursor to D.) To finish our triangle, create segment $\overline{BD}$.

By the Pythagorean Theorem we know that the length of $\overline{AD}$ will be $\sqrt{5}$. So we have constructed all the numbers that appear in the fraction for the golden ratio, but we have not actually found a point on $\overline{AB}$ that will subdivide this segment in this ratio.

Let's experiment a bit with what we are trying to find. Suppose G is a point between A and B that subdivides the segment into the golden ratio; that is $\frac{AB}{AG} = \frac{1+\sqrt{5}}{2}$. Let x be the length of $\overline{AG}$. (Draw a picture on a scrap piece of paper—all good mathematicians always have paper on hand when reading math books!)

Exercise 1.3.1. Given that the length of $\overline{AB}$ is 2, argue that x must be equal to $\sqrt{5}-1$.

From this exercise, we see that to finish our construction we need to find a point G on $\overline{AB}$ such that the length of $\overline{AG}$ is $\sqrt{5}-1$. Explore (on paper) for a couple minutes how you might find such a point and then continue on with the project.

We make use of the fact that we already have a length of $\sqrt{5}$ in the hypotenuse of our right triangle. To cut off a length of 1 from the hypotenuse, create a circle centered at D with radius point B (using the click and drag technique as described above). Then, multi-select $\overline{AD}$ and this new circle and construct the intersection point F (using the Intersect button).

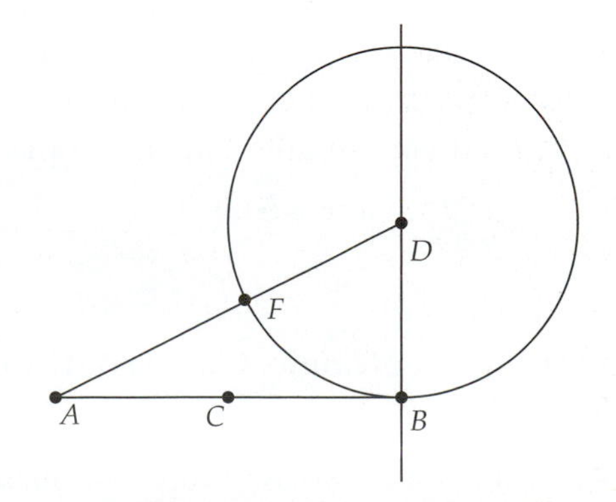

To construct G, we just need to transfer $\overline{AF}$ to $\overline{AB}$. Create a circle with center A and radius point F and construct the intersection point G of this new circle with $\overline{AB}$. (Use the Intersect button again.) Multi-select A, G, and B and choose **Ratio** (**Measure** menu) to compute the ratio of the length of $\overline{AG}$ to the length of $\overline{GB}$.

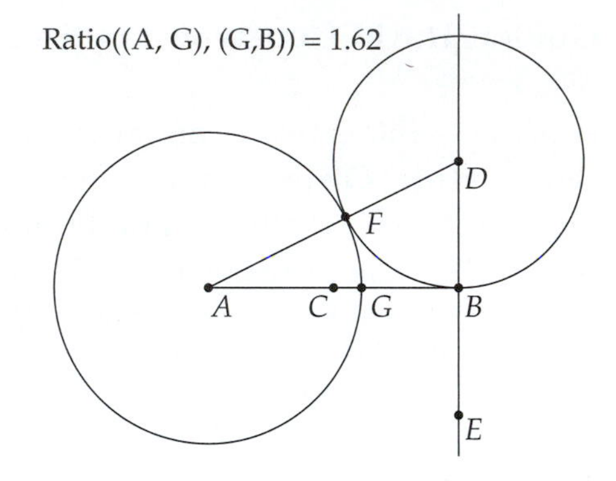

It looks like our analysis was correct. We have constructed a golden section for $\overline{AB}$ at G!

1.3.2 Golden Rectangles

The golden ratio has been used extensively in art and architecture—not in the subdivision of a single segment but in the creation of rectangular shapes called golden rectangles. A *Golden Rectangle* is a rectangle where the ratio of the long side to the short side is exactly the golden ratio.

To construct a golden rectangle, we will again need to construct the numbers 1, 2, and $\sqrt{5}$. But this time we can interpret the numerator and denominator of the golden ratio fraction $\frac{1+\sqrt{5}}{2}$ as the separate side lengths of a rectangle. It makes sense to start out with a square of side length 2, as we can take one of its sides as the denominator and can split an adjacent side in two to get the 1 term in the numerator. To get the $\sqrt{5}$ term, we need to extend this smaller side appropriately.

Golden Rectangle Construction Step 1

To start the construction, clear the screen (**Clear** (**Edit** menu)) and create segment $\overline{AB}$. We will assume this segment is of length 2.

To construct a square on $\overline{AB}$, first construct the perpendicular to $\overline{AB}$ at A. (Multi-select the segment and A and click on the Perpendicular button in the Construct panel.) Then *construct* a circle at A with radius $\overline{AB}$ by multi-selecting point A and segment $\overline{AB}$ and clicking on the Circle Construction button (second from left in second row of Construct panel). Next, multi-select the circle and the perpendicular and construct the intersection point C. Do a similar series of constructions at B to get point E as shown.

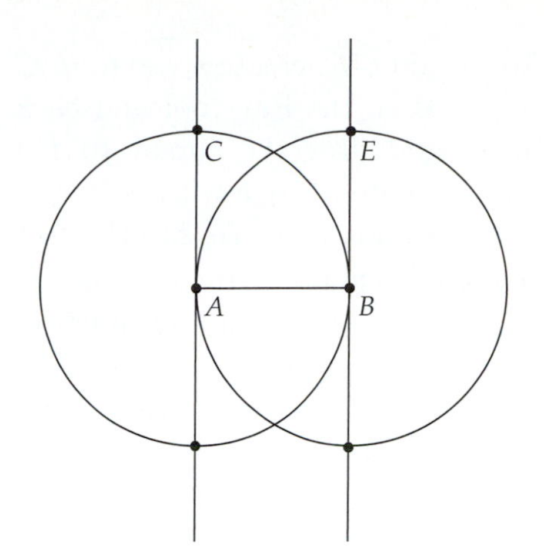

Points A, B, C, and E will form a square. Hide all of the perpendiculars and circles (select the objects and choose **Hide Object** (**View** menu)) and create the segments $\overline{AC}$, $\overline{CE}$, and $\overline{EB}$ to form a square.

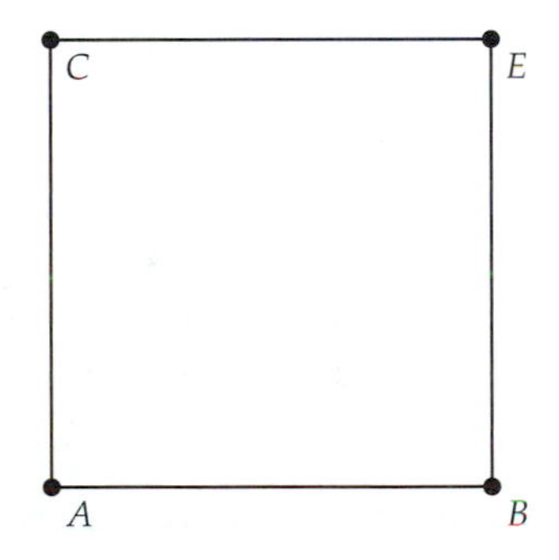

Golden Rectangle Construction Step 2

Select $\overline{CE}$ and construct the midpoint G of $\overline{CE}$. If we can extend $\overline{CE}$ to a point H so that $\overline{EH}$ has length $\sqrt{5} - 1$, then segment $\overline{CH}$ will have length $\sqrt{5} + 1$ and we will have the length ratio of $\overline{CH}$ to $\overline{AB}$ equal to the golden ratio.

To extend $\overline{CE}$, create a ray from C to E. (Use the Ray tool and click on C and drag the cursor to E.) Then, create a circle with center at G and radius point B. Multi-select this circle, and the ray through C and E, and construct the intersection point H.

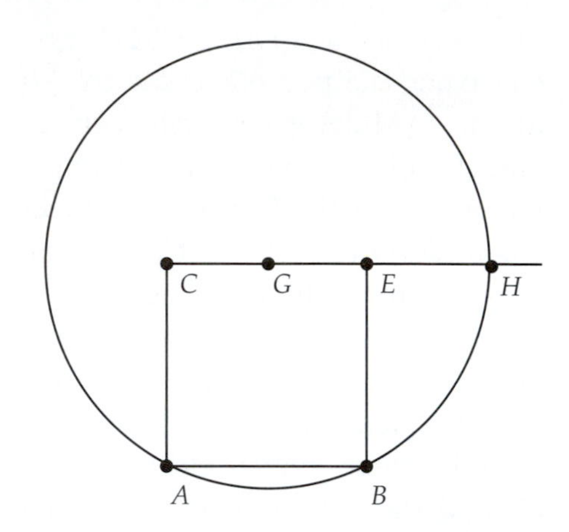

Exercise 1.3.2. Use a right triangle to argue that $\overline{EH}$ has length $\sqrt{5}-1$, given the construction of H done in the step above.

Golden Rectangle Construction Step 3

Finally, to finish off the rectangle partially defined by B, E, and H, we construct a perpendicular to ray $\overline{GE}$ at H, then create a ray from A to B, and construct the intersection point J of the perpendicular with this ray. Multi-select A, B, and J and choose **Ratio (Measure** menu). We have created a segment ($\overline{AJ}$) that is cut in the golden ratio at B, as well as a rectangle ($EHJB$) that is "golden."

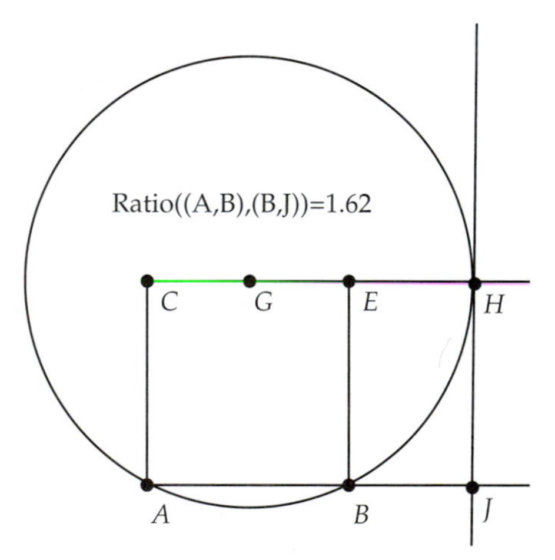

Exercise 1.3.3. Grab a ruler and a few of your friends and measure the proportions of the rectangles enclosing your friends' faces. Measure the ratio of the distance from the bottom of the chin to the top of the head to the distance between the ears. Make a table of these distances and find the average "face ratio." Does the average come close to the golden ratio? Find some magazine photos of actors and actresses considered beautiful. Are their faces "golden"?

The Fibonacci sequence has a fascinating connection with the golden ratio. The sequence is defined as a sequence of numbers $u_1, u_2, u_3, \ldots$ where $u_1 = 1, u_2 = 1$, and $u_n = u_{n-1} + u_{n-2}$. Thus, $u_3 = 2, u_4 = 3, u_5 = 5$, and so on. The first ten terms of this sequence are 1, 1, 2, 3, 5, 8, 13, 21, 34, 55, 89.

These numbers come up in surprising places. They appear in the branching patterns of plants, in the numbers of rows of kernels in a sunflower head, and in many spiral patterns, such as the spiral found in the Nautilus shell. The ratio of successive Fibonacci numbers is also related to the golden ratio. Consider the first few ratios of terms in the sequence

$$\begin{aligned} \frac{2}{1} &= 2 \\ \frac{3}{2} &= 1.5 \\ \frac{5}{3} &= 1.666... \\ \frac{8}{5} &= 1.6 \\ \frac{13}{8} &= 1.625 \\ \frac{21}{13} &= 1.61538... \end{aligned}$$

It appears that these ratios are approaching the golden ratio. In fact, these ratios actually do converge *exactly* to the golden ratio (see [25] for a proof).

Exercise 1.3.4. Find five objects in your environment that have dimensions given by two succeeding terms in the Fibonacci sequence. For example, a simple 3x5 index card is close to being a golden rectangle as $\frac{5}{3}$ is one of our Fibonacci ratios. Once you start looking, you will be amazed at how many simple, everyday objects are nearly golden.

Project Report

This ends the exploratory phase of this project. Now it is time for the explanation phase—the final project report. The report should be carefully written and should include three main components: an introduction, a section discussing results, and a conclusion.

The introduction should describe what was to be accomplished in the project. It should focus on one or two major themes or ideas that were explored.

The main body of the report will necessarily be composed of the results of your investigation. This will include a summary of the constructions you carried out. This should not be a verbatim list or recipe of what you did, but rather a general discussion of what you were asked to construct and what you discovered in the process. Also, all results from exercises should be included in this section.

The conclusion of the report should document what you learned by doing the project and also include any interesting ideas or conjectures that you came up with while doing the project.

1.4 The Rise of the Axiomatic Method

In the last section we looked at the ancient Greeks' search for the perfect harmony of proportion. In this section we consider another example of the Greeks' quest for perfection—that of perfection of *reasoning*. This quest culminated in the creation of a pattern of reasoning called the *axiomatic method*.

The axiomatic method is based on a system of *deductive* reasoning. In a deductive system, statements used in an argument must be derived, or based upon, prior statements used in the argument. These prior statements must themselves be derived from even earlier statements, and so on. There are two logical traps that one can fall into in such a system.

First, there is the trap of producing a never-ending stream of prior statements. Consider, for example, the definition of a word in the dictionary. If we want to define a word like *orange* we need to use other words such as *fruit* and *round*. To define *fruit* we need to use *seed*, and so forth. If every word required a different word in its definition, then we would need an infinite number of words in the English language!

Second, there is the trap of circular reasoning. In some dictionaries a *line* is defined as a type of curve, and a *curve* is defined as a line that deviates from being straight.

The Greeks recognized these traps and realized that the only way out of these logical dilemmas was to establish a base of concepts and statements that would be used *without* proof. This base consisted of *undefined terms* and *postulates/axioms*.

Undefined terms were those terms that would be accepted without any further definition. For example, in the original formulation of Euclid's geometry, the terms *breadth* and *length* are undefined, but a *line* is defined as "breadth-less length." One may argue whether this is a useful definition of a line, but it does allow Euclid to avoid an infinite, or circular, regression of definitions. In modern treatments of Euclidean geometry, the terms *point* and *line* are typically left undefined.

A postulate or axiom is a logical statement about terms that is accepted as true without proof. To the Greeks, postulates referred to statements about the specific objects under study and axioms (or *common notions*) referred to more universal statements about general logical systems. For example, the statement "A straight line can be drawn from any point to any point" is the first of Euclid's five postulates; whereas the statement "If equals be added to equals, the wholes are equal" is one of Euclid's axioms. The first is a statement about the specifics of a geometric system and the second is a general logical statement. In modern mathematical axiomatic systems, there is no distinction between these two types of mutually accepted statements, and the terms *axiom* and *postulate* are used interchangeably to refer to a statement accepted without proof.

Starting from a base of undefined terms and agreed upon axioms, we can define other terms and use our axioms to argue the truth of other statements. These other statements are called the *theorems* of the system. Thus, our deductive system consists of four components:

1. Undefined Terms

2. Axioms (or Postulates)

3. Defined Terms

4. Theorems

A system comprising these four components, along with some basic rules of logic, is an axiomatic system. (In Appendix A there is a complete listing of the axiomatic system Euclid used at the start of his first book on plane geometry.)

One way to think about an axiomatic system is by analogy with playing a game like chess. We could consider the playing pieces (as black and white objects) and the chessboard as undefined parts of the game. They just exist and we use them. A particular playing piece, for example the bishop, would be a defined term, as it would be a special kind of playing piece. The *rules* of chess would be axioms. The rules are the final say in what is allowed and what is not allowed in playing the game. Everyone (hopefully) agrees to play by the rules. Once the game starts, a player moves about the board, capturing his or her opponent's pieces. A particular configuration of the game, for example with one player holding another player in check, would be like a theorem in the game in that it is derived from the axioms (rules), using the defined and undefined terms (pieces) of the game, and it is a configuration of the game that can be verified as legal or not, using the rules. For example, if we came upon a chessboard set up in the starting position, but with all of a player's pawns behind all of the other pieces, we could logically conclude that this was not a legal configuration of the game.

It is actually very useful to have a game analogy in mind when working in an axiomatic system. Thinking about mathematics and proving theorems is really a grand game and can be not only challenging and thought-provoking like chess, but equally as enjoyable and satisfying.

As an example of playing the axiomatic game, suppose that we had a situation where students enrolled in classes. *Students* and *classes* will be left as undefined terms as it is not important for this game what they actually mean. Suppose we have the following rules (axioms) about students and classes.

A1 There are exactly three students.

A2 For every pair of students, there is exactly one class in which they are enrolled.

A3 Not all of the students belong to the same class.

A4 Two separate classes share at least one student in common.

What can be deduced from this set of axioms? Suppose that two classes shared more than one student. Let's call these classes C_1 and C_2. If they share more than one student—say students A and B are in both classes—then we would have a situation where A and B are in more than one class. This clearly contradicts the rule we agreed to in Axiom 2, that two students are in one and only one class. We will use a rule of logic that says that an assumption that contradicts a known result or an axiom cannot be true. The assumption that we made was that two classes could share more than one student. The conclusion we must make is that this assumption is false, and so two classes cannot share more than one student.

By Axiom 4 we also know that two separate classes share at least one student. Thus, two separate classes must have one and only one student in common. We can write these results down as a *theorem*. The set of explanations given above is called a *proof* of the theorem.

Theorem 1.1. *Two separate classes share one and only one student in common.*

Here are a couple more results using our axiomatic system:

Theorem 1.2. *There are exactly three classes in our system.*

Proof: By Axiom 2 we know that for each pair of students, there is a class. By Axiom 3 all three students cannot be in a common class. Thus, there must be at least three classes, say C_1, C_2, C_3, as there are three different pairs of students. Suppose there is a fourth class, say C_4. By the theorem just proved, there must be a student shared by each pairing of C_4 with one of the other three classes. So C_4 has at least one student. It cannot contain all three students by Axiom 3. Also, it cannot have just one student since if it did, then classes C_1, C_2, and C_3 would be forced by Axiom 4 to share this student and, in addition, to have three other, different students among them, because the three classes must have different pairs of students. This would mean that there are at least four students and would contradict Axiom 1. Thus, C_4 must have exactly two students. But, since this pair of students must already be in one of the other three classes, we have a contradiction to Axiom 2. Thus, there cannot be a fourth class. □

Theorem 1.3. *Each class has exactly two students.*

Proof: By the previous theorem we know that there are exactly three classes. By Axiom 4 we know that there is at least one student in a class. Suppose a class had just one student, call this student S. All classes would then have student S by Axiom 4. The other two students are in some class, call it class X, by Axiom 2. Class X must then have all three students as it also needs to have student S, the student common to all three classes. But, this contradicts Axiom 3. Thus, all classes must have at least two students and by Axiom 3 must have exactly two students. □

It is important to point out that the precise meaning of the terms *students* and *classes* is not important in this system. We could just as well have used the following axiom set:

A1 There are exactly three snarks.

A2 For every pair of snarks, there is exactly one bittle.

A3 Not all of the snarks belong to the same bittle.

A4 Two separate bittles have at least one snark in common.

By changing the labels in the theorems above, we would get equivalent theorems about snarks and bittles. The point of this silly little aside is that we are concerned about the *relationships* and *patterns* among the objects in an abstract axiomatic system and not about the objects themselves. This is the great insight of the Greeks—that it is the relationships that matter and not how we apply those relationships to objects.

The insistence on proofs in formal axiomatic mathematics can seem, at times, to be a tedious exercise in belaboring the obvious. This sentiment is actually as old as Greek geometry itself. J. L. Heilbron, in [20] describes how the Greek philosopher Epicurus (341–270 BC) criticized Euclid's proof that no side of a triangle can be longer than the sum of the other two sides. As Epicurus stated, "It is evident even to an ass." For if a donkey wanted to travel to a bale of hay, it would go directly there along a line and not go through any point not on that line (Fig. 1.5).

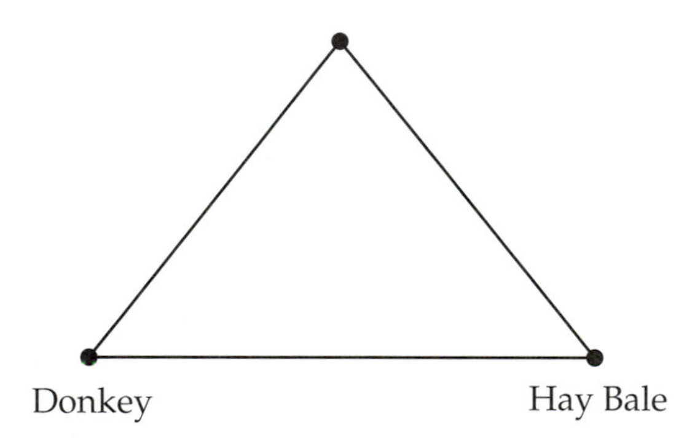

Fig. 1.5 Donkey Geometry

But, the Greek geometer Proclus (411–485 AD) refuted this criticism by arguing that something that seems evident to our senses cannot be relied on for scientific investigation. By training our minds in the most careful and rigorous forms of reasoning abstracted from the real world, we are preparing our minds for the harder task of reasoning about things that we cannot perceive. For Proclus, this type of reasoning "arouses our innate knowledge, awakens our intellect, purges our understanding, brings to light the concepts that belong essentially to us, takes away the forgetfulness and ignorance that we have from birth, and sets us free from the bonds of unreason" [20, page 8].

One clear illustration of Proclus' point is the development of modern views on the nature of the universe. From the time of Euclid, mathematicians believed that the universe was *flat*, that it was essentially a three-dimensional version of the flat geometry Euclid developed in the plane. In fact, Euclidean geometry was considered the only possible axiomatic geometric system. This was a reasonable extrapolation from these mathematicians' experience of the world. However, in the nineteenth century, a revolution occurred in the way mathematicians viewed geometry. *Non-Euclidean* geometries were developed that were just as logically valid as Euclidean geometry. The notion of a *curved* universe was now mathematically possible, and in 1854 George Frederich Riemann (1826–1866) set out the basic mathematical principles for analyzing curved spaces of not just three dimensions, but *arbitrary* dimensions. This mathematical theory must have seemed incredibly wild and impractical to non-mathematicians. However, the groundwork laid by Riemann was fundamental in Einstein's development of the theory of relativity and in his view of space-time as a four-dimensional object.

This revolution in the axiomatic basis of geometry laid the groundwork for a movement to formalize the foundations of *all* of mathematics. This movement reached its peak in the *formalist* school of the late 1800s and early 1900s. This school was led by David Hilbert (1862–1943) and had as its goal the axiomatic development of all of mathematics from first principles. Hilbert's *Grundlagen der Geometrie*, published in 1899, was a careful development of Euclidean geometry based on a set of 21 axioms.

Exercise 1.4.1. Trace the following words through a dictionary until a circular chain of definitions has been formed for each: power, straight, real (e.g., power - strength - force - power). Why must dictionary definitions be inherently circular?

Exercise 1.4.2. In the game of Nim, two players are given several piles of coins, each pile having a finite number of elements. On each turn a player picks a pile and removes as many coins as he or she wants from that pile but must remove at least one coin. The player who picks up the last coin wins. (Equivalently, the player who no longer has coins to pick up loses.) Suppose that there are two piles with one pile having more coins than the other. Show that the first player to move can always win the game.

Exercise 1.4.3. Consider a system where we have children in a classroom choosing different flavors of ice cream. Suppose we have the following axioms:

A1 There are exactly five flavors of ice cream: vanilla, chocolate, strawberry, cookie dough, and bubble gum.

A2 Given any two different flavors, there is exactly one child who likes these two flavors.

A3 Every child likes exactly two different flavors among the five.

How many children are there in this classroom? Prove your result.

Exercise 1.4.4. Using the ice cream axiom system, show that any pair of children likes at most one common flavor.

Exercise 1.4.5. In the ice cream system, show that for each flavor there are exactly four children who like that flavor.

One of the most universal of abstract mathematical structures is that of a *group*. A group G consists of a set of undefined objects called *elements* and a *binary operation* "$\circ$" that relates two elements to a third. The axioms for a group are

A1 For all elements x and y, the binary operation on x, y is again an element of G. That is, for all $x, y \in G$, $x \circ y \in G$.

A2 For all $x, y, z \in G$, $(x \circ y) \circ z = x \circ (y \circ z)$. That is, the binary operation is *associative*.

A3 There is a special element $e \in G$ such that $x \circ e = x$ for all $x \in G$. The element e is called the *identity* of G.

A4 Given $x \in G$, there is an element $x^{-1} \in G$ such that $x \circ x^{-1} = e$. The element x^{-1} is called the *inverse* to x.

Exercise 1.4.6. Show that if $x, y, z \in G$ and $x \circ z = y \circ z$, then $x = y$.

Exercise 1.4.7. Show that the binary operation is *commutative* with the identity. That is, show that for all $x \in G$, we have $x \circ e = e \circ x$. [Hint: Be careful—you may use only the four axioms and the previous exercise.]

Exercise 1.4.8. Show that a group G can have only one identity.

In 1889 Giuseppe Peano (1858–1932) published a set of axioms, now known as the *Peano axioms*, as a formal axiomatic basis for the natural numbers. Peano was one of the first contributors to the modern axiomatic and formalist view, and his system included five axioms based on the undefined terms *natural number* and *successor*. We will let N stand for the set of all natural numbers in the following listing of Peano's axioms.

A1 1 is a natural number.

A2 Every natural number x has a successor (which we will call x') that is a natural number, and this successor is unique.

A3 No two natural numbers have the same successor. That is, if x, y are natural numbers, with $x \neq y$, then $x' \neq y'$.

A4 1 is not the successor of any natural number.

A5 Let M be a subset of natural numbers with the property that

(i) 1 is in M.

(ii) Whenever a natural number x is in M, then x' is in M.

Then, M must contain all natural numbers, that is, $M = N$.

Exercise 1.4.9. The fifth Peano axiom is called the *Axiom of Induction.* Let M be the set of all natural numbers x for which the statement $x' \neq x$ holds. Use the axiom of induction to show that $M = N$, that is, $x' \neq x$ for all natural numbers x.

Exercise 1.4.10. Must every natural number be the successor of some number? Clearly, this is not the case for 1 (why not?), but what about other numbers? Consider the statement "If $x \neq 1$, then there is a number u with $u' = x$." Let M be the set consisting of the number 1 plus all other x for which this statement is true. Use the axiom of induction to show that $M = N$ and, thus, that every number other than 1 is the successor of some number.

Exercise 1.4.11. Define *addition* of natural numbers as follows:

- For every x, define $x + 1 = x'$.
- For every x and y, define $x + y' = (x + y)'$.

Show that this addition is well-defined. That is, show that for all x and w, the value of $x + w$ is defined. [Hint: You may want to use induction.]

Exercise 1.4.12. Think about how you might define multiplication in the Peano system. Come up with a two-part definition of multiplication of natural numbers.

Exercise 1.4.13. Hilbert once said that "mathematics is a game played according to certain simple rules with meaningless marks on paper." Lobachevsky, one of the founders of non-Euclidean geometry, once said, "There is no branch of mathematics, however abstract, which may not someday be applied to phenomena of the real world." Discuss these two views of the role of mathematics. With which do you most agree? Why?

Exercise 1.4.14. Historically, geometry had its origins in the empirical exploration of figures that were subsequently abstracted and proven deductively. Imagine that you are designing the text for a course on geometry. What emphasis would you place on discovery and empirical testing? What emphasis would you place on the development of proofs and axiomatic reasoning? How would you balance these two ways of exploring geometry so that students would become confident in their ability to reason logically, yet also gain an intuitive understanding of the material?

1.5 Properties of Axiomatic Systems

In this section we will look a bit deeper into the properties that characterize axiomatic systems. We will look at the axiomatic structure itself and at its properties.

In the last section, we looked at an axiomatic system consisting of classes and students. We saw that we could re-label these terms as snarks and bittles without any real change in the structure of the system or in the relationships between the terms of the system. In an axiomatic system, it does not matter what the terms represent. The only thing that matters is how the terms are related to each other. By giving the terms a specific meaning, we are creating an *interpretation* of the axiomatic system.

Suppose in our example of the last section that we replaced *class* and *student* by *line* and *point*. Also, suppose that we interpret *line* to be one of the three segments shown below, and *point* to be any one of the three endpoints of the segments, as shown in Fig. 1.6.

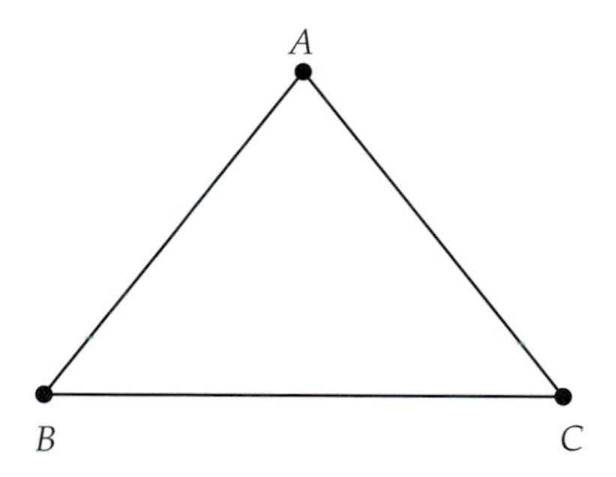

Fig. 1.6 Three-Point Geometry

Then, our axioms would read as follows:

A1 There are exactly three points.

A2 Two distinct points belong to one and only one line.

A3 Not all of the points belong to the same line.

A4 Two separate lines have at least one point in common.

We should check that our axioms still make correct sense with this new interpretation and they do. In our new interpretation, the theorems above now say that there are exactly three lines, each pair of lines intersects in exactly one point, and each line has exactly two points, as can be seen in Fig. 1.6. This system is called *Three-Point geometry.*

This new interpretation of our original axiomatic system is called a *model* of the system.

Definition 1.1. A *model* is an interpretation of the undefined terms of an axiomatic system such that all of the axioms are true statements in this new interpretation.

In any model of an axiomatic system, all theorems in that system are true when interpreted as statements about the model. This is the great power of the axiomatic method. A theorem of the abstract system needs to be proved once, but can be interpreted in a wide variety of models.

Three important properties of an axiomatic system are *consistency*, *independence*, and *completeness*.

1.5.1 Consistency

Consider the following axiomatic system:

A1 There are exactly three points.

A2 There are exactly two points.

One may consider this a rather stupid system, and rightly so. It is basically useless as Axiom 1 contradicts Axiom 2. It would be impossible to logically deduce theorems in this system as we start out with a fundamental contradiction. The problem here is that this system is *inconsistent.* One may argue that to avoid creating an inconsistent system, we just need to choose axioms that are not self-contradictory. This is not always easy to do. It may be the case that the axioms look fine, but there is some theorem that contradicts one of the axioms. Or, perhaps two theorems contradict one another. Any of these situations would be disastrous for the system.

Definition 1.2. An axiomatic system is *consistent* if no two statements (these could be two axioms, an axiom and theorem, or two theorems) contradict each other.

To determine whether an axiomatic system is consistent, we would have to examine every possible pair of axioms and/or theorems in the system. In systems such as Euclidean geometry, this is not possible. The best we can do is to show *relative* consistency. If we can find a model for a system—let's call it system A—that is embedded in another axiomatic system B, and if we know that system B is consistent, then system A must itself be consistent. For, if there were two statements in A that were contradictory, then this would be a contradiction in system B, when interpreted in the language of B. But, B is assumed consistent, so there can be no contradictory statements in B.

If we believe that Euclidean geometry is consistent, then the axiomatic system for classes and students we have described must be consistent as it has a model (Three-Point geometry) that is embedded in Euclidean geometry.

1.5.2 Independence

Definition 1.3. An individual axiom in an axiomatic system is called *independent* if it cannot be proved from the other axioms.

For example, consider the following system for points and lines:

A1 There are exactly three points.

A2 Two distinct points belong to one and only one line.

A3 Not all of the points belong to the same line.

A4 Two separate lines have at least one point in common.

A5 A line has exactly two points.

In the preceding section, we saw that Axiom 5 could be *proved* from the first four. Thus, Axiom 5 is not independent. While this axiomatic system would be perfectly fine to use (it is just as consistent as the original), it is not as economical as it could be. A system with numerous axioms is difficult to remember and confusing to use as a basis for proving theorems. Just as physicists search for the most simple and elegant theories to describe the structure of matter and the evolution of the universe, so too mathematicians search for the most concise and elegant basis for the foundations of their subject. However, it is often the case that the more compact an axiomatic system is, the more work has to be done to get beyond elementary results.

How do we show an axiom is independent of the other axioms in an axiomatic system? Suppose that we could find a model for a system including Axiom X. Suppose additionally that we could find a model for the system where we replaced Axiom X by its logical negation. If both models are consistent then Axiom X must be independent of the other axioms. For if it were dependent, then it would have to be provably true from the other axioms and its negation would have to be provably false. But, if consistent models exist with both X and the negation of X valid, then X must be independent of the other axioms.

As an example, consider the system for Three-Point geometry. Suppose we replaced Axiom 3 with "All of the points belong to the same line." A model for this would be just a line with three points. This would clearly be consistent with the first two axioms. Since there is only one line in the system, Axiom 4 would be vacuously true. (Here we use the logical rule that a statement is true if the hypothesis, the existence of two lines in this case, is false.) Thus, we can say that Axiom 3 is independent of Axioms 1, 2, and 4.

1.5.3 Completeness

The last property we will consider is that of completeness.

Definition 1.4. An axiomatic system is called *complete* if it is impossible to add a new consistent and independent axiom to the system. The new axiom can use only defined and undefined terms of the original system.

An equivalent definition would be that every statement involving defined and undefined terms is provably true or false in the system. For if there were a statement that could not be proved true or false within the system, then it would be independent of the system, and we could add it to get an additional axiom.

As an example consider the following system for points and lines:

A1 There are exactly three points.

A2 Two distinct points belong to one and only one line.

Now consider the additional statement "Not all of the points belong to the same line." (This is our original Axiom 3 from the last section.) We have already looked at a consistent model for this larger system. Now, consider the logical opposite of this statement: "All of the points belong to the same line." A consistent model for a system with axioms A1, A2, and this additional statement is that of three points on a line. Thus, the statement "Not all of the points belong to the same line" is not provably true or false within the smaller system consisting of just axioms A1 and A2, and this smaller system cannot be complete.

What of the four axioms for Three-Point geometry? Do they form a consistent system? Suppose we label the points A, B, C. We know from Theorem 1.2 and Theorem 1.3 in section 1.4 that there must be exactly three lines in Three-Point geometry and that each line has exactly two points. The three lines can then be symbolized by pairs of points: $(A, B), (A, C), (B, C)$.

Suppose we had another model of Three-Point geometry with points L, M, N. Then, we could create a one-to-one correspondence between these points and the points A, B, C so that the lines in the two models could be put into one-to-one correspondence as well. Thus, any two models of this system are essentially the same, or *isomorphic*; one is just a re-labeling of the other. This must mean that the axiomatic system is complete. For if it were not complete, then a statement about points and lines that is independent of this system could be added as a new axiom. If this were the case, we could find a consistent model not only for this new augmented system, but also for the system we get by adding the *negation* of this new axiom, and obviously these two models could not be isomorphic.

An axiomatic system is called *categorical* if all models of that system are isomorphic to one other. Thus, Three-Point geometry is a categorical system. In Chapters 7 and 8 we will see a much more powerful display of this idea of isomorphism of systems when we look in detail at the system of non-Euclidean geometry called *Hyperbolic Geometry.*

1.5.4 Gödel's Incompleteness Theorem

We cannot leave the subject of completeness without taking a side-trip into one of the most amazing results in modern mathematics—the Incompleteness Theorem of Kurt Gödel.

At the second International Congress of Mathematics, held in Paris in 1900, David Hilbert presented a lecture entitled "Mathematical Problems" in which he listed 23 open problems that he considered critical for the development of mathematics. Hilbert was a champion of the *formalist* school of thinking in mathematics, which held that all of mathematics could be built as a logical axiomatic system.

Hilbert's second problem was to show that the simplest axiomatic system, the natural numbers as defined by Peano, was consistent; that is, the system could not contain theorems yielding contradictory results.

Ten years after Hilbert posed this challenge to the mathematical community, Bertrand Russell (1872–1970) and Alfred North Whitehead (1861–1947) published the first volume of the *Principia Mathematica*, an ambitious project to recast all of mathematics as an expression of formal logic. This line of reasoning has become known as *logicism*. Russell and Whitehead succeeded in deriving set theory and natural number arithmetic from a formal logical base, yet their development still begged the question of consistency. However, the *Principia* was very influential in that it showed the power of formal logical reasoning about mathematics.

In 1931, an Austrian mathematician named Kurt Gödel (1906–1978) published a paper titled "On Formally Undecidable Propositions of Principia Mathematica and Related Systems" in the journal *Monatshefte für Mathematik und Physik*. In this paper Gödel used the machinery of formal logic to show the impossibility of the universalist approach taken by the formalists and the logicists. This approach had as its goal the development of *consistent* and *complete* systems for mathematics.

To address the question of consistency, Gödel showed the following result:

Theorem 1.4. *Given a consistent axiomatic system that contains the natural number system, the consistency of such a system cannot be proved from within the system.*

Thus, Hilbert's second problem is impossible! If one built a formal system, even one based on pure logic like Whitehead and Russell, then Gödel's theorem implies that it is impossible to show such a system has no internal contradictions.

Even more profound was the following result of Gödel's on the completeness of systems:

Theorem 1.5 (Incompleteness Theorem). *Given a consistent axiomatic system that contains the natural number system, the system must contain* undecidable *statements, that is, statements about the terms of the system that are neither provable nor disprovable.*

In other words, a system sufficiently powerful to handle ordinary arithmetic will necessarily be incomplete! This breathtaking result was revolutionary in its scope and implication. It implied that there might be a simple statement in number theory that is undecidable, that is impossible to prove true or false. Even with this disturbing possibility, most mathematicians have accepted the fact of incompleteness and continue to prove those theorems that are decidable.

Now that we have reviewed most of the important features of axiomatic systems, we will take a closer look in the next section at the first, and for most of history the most important, axiomatic system—that of Euclid's geometry.

Exercise 1.5.1. Consider sets as collections of objects. If we allow a set to contain objects that are themselves sets, then a set can be an element of another set. For example, if sets A, B are defined as $A = \{\{a, b\}, c\}$ and $B = \{a, b\}$, then B is an element of A. Now, it may happen that a set is an element of itself! For example, the set of all mathematical concepts is itself a mathematical concept.

Consider the set S which consists only of those sets which are not elements of themselves. Can an axiomatic system for set theory, that allows the existence of S, be a consistent system? [Hint: Consider the statement "S is an element of itself."] The set S has become known as the *Russell Set* in honor of Bertrand Russell.

Exercise 1.5.2. Two important philosophies of mathematics are those of the *platonists* and the *intuitionists.* Platonists believe that mathematical ideas actually have an independent existence and that a mathematician only discovers what is already there. Intuitionists believe that we have an intuitive understanding of the natural numbers and that mathematical results can and should be derived by constructive methods based on the natural numbers, and such constructions must be *finite.*

Of the four main philosophies of mathematics—platonism, formalism, logicism, and intuitionism—which most closely matches the way you view mathematics? Which is the prevalent viewpoint taken in the teaching of mathematics?

Exercise 1.5.3. Gödel's Incompleteness Theorem created a revolution in the foundations of mathematics. It has often been compared to the discovery of incommensurable magnitudes by the Pythagoreans in ancient Greece. Research the discovery of incommensurables by the Pythagoreans and compare its effect on Greek mathematics with the effect of Gödel's theorem on modern mathematics.

The exercises that follow deal with various simple axiomatic systems. To gain facility with the abstract notions of this section, carefully work out these examples.

In Four-Point geometry we have the same types of undefined terms as in Three-Point geometry, but the following axioms are used:

A1 There are exactly four points.

A2 Two distinct points belong to one and only one line.

A3 Each line has exactly two points belonging to it.

Exercise 1.5.4. Show that there are exactly six lines in Four-Point geometry.

Exercise 1.5.5. Show that each point belongs to exactly three lines.

Exercise 1.5.6. Show that Four-Point geometry is relatively consistent to Euclidean geometry. [Hint: Find a model.]

Exercise 1.5.7. A regular *tetrahedron* is a polyhedron with four sides being equilateral triangles (pyramid-shaped). If we define a point to be a vertex of the tetrahedron and a line to be an edge, will the tetrahedron be a model for Four-Point geometry? Why or why not?

Exercise 1.5.8. Consider an axiomatic system that consists of elements in a set S and a set P of pairings of elements (a, b) that satisfy the following axioms:

A1 If (a, b) is in P, then (b, a) is not in P.

A2 If (a, b) is in P and (b, c) is in P, then (a, c) is in P.

Let $S = \{1, 2, 3, 4\}$ and $P = \{(1, 2), (2, 3), (1, 3)\}$. Is this a model for the axiomatic system? Why or why not?

Exercise 1.5.9. In the previous problem, let S be the set of real numbers and let P consist of all pairs (x, y) where $x < y$. Is this a model for the system? Why or why not?

Exercise 1.5.10. Use the results of the previous two exercises to argue that the axiomatic system with sets S and P is not complete. Think of another independent axiom that could be added to the axioms in Exercise 1.5.8, for which S and P from Exercise 1.5.8 is still a model, but S and P from Exercise 1.5.9 is not a model.

1.6 Euclid's Axiomatic Geometry

The system of deductive reasoning begun in the schools of Thales and Pythagoras was codified and put into definitive form by Euclid (ca. 325–265 BC) around 300 BC in his 13-volume work *Elements*. Euclid was a scholar at one of the great schools of the ancient world, the Museum of Alexandria, and was noted for his lucid exposition and great teaching ability. The *Elements* were so comprehensive in scope that this work superseded all previous textbooks in geometry.

What makes Euclid's exposition so important is its clarity. Euclid begins Book I of the *Elements* with a few definitions, a few rules of logic, and ten statements that are axiomatic in nature. Euclid divides these into five geometric statements which he calls *postulates*, and five *common notions*. These are listed in their entirety in Appendix A. We will keep Euclid's terminology and refer to the first five axiomatic statements—those that specifically deal with the geometric basis of his exposition—as postulates.

1.6.1 Euclid's Postulates

Euclid I To draw a straight line from any point to any point.

Euclid II To produce a finite straight line continuously in a straight line.

Euclid III To describe a circle with any center and distance (i.e., radius).

Euclid IV That all right angles are equal to each other.

Euclid V If a straight line falling on two straight lines makes the interior angles on the same side less than two right angles, the two straight lines, if produced indefinitely, meet on that side on which are the angles less than the two right angles.

The first three postulates provide the theoretical foundation for constructing figures based on a hypothetically *perfect* straightedge and compass. We will consider each of these three in more detail, taking care to point out exactly what they say and what they do *not* say.

The first postulate states that given two points, one can construct a line connecting these points. Note, however, that it does not say that there is *only* one line joining two points.

The second postulate says that finite portions of lines (i.e., segments) can be extended. It does not say that lines are *infinite* in extent, however.

The third postulate says that given a point and a distance from that point, we can construct a circle with the point as center and the distance as radius. Here, again, the postulate does not say anything about other properties of circles, such as continuity of circles.

As an example of Euclid's beautiful exposition, consider the very first theorem of Book I of the *Elements*. This is the construction of an equilateral triangle on a segment $\overline{AB}$ (Fig. 1.7). The construction goes as follows:

1. Given segment $\overline{AB}$.

2. Given center A and distance equal to the length of $\overline{AB}$, construct circle $c1$. (This is justified by Postulate 3.)

3. Given center B and distance equal to the length of $\overline{BA}$, construct circle $c2$. (This is justified by Postulate 3.)

4. Let C be a point of intersection of circles $c1$ and $c2$.

5. Construct segments from A to C, from C to B, and from B to A. (This is justified by Postulate 1.)

6. Since $\overline{AC}$ and $\overline{AB}$ are radii of circle $c1$, and $\overline{CB}$ and $\overline{AB}$ are radii of $c2$, then $\overline{AC}$ and $\overline{CB}$ are both equal to $\overline{AB}$ and, thus, must be equal to one another. (This is justified by use of the Common Notions.)

7. By definition, then, triangle ABC is equilateral.

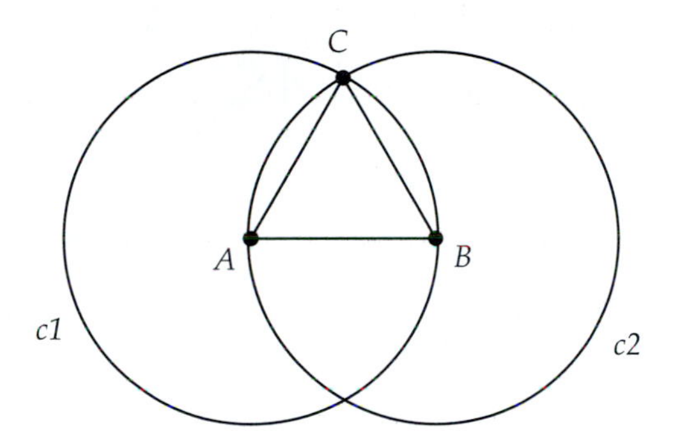

Fig. 1.7 Euclid Book I, Proposition I

The proof is an *almost* perfect example of a well-written mathematical argument. Euclid takes care to justify the steps in the construction in terms of the initial set of five postulates and five common notions, *except* for the step where the intersection point of the two circles is found. Euclid here assumed a principle of continuity of circles. That is, since circle $c1$ has points inside *and* outside of another circle ($c2$), then the two circles must intersect somewhere. In other words, there are no *holes* in the circles. To make this proof rigorous, we would have to add a postulate on circle continuity, or prove circle continuity as a consequence of the other five postulates.

We can see that from the very first proposition, Euclid was not logically perfect. Let's look at the last two postulates of Euclid.

Whereas the first three postulates are ruler and compass statements, the fourth postulate says that the rules of the game we are playing do not change as we move from place to place. In many of Euclid's theorems, he moves parts of figures on top of other figures. Euclid wants an axiomatic basis by which he can assume that segment lengths and angles remain unchanged when moving a geometric figure. In the fourth postulate, Euclid is saying that, at least, right angles are always equal no matter what configuration they are in.

In many of his proofs, Euclid assumes much more than Postulate 4 guarantees. For example, in his proof of the Side-Angle-Side Congruence Theorem for triangles, Euclid starts with two triangles ABC and DEF with $\overline{AB}$ congruent to $\overline{DE}$, $\overline{AC}$ congruent to $\overline{DF}$, and angle A congruent to angle D, as shown in Fig. 1.8. (For the time being, assume *congruent* to mean equal in magnitude.)

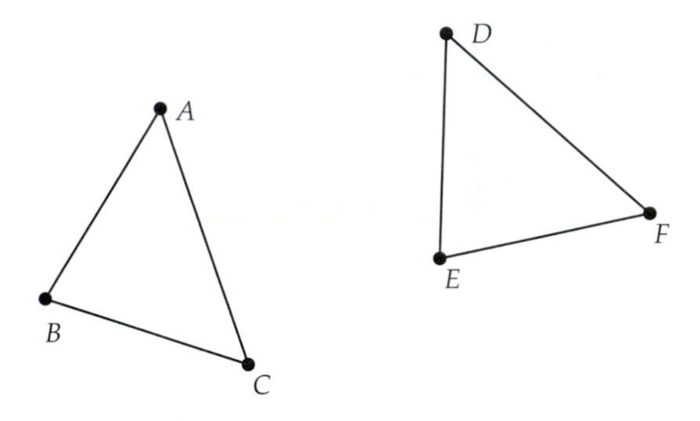

Fig. 1.8 Euclid SAS Theorem

Euclid proceeds by moving angle A on top of angle D. Then, he says that since two sides are congruent and the angles at A and D are congruent, then point B must move to be on top of point E and point C must move to be on top of point F. Thus, the two triangles have to be congruent.

The movement or *transformation* of objects was not on a solid logical basis in Euclid's geometry. Felix Klein, in the late 1800s, developed an axiomatic basis for Euclidean geometry that *started* with the notion of an existing set of transformations and then constructed geometry as the set of figures that are unchanged by those transformations. We will look at transformations and transformational geometry in greater detail in Chapter 5.

The final postulate seems very different from the first four. Euclid himself waited until Proposition 29 of Book I before using the fifth postulate as a justification step in the proof of a theorem. Euclid's fifth postulate is often referred to as the *parallel postulate*, even though it is actually more of an "anti-parallel" postulate as it specifies a sufficient condition for two lines to intersect.

From the time of Euclid, the fifth postulate's axiomatic status has been questioned. An axiom should be a statement so obvious that it can be accepted without proof. The first four of Euclid's postulates are simple statements about the construction of figures, statements that resonate with our practical experience. We can draw lines and circles and can measure angles. However, we cannot follow a line indefinitely to see if it intersects another line.

Many mathematicians tried to find simpler postulates, ones that were more intuitively believable, to replace Euclid's fifth postulate, and then prove the postulate as a *theorem*. Others tried to prove Euclid's fifth postulate *directly* from the first four. That is, they tried to show that the fifth postulate was not *independent* of the other four.

One of the attempts to find a simpler postulate for Euclid's fifth was that of John Playfair (1748–1819). His substitute can be expressed as follows:

> Given a line and a point not on the line, it is possible to construct one and only one line through the given point parallel to the line.

This statement, which has become known as *Playfair's Postulate*, is certainly easier to read and understand when compared to Euclid's fifth postulate. However, it is not hard to show that this statement is logically equivalent to Euclid's fifth postulate and, thus, does not really *simplify* Euclid's system at all. A full discussion of this and other statements about parallelism can be found in the next chapter.

Other mathematicians tried to prove that Euclid's fifth postulate was actually a *theorem* that could be derived from the first four postulates. A popular method of attack was to assume the logical opposite of Euclid's fifth postulate and try to prove this new statement false, or find a contradiction to an already accepted result. Amazingly, no one could prove that the negation of the fifth postulate was false or produced a contradiction.

In hindsight, this persistent lack of success would seem to imply that there could be consistent *non-Euclidean* geometries, obtained by replacing the fifth postulate by its opposite. However, this possibility ran counter to the generally held belief that Euclidean geometry was the only consistent geometry possible. This belief had been a bedrock of philosophy from the time of Euclid, and it was not until the 1800s that non-Euclidean geometry was fully explored. Even then, Carl Frederich Gauss (1777–1855), who was undoubtedly the first to recognize the consistency of non-Euclidean geometry, refrained from publishing results in this area. It wasn't until the work of Janos Bolyai (1802–1860) and Nicolai Lobachevsky (1793–1856) that theorems resulting from non-Euclidean replacements for Euclid's fifth postulate appeared in print. We will look at the fascinating development of non-Euclidean geometry in more detail in Chapter 7.

To end this discussion of Euclid's original axiomatic system, let's consider the system in terms of the properties discussed in the last section. Perhaps the most important question is whether this system is complete. A careful reading of the postulates shows that Euclid *assumed* the existence of points with which to construct his geometry. However, none of the postulates implies that points exist! For example, the statement "There exists at least two points" cannot be proven true or false from the other postulates. Thus, Euclid's axiomatic system is not complete.

Euclid's work contains quite a few *hidden* assumptions that make his axiomatic system far from complete. However, we should not be too hard on Euclid. It was only in the late 1800s that our modern understanding of abstract axiomatic systems and their properties was developed. No theorem that Euclid included in the *Elements* has ever been shown to be false, and various attempts to provide better axioms for Euclidean geometry have served to shore up the *foundations* of the subject, but the results that Euclid proved thousands of years ago are just as true today.

The first geometric axiomatic system to be shown logically consistent (at least relative to the consistency of natural number arithmetic) was the system developed by David Hilbert in 1899 [21]. Other modern axiomatic systems for geometry have been developed since then and it would be interesting to compare and contrast these. However, having a good axiomatic foundation is only the first step in exploring geometry. One must also *do* geometry, that is, one must experiment with geometric concepts and their patterns and properties. In the next chapter we will start with a basic set of geometric results and assumptions that will get us up and running in the grand game of Euclidean geometry.

Exercise 1.6.1. Euclid's *Elements* came to be known in Europe through a most circuitous history. Research this history and describe the importance of Arabic scholars to the development of Western mathematics of the late Middle Ages.

Exercise 1.6.2. The 13 books of Euclid's *Elements* contain a wealth of results concerning not only plane geometry, but also algebra, number theory, and solid geometry. Book II is concerned with *geometric algebra*, algebra based on geometric figures. For example, the first proposition of Book II is essentially the distributive law of algebra, but couched in terms of areas. Show, using rectangular areas, how the distributive law ($a(b+c) = ab + ac$) can be interpreted geometrically.

Exercise 1.6.3. Show how the algebraic identity $(a+b)^2 = a^2 + 2ab + b^2$ can be established geometrically.

Exercise 1.6.4. Euclid's *Elements* culminates in Book XIII with an exploration of the five *Platonic Solids*. Research the following questions:

- What are the Platonic Solids? Briefly discuss why there are only five such solids.
- The solids have been known for thousands of years as *Platonic* solids. What is the relationship to Plato and how did Plato use these solids in his explanation of how the universe was constructed?

Exercise 1.6.5. The *Elements* is most known for its development of geometry, but the 13 books also contain significant non-geometric material. In Book VII of the *Elements*, Euclid explores basic number theory. This book starts out with one of the most useful algorithms in all of mathematics, the *Euclidean Algorithm.* This algorithm computes the *greatest common divisor* (*gcd*) of two integers. A *common divisor* of two integers is an integer that divides both. For example, 4 is a common divisor of 16 and 24. The *greatest common divisor* is the largest common divisor. The Euclidean Algorithm works as follows:

- Given two positive integers a and b, assume $a \geq b$.
- Compute the quotient q_1 and remainder r_1 of dividing a by b. That is, find integers q_1, r_1 with $a = q_1 b + r_1$ and $0 \leq r_1 < b$. If $r_1 = 0$, we stop and b is the $gcd(a, b)$.
- Otherwise, we find the quotient of dividing b by r_1, that is, find q_2, r_2 with $b = q_2 r_1 + r_2$ and $0 \leq r_2 < r_1$. If $r_2 = 0$, we stop and r_1 is the $gcd(a, b)$.
- Otherwise, we iterate this process of dividing each new divisor (r_1, r_2, etc.) by the last remainder (r_2, r_3, etc.), until we finally reach a new remainder of 0. The last non-zero remainder is then the gcd of a, b.

Use this algorithm to show that the $gcd(36, 123) = 3$.

Exercise 1.6.6. Prove that the Euclidean Algorithm works, that it does find the gcd of two positive integers.

Exercise 1.6.7. Suppose we re-interpreted the term *point* in Euclid's postulates to mean a point on a sphere and a *line* to be a part of a great circle on the sphere. A *great circle* is a circle on the sphere cut by a plane passing through the center of the sphere.

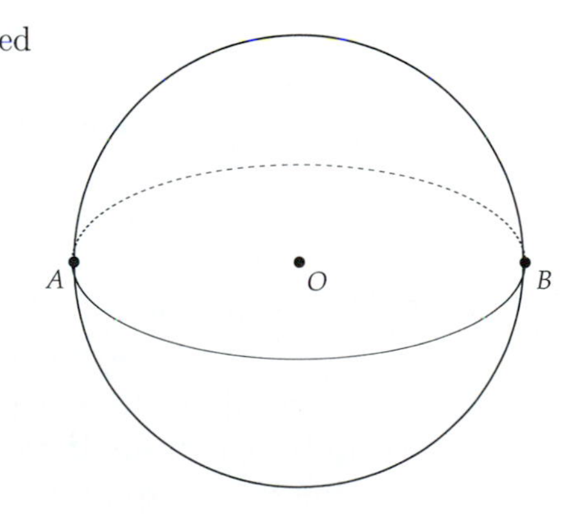

Find a ball (beach ball, bowling ball, soccer ball, etc.) and a string and experiment with constructing lines on the sphere. Now, think about how you would define a circle on the sphere. Experiment with the string and come up with a definition of a circle. How would you define an angle?

Show that Euclid's first four postulates hold in this new *spherical geometry*, but that the fifth postulate (or equivalently Playfair's Postulate) does *not* hold.

In the following exercises, experiment with spherical geometry to determine if each statement is most likely true or false in spherical geometry. Give short explanations (not proofs—we do not yet have the resources to do proofs in spherical geometry) for your answers.

Exercise 1.6.8. The sum of the angles of a triangle is 180 degrees.

Exercise 1.6.9. Given a line and a point not on the line, there is a perpendicular to the line through that point.

Exercise 1.6.10. A four-sided figure with three right angles must be a rectangle.

Exercise 1.6.11. One can construct a triangle with three right angles.

1.7 Project 2 - A Concrete Axiomatic System

We have seen that the axioms in an axiomatic system are the statements about the terms of the system that must be accepted without proof. Axioms are generally statements that are written down at the start of the development of some mathematical area.

When exploring geometry using *Geometry Explorer*, there are similarly constructions and tools that one uses *without proof.* That is, we draw lines and circles, find intersections, and construct perpendiculars using the tools built into the software, assuming that these constructions are valid. We might say that the abstract Euclidean axiomatic system has been made *concrete* within the software world of the program.

When one first starts up the *Geometry Explorer* program, constructions are based on the Euclidean axiomatic system. However, there is a second axiomatic system available for exploration in the program. To explore this system, start up the program and choose **New** (**File** menu). A dialog box will pop up asking you to specify which type of geometry environment you wish to work in. Choose "Hyperbolic" and hit "Okay." A new window will pop up with a large circle inside. Note that the tools and buttons in this window are the same as we had in the Euclidean main window (Fig. 1.9).

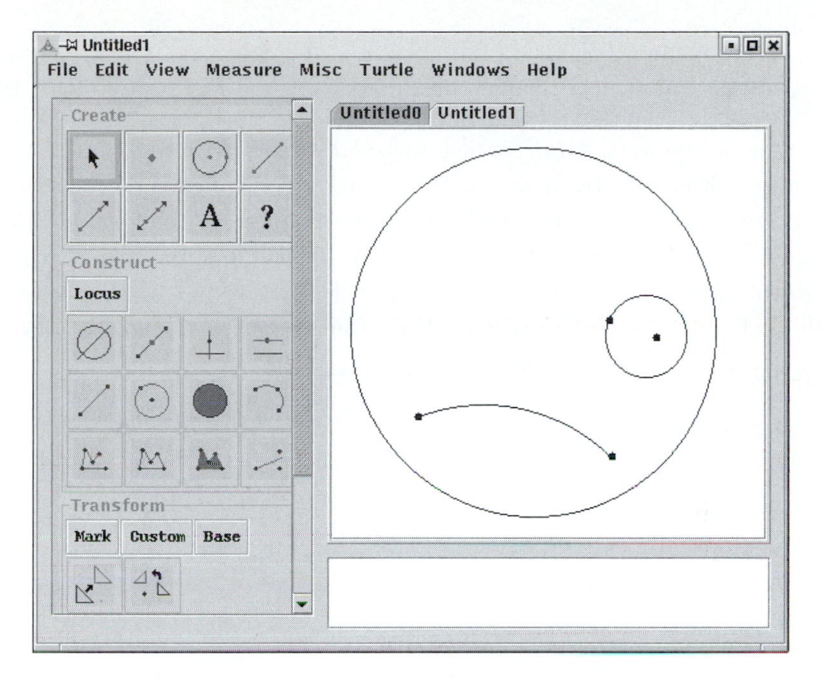

Fig. 1.9 A Concrete Axiomatic System

Try creating some segments, lines, rays, and circles using the tools in the Create panel of buttons.

How is this system like Euclidean geometry and how is it different? Let's review Euclid's first postulate—that between any two points we can draw a segment. As we experiment with creating segments on the screen, we see that points in this new geometry are restricted to points within the large disk. However, it does appear that given any two points in this disk, we can always create a *segment* connecting those points. It is true that segments do not look "normal" in the sense that they are not straight. However, we must use the axiomatic view in this new geometry. That is, a segment is whatever gets created when we use the Segment tool. Thus, it appears that Euclid's first postulate holds in this new geometry.

Euclid's second postulate states that segments can always be extended. Before we explore the validity of this postulate in our new geometry, let's consider what is meant by *extending* segments. In Euclidean geometry we can extend a segment $\overline{AB}$ by *moving* one of the endpoints, say B, a certain distance in the direction determined by the segment $\overline{AB}$. (Think of the direction as an arrow pointing from A to B.) As has been mentioned before, the movement of objects in Euclidean geometry is not specified in the postulates, but is nevertheless assumed by Euclid in many of his proofs.

We will define a *translation* of an object to be movement of the object in a certain direction by a specified distance (or length). The combined effect of distance along a specified direction is so useful that we will define a term for it; we will define a *vector* to be a quantity encapsulating a direction and length. Thus, a translation is defined by a vector. Also, a vector can be defined by a segment, with the direction given by the endpoints of the segment and the length given by the length of the segment.

For example, here we have a segment $\overline{AB}$ and another segment $\overline{CD}$. We have translated $\overline{CD}$ by the vector determined by $\overline{AB}$ (i.e., moved all of the points on $\overline{CD}$ in the direction from A to B, through a distance equal to the length of $\overline{AB}$).

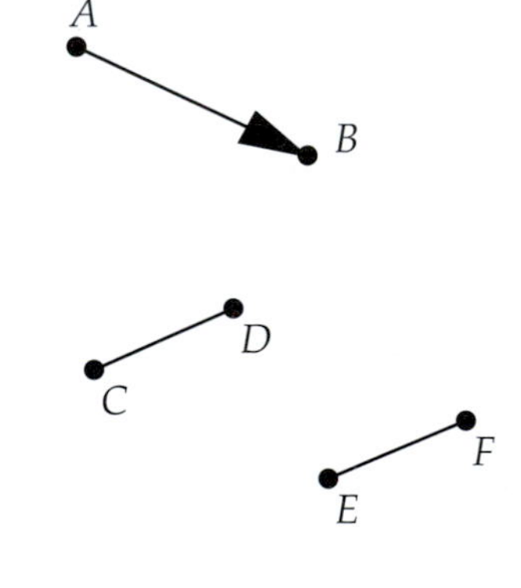

Can we do this in our new geometry? Clear the screen (**Clear** (**Edit** menu)) and create two segments $\overline{AB}$ and $\overline{CD}$ on the screen.

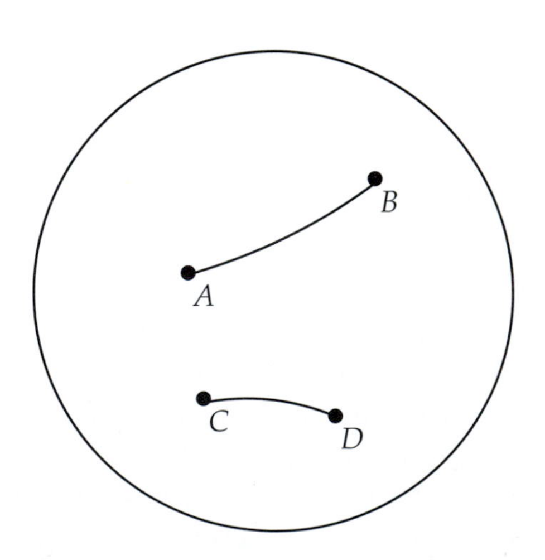

We will now define $\overline{AB}$ as a vector. Multi-select points A and B (in that order) and click and hold the mouse on the button labeled **Mark** in the Transform panel of buttons. A menu will pop down. Choose **Vector** to define $\overline{AB}$ as a vector, that is a direction and length. A dialog box will pop up asking for "Translate Type." Choose "By two points."

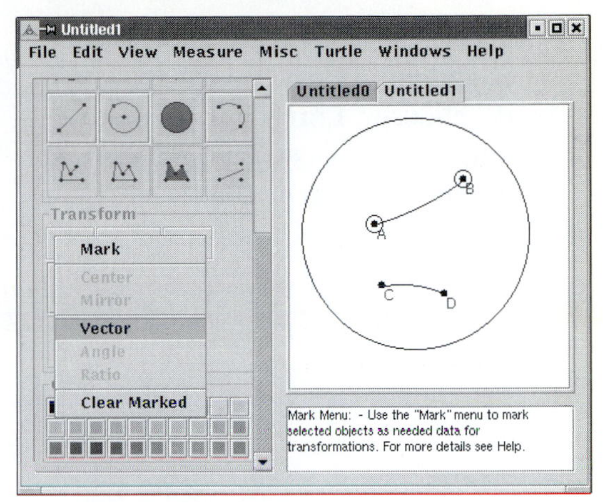

Now, we will translate $\overline{CD}$ by this vector. Select the segment $\overline{CD}$ (click somewhere between C and D) and then click on the Translate button (first button in first row of Transform panel) to translate $\overline{CD}$ to a new segment $\overline{EF}$. Note that this translation leaves $\overline{CD}$ unchanged and translates $\overline{CD}$ as a *copied* segment to its new position.

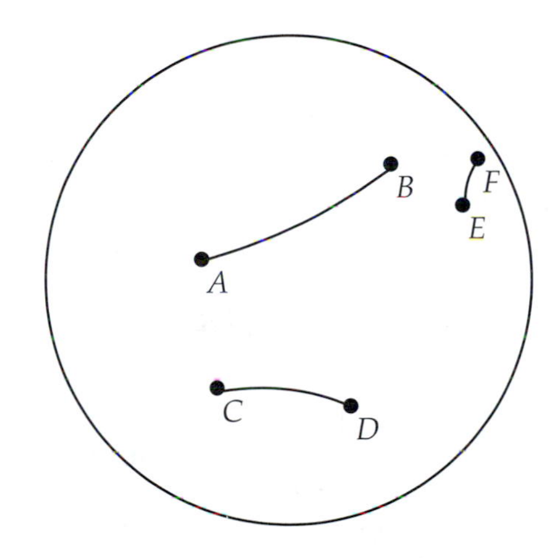

You may be surprised by the appearance of the translated segment $\overline{EF}$. Its length should be the same as the length of $\overline{CD}$ (we just picked it up and moved it), but it looks a lot smaller in the picture above. In your construction, it may look smaller, the same size, or perhaps larger. But, is it really different in length? Remember that we have to think about this system as its own "universe," where the axioms and definitions of geometric quantities like length are determined *within* the system, not by imposing definitions from outside the system.

Let's check the lengths from within the system. Select segment $\overline{CD}$ and choose **Length** (**Measure** menu). Then, select $\overline{EF}$ and choose **Length** (**Measure** menu).

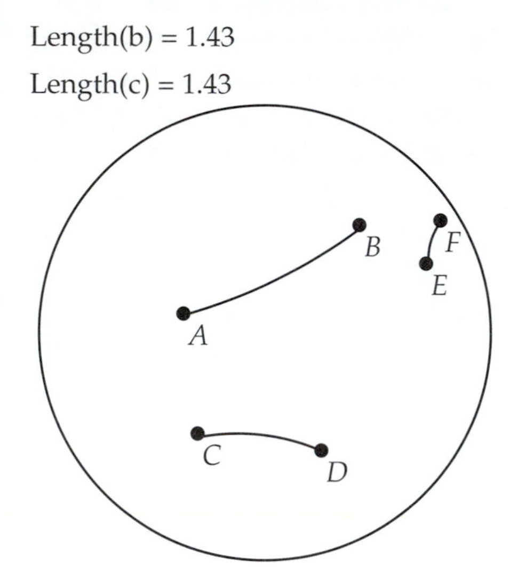

Interesting! The length of segment $\overline{CD}$ is preserved under translation. Try moving $\overline{CD}$ around and verify that this property persists. Here is a configuration where the translation looks almost like it would in Euclidean geometry, with movement from A to B taking D to F and C to E. However, other configurations look completely alien to our usual notions of translation.

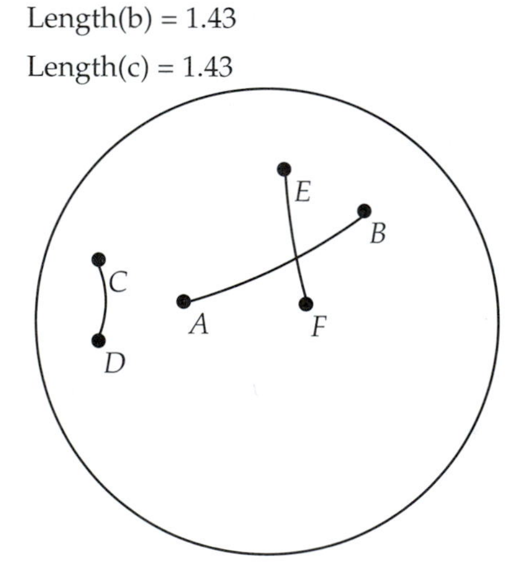

The point of this example is that translation in this strange new geometry has some properties, like preservation of length, that are just like Euclidean translations, but in other respects behaves quite differently from Euclidean translations.

Let's return to the question of Euclid's second postulate and whether it holds in this new geometry.

Clear the screen (**Clear** (**Edit** menu)) and create a segment $\overline{AB}$. Define $\overline{AB}$ as a vector, as you did previously. Then, select B and translate it to point C by clicking the Translate button in the Transform panel.

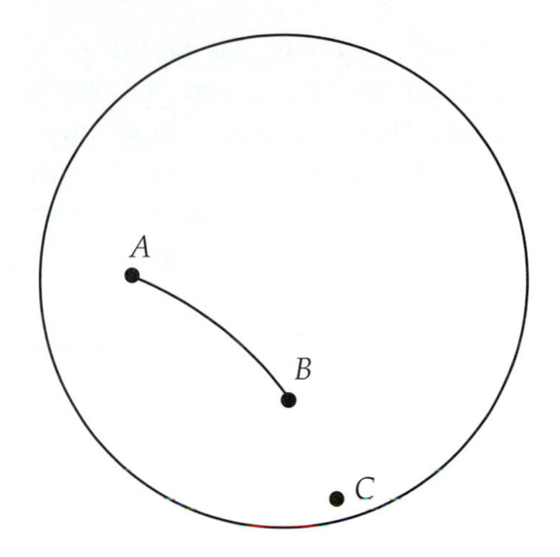

Measure the distance from A to B and then the distance from B to C (using **Distance** (**Measure** menu)). Since the translation has moved point C a certain distance along the direction from A to B, we have successfully *extended* $\overline{AB}$.

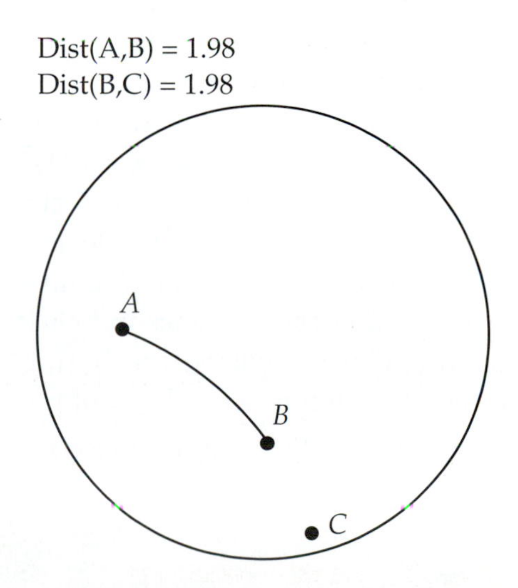

Select segment $\overline{AB}$ and drag it to the boundary. Even though it appears that the points have piled up at the boundary, point C still extends beyond $\overline{AB}$, as is evident by the distance measurements.

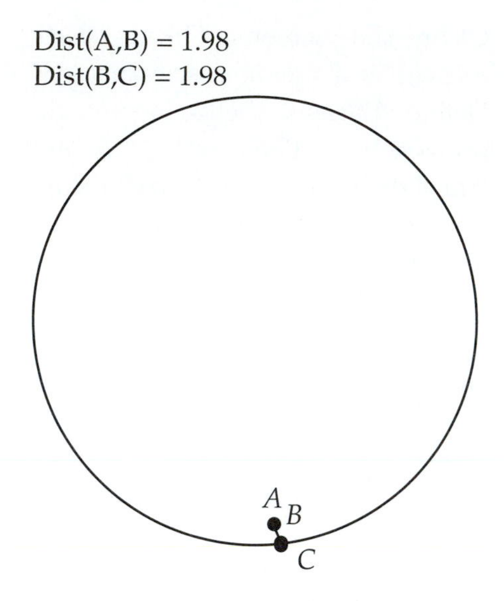

Now, let's consider Euclid's third postulate on the constructibility of circles.

Clear the screen (**Clear** (**Edit** menu)) and create a segment $\overline{AB}$ and a point C. To construct the circle of radius equal to the length of $\overline{AB}$ having center point C, multi-select C and the segment and click on the Circle Constructor button (second button in second row of the Construct panel).

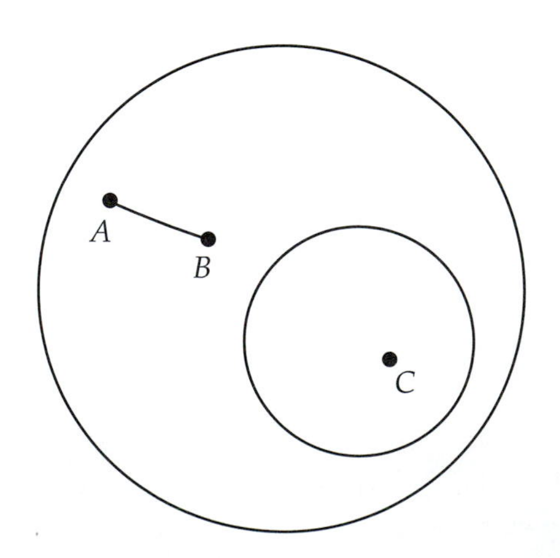

If we drag point C toward the boundary, we see that the circle with radius $\overline{AB}$ still exists, although it doesn't look like we would expect it to look.

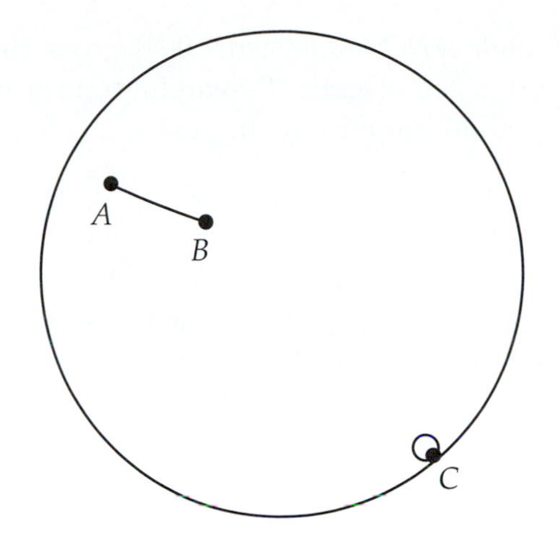

Thus, it appears that circles with any center and any specified length can be constructed in this geometry.

Let's now consider the fourth Euclidean axiom—that all right angles are congruent. Let's consider not just a right angle, but *any* angle.

Clear the screen and create a segment $\overline{AB}$ near the center of the disk. Then, create another segment $\overline{AC}$ originating from A. Next, multi-select the points B, A, and C (in that order) and choose **Angle** (**Measure** menu).

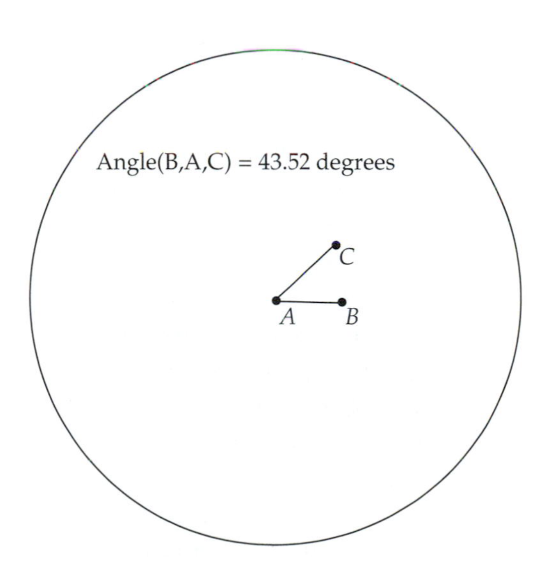

Note that if you select the points in the reverse order (try it), you will not have an equivalent angle measure. The reason is that *Geometry Explorer* measures angles in an *oriented* fashion, going counterclockwise. Thus, if angle ABC has measure 90 degrees, then angle CBA will have measure $360 - 90 = 270$ degrees. Be careful to measure angles in the correct orientation as you do the labs in the text.

Multi-select the three points A, B, and C and drag the mouse, moving the angle around the screen. Verify that movement of the angle has no effect on the angle measurement.

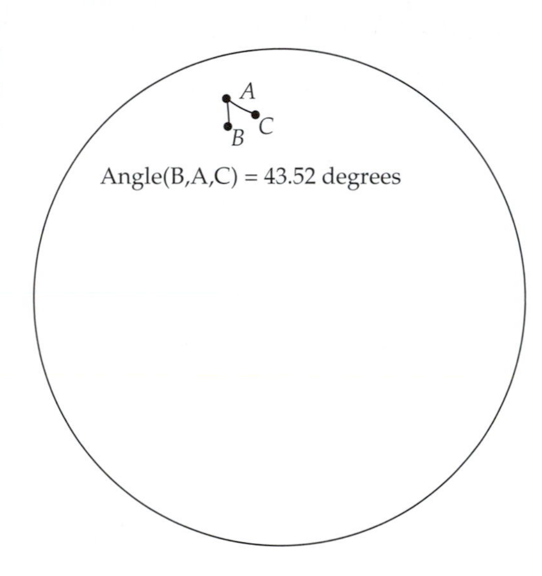

We see that the first four of Euclid's axioms still hold in this strange new geometry. What about the parallel axiom?

Clear the screen and create a line l near the bottom. Then, create a point P above l. As shown in the figure at the right, one can create many lines through P that do not intersect l, that is, are *parallel* to l. In fact, there should be two unique parallels through P that are *just* parallel to l. These are lines n and m in the figure.

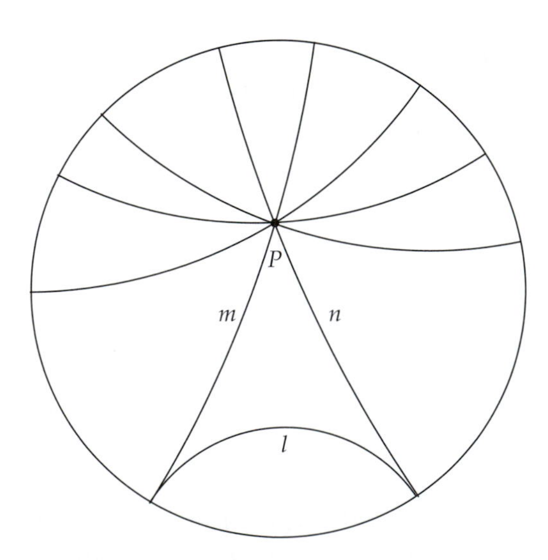

The geometry we have been discussing in this project is called *hyperbolic* geometry. From the last construction above on parallel lines we see that Euclid's fifth postulate does *not* hold in this geometry, and so we have had our first exploration of a *non-Euclidean* geometry. We will cover this geometry in more depth in Chapter 7.

Exercise 1.7.1. Determine which of the following Euclidean properties holds in the hyperbolic geometry of *Geometry Explorer*.

- Rectangles can be constructed. (Try any construction you remember from high school geometry.)
- The sum of the angles of a triangle is 180 degrees.
- Euclid's construction of an equilateral triangle.
- Given a line and a point not on the line there is a perpendicular to the line through that point.

Project Report

Your report should include a general discussion of how this new system of geometry is similar to Euclidean geometry and how it is different. This discussion should reference the five postulates as we have explored them in this chapter, and should include your analysis from Exercise 1.7.1.

Chapter 2

Euclidean Geometry

Euclid alone has looked on Beauty bare.
Let all who prate of Beauty hold their peace,
And lay them prone upon the earth and cease
To ponder on themselves, the while they stare
At nothing, intricately drawn nowhere
In shapes of shifting lineage; let geese
Gabble and hiss, but heroes seek release
From dusty bondage into luminous air.
O blinding hour, O holy, terrible day,
When first the shaft into his vision shone
Of light anatomized! Euclid alone
Has looked on Beauty bare. Fortunate they
Who, though once only and then but far away,
Have heard her massive sandal set on stone.

—Edna St. Vincent Millay (1892–1950)

In the last chapter we saw that the origins of our modern views on geometry can be traced back to the work of Euclid and earlier Greek mathematicians. However, Euclid's axiomatic system is not a *complete* axiomatic system. Euclid makes many assumptions that are never formalized as axioms or theorems. These include assumptions about the existence of points, the unboundedness and continuity of space, and the transformation of figures.

Since the time of Euclid, mathematicians have attempted to "complete" Euclid's system by either supplementing Euclid's axioms with additional axioms or scrapping the whole set and replacing them with an entirely new set of axioms for Euclidean geometry.

Hilbert's *Grundlagen der Geometrie* [21] is the best example of a consistent formal development of geometry in the spirit of Euclid. Hilbert uses six undefined terms (*point*, *line*, *plane*, *on*, *between*, *congruent*) and a set of 21 axioms as a foundation for Euclidean geometry. Hilbert's axioms for plane geometry are included in Appendix D. Some of Euclid's axioms are repeated in Hilbert's axioms and others are provable as theorems. Is Hilbert's system complete? Is it consistent? Hilbert developed a model for his geometry based on the real numbers. Assuming real arithmetic is consistent, then Hilbert's system is consistent as well [21], [pages 29-30]. Hilbert had hoped to show his system *complete*, but as we saw in the last chapter, Gödel's Incompleteness Theorem implies the *incompleteness* of Hilbert's system, as it does any system containing ordinary arithmetic.

A very different axiomatic development of Euclidean geometry was created in 1932 by G. D. Birkhoff [6]. This system departs significantly from Euclid's style of using purely *geometric* constructions in axioms and theorems and assumes at the outset an arithmetic structure of the real numbers. The system has only four axioms and two of these postulate a correspondence between numbers and geometric angles and lengths. From this minimal set of axioms (fewer than Euclid!), Birkhoff develops planar Euclidean geometry, and his development is as logically consistent as Hilbert's. Birkhoff achieves his economy of axioms by assuming the existence of the real numbers and the vast array of properties that entails, whereas Hilbert has to construct the reals from purely geometric principles.

We will not take the time at this point to go through the sequence of proofs and definitions needed to base Euclidean geometry firmly and carefully on Hilbert's axioms or Birkhoff's axioms. Appendix D contains an outline of Hilbert's approach to geometry, while Birkhoff's system is described in Appendix C and also discussed at the end of Chapter 3. We will, however, review some of the basic results from classical Euclidean geometry as a basis for considering some interesting results not often covered in high school geometry. We will take care to provide precise definitions of terms and to point out areas where the work of Hilbert and Birkhoff has filled in the "gaps" of Euclid described above.

2.1 Angles, Lines, and Parallels

The simplest figures in geometry are points, lines, and angles. The terms *point* and *line* will remain undefined, and an *angle* will be defined in terms of *rays*, which are themselves defined in terms of *segments*. We used many of these terms in the last chapter without really giving them a precise definition. To begin our review of Euclidean geometry, it is critical to have a solid set of definitions for these basic terms.

Definition 2.1. A *segment* $\overline{AB}$ consists of points A and B and all those points C on the line through A and B such that C is between A and B.

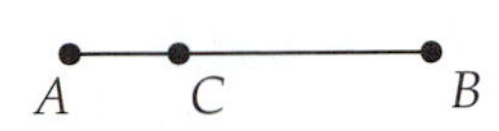

Note that this definition relies on the first of Euclid's postulates, the postulate on the existence of lines. Also, this definition uses an undefined term—*between*. The notion of the ordering of points on a line is a critical one, but one which is very hard to define. Hilbert, in his axiomatic system, leaves the term *between* undefined and establishes a set of axioms on how the quality of *betweenness* works on points. We will assume that *betweenness* works in the way our intuition tells us it should.

Also, note that this definition is not very useful unless we actually have points to work with and can tell when they lie on a line. Here, again, Hilbert introduces a set of axioms about the existence of points and how the property of being *on* a line works, the term *on* being left undefined.

Definition 2.2. A *ray* $\overrightarrow{AB}$ consists of the segment $\overline{AB}$ together with those points C on the line through A and B such that B is between A and C.

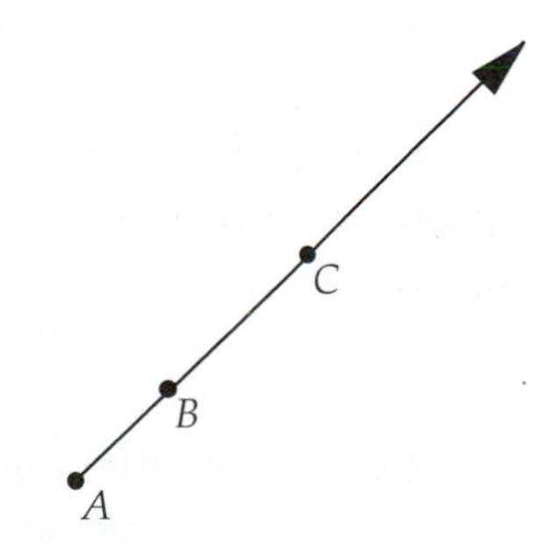

Definition 2.3. The *angle with vertex A* consists of the point A together with two rays $\overrightarrow{AB}$ and $\overrightarrow{AC}$ (the *sides* of the angle). We denote an angle with vertex A and sides $\overrightarrow{AB}$ and $\overrightarrow{AC}$ by $\angle BAC$.

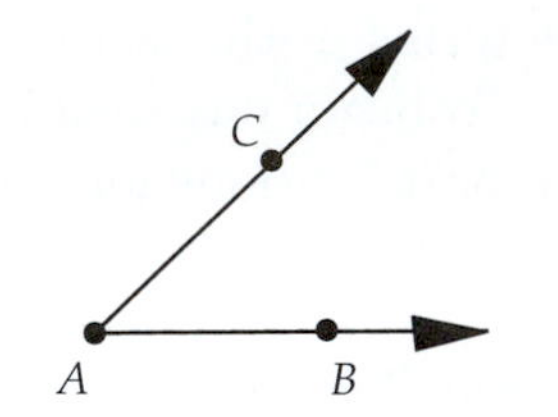

Definition 2.4. Two angles with a common vertex and whose sides form two lines are called *vertical* angles.

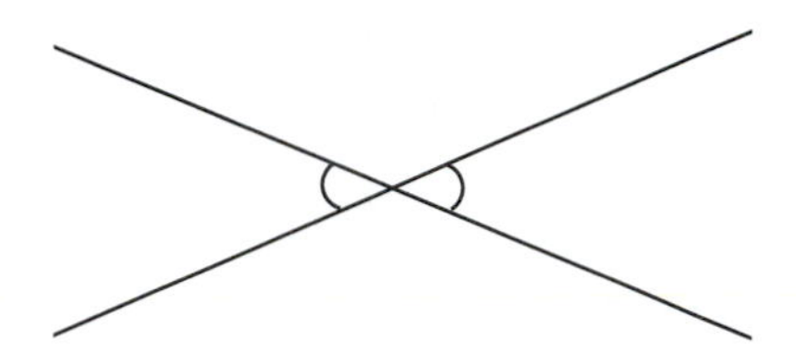

Definition 2.5. Two angles are *supplementary* if they share a common side and the other sides lie in opposite directions on the same line. (Here $\angle BAC$ and $\angle CAD$ are supplementary.)

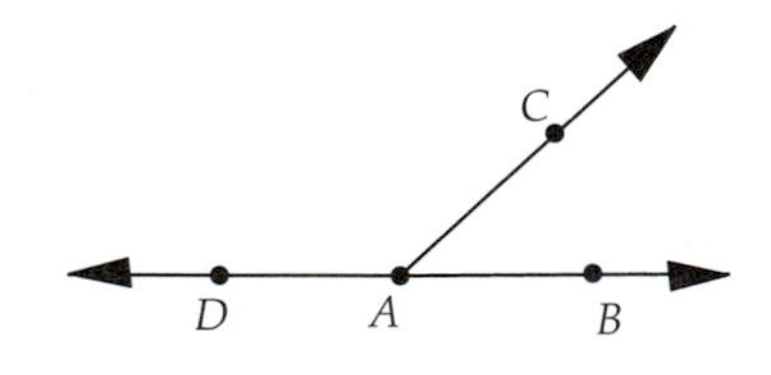

We will often talk of *congruent* angles and segments. Congruence is another of those terms that is difficult to define explicitly. For Euclid, congruence of angles or segments meant the ability to move an angle or segment on top of another angle or segment so that the figures coincided. This transformational geometry pre-supposes an existing set of transformations that would itself need an axiomatic basis. Hilbert leaves the term *congruence* undefined and provides a set of axioms as to how the quality of *congruence* works for angles, as well as for segments.

When speaking of two angles being congruent, we will assume that we have a way of associating an angle with a real number, called the *degree measure* of the angle, in such a way that

- The measure of an angle is a real number between 0 and 180 degrees.

- Congruent angles have the same angle measure.
- The measure of a right angle is 90.
- If two angles are supplementary, then the sum of the measures of the angles is 180.

Definition 2.6. The *degree measure* of $\angle ABC$ will be denoted by $m\angle ABC$.

Similarly, for any segment $\overline{AB}$ we will assume that we have a way of associating a real number, called the *length* of $\overline{AB}$, satisfying the usual notions of Euclidean length. For example, congruent segments will have the same *length*, and if B is between A and C, then the length of $\overline{AC}$ is equal to the sum of the lengths of $\overline{AB}$ and $\overline{BC}$.

Definition 2.7. The *length* of $\overline{AB}$ will be denoted by AB.

In classical Euclidean geometry, the *orientation* of an angle is not significant. Thus, $\angle ABC$ is congruent to $\angle CBA$, and both angles have the same measure, which is less than or equal to 180 degrees. This is distinctly different from the notion of angle commonly found in trigonometry, where orientation is quite significant. Be careful of this distinction between oriented and unoriented angles, particularly when using *Geometry Explorer*, as the program measures angles as *oriented* angles.

Definition 2.8. A *right angle* is an angle that has a supplementary angle to which it is congruent.

Definition 2.9. Two lines that intersect are *perpendicular* if one of the angles made at the intersection is a right angle.

Definition 2.10. Two lines are *parallel* if they do not intersect.

Definition 2.11. A *bisector* of a segment $\overline{AB}$ is a point C on the segment such that $\overline{AC}$ is congruent to $\overline{CB}$. A bisector of an angle $\angle BAC$ is a ray $\overrightarrow{AD}$ such that $\angle BAD$ is congruent to $\angle DAC$.

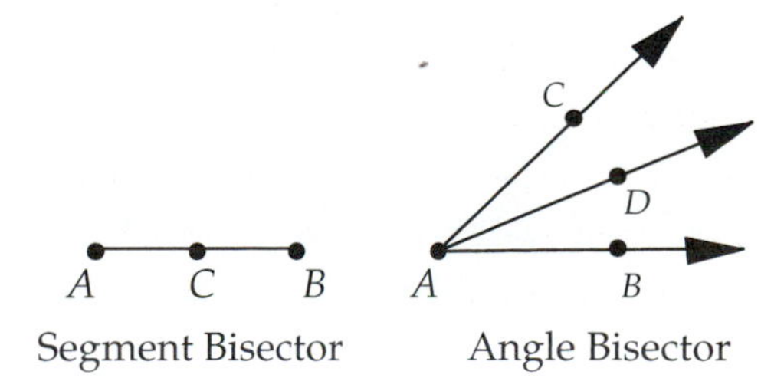

Now that we have a good base of precise definitions, let's review some of the basic results on points, lines, and angles from the first 28 propositions of Book I of Euclid's *Elements*. We will review these propositions without proof. Proofs of these results can be found in [17].

Theorem 2.1. *(Prop. 9, 10 of Book I) Given an angle or segment, the bisector of that angle or segment can be constructed.*

We point out that to *construct* a geometric figure we mean a very specific geometric process, a sequence of steps carried out using only a straightedge and compass. Why just these two types of steps? Clearly, if one does geometry on paper (or sand), the most basic tools would be a straightedge, for drawing segments, and a compass, for drawing circles. Euclid, in his first three axioms, makes the assumption that there are *ideal* versions of these tools that will construct *perfect* segments and circles. Euclid is making an abstraction of the concrete process we carry out to draw geometric figures. By doing so, he can prove, without a shadow of a doubt, that properties such as the apparent equality of vertical angles are *universal* properties and not just an artifact of how we draw figures.

For example, how would we use a straightedge and compass to bisect an angle? The following is a list of steps that will bisect $\angle BAC$ (see Fig. 2.1). The proof is left to the reader as an exercise at the end of the next section.

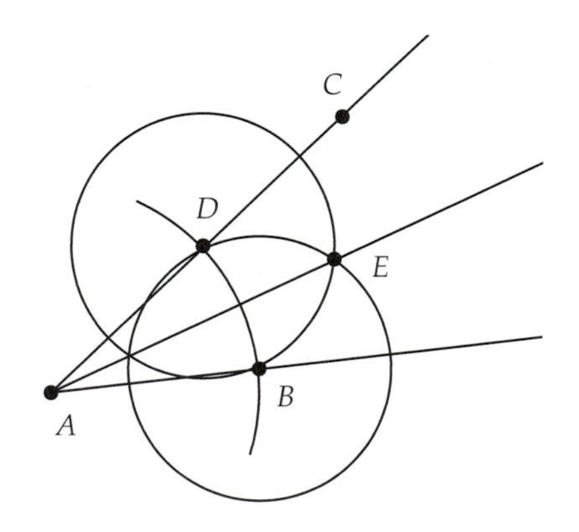

Fig. 2.1 Bisector of an Angle

- At A construct a circle of radius $\overline{AB}$, intersecting $\overrightarrow{AC}$ at D.
- At B construct a circle of radius $\overline{BD}$ and at D construct a circle of radius $\overline{BD}$. These will intersect at E.

- Construct ray $\overrightarrow{AE}$. This will be the bisector of the angle.

A more complete discussion of constructions can be found in Chapter 4. You may wish to review some of the simple constructions in the first section of that chapter before continuing.

Theorem 2.2. *(Prop. 11, 12 of Book I) Given a line and a point either on or off the line, the perpendicular to the line through the point can be constructed.*

Theorem 2.3. *(Prop. 13, 15 of Book I) If two straight lines cross one another they make the vertical angles congruent. Also, the angles formed by one of the lines crossing the other either make two right angles or will have angle measures adding to two right angles. In Fig. 2.2,* $\angle BEC$ *and* $\angle AED$ *are congruent and* $m\angle BEC + m\angle CEA = 180$.

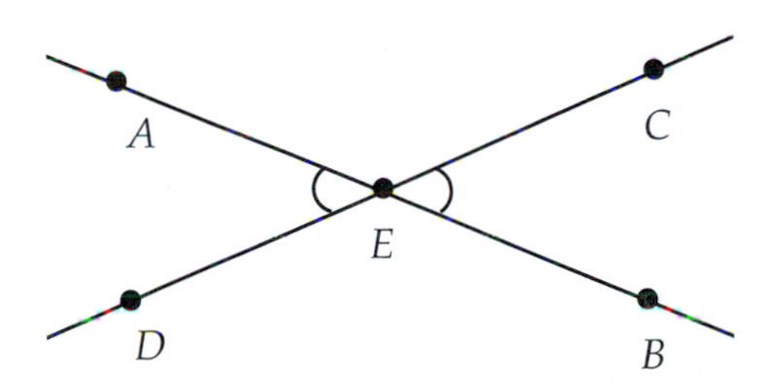

Fig. 2.2 Angles Made by Lines

Theorem 2.4. *(Prop. 14 of Book I) If the measures of two adjacent angles add to two right angles (180 degrees), then the angles form a line.*

Theorem 2.5. *(Prop. 23 of Book I) Given an angle, a line (or part of a line), and a point on the line, an angle whose vertex is at the point, and whose measure is equal to the measure of the given angle, can be constructed.*

For the next set of theorems, we need a way of specifying types of angles created by a line crossing two other lines.

Definition 2.12. Let t be a line crossing lines l and m and meeting l at A and m at A', with $A \neq A'$ (Fig. 2.3). Choose points B and C on either side of A on l and B' and C' on either side of A' on m, with C and C' on the same side of t.

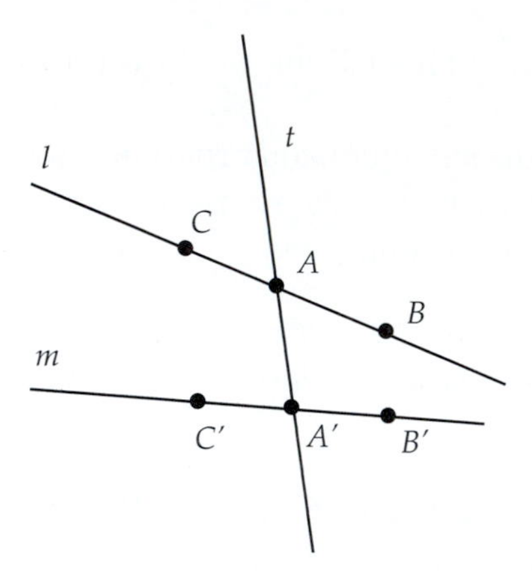

Fig. 2.3 Interior-Exterior Angles

Then $\angle CAA'$, $\angle C'A'A$, $\angle BAA'$, and $\angle B'A'A$ are called *interior* angles (the angles having $\overrightarrow{AA'}$ as a side). Also, $\angle CAA'$ and $\angle B'A'A$ are called *alternate interior angles*, as are $\angle C'A'A$ and $\angle BAA'$. All other angles formed are called *exterior* angles. Pairs of nonadjacent angles, one interior and one exterior, on the same side of the crossing line t are called *corresponding* angles.

Definition 2.13. We will say an angle is *greater* than another angle if its angle measure is greater than the other angle's measure.

Theorem 2.6. *(The Exterior Angle Theorem, Prop. 16 of Book I) Given* ΔABC*, if one of the sides* ($\overline{AC}$) *is extended, then the exterior angle produced* ($\angle DAB$) *is greater than either of the two interior and opposite angles* ($\angle BCA$ *or* $\angle ABC$) *as shown in Fig. 2.4.*

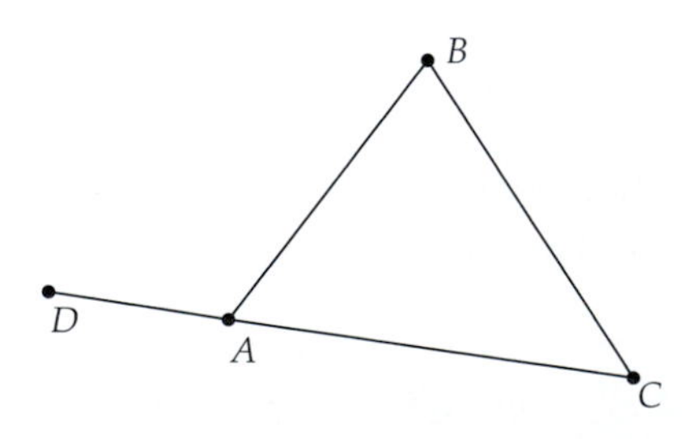

Fig. 2.4

Theorem 2.7. *(Prop. 27 of Book I) If a line n falling on two lines l and m makes the alternate interior angles congruent to one another, then the two lines l and m must be parallel (Fig. 2.5).*

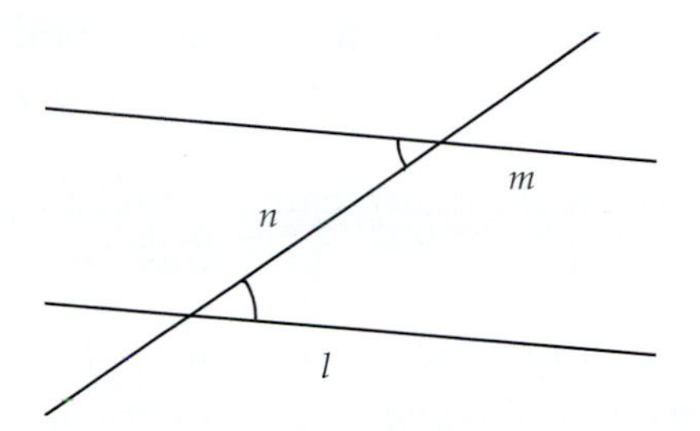

Fig. 2.5 Alternate Interior Angles

Theorem 2.8. *(Prop. 28 of Book I) If a line n falling on two lines l and m makes corresponding angles congruent, or if the sum of the measures of the interior angles on the same side equal two right angles, then l and m are parallel (Fig. 2.6).*

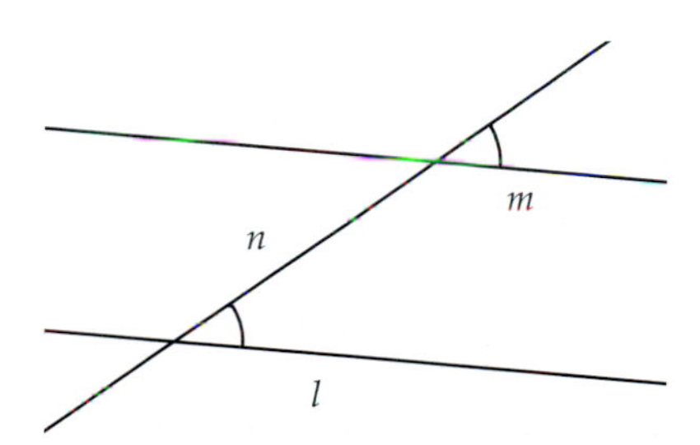

Fig. 2.6 Angles and Parallels

We note here that all of the preceding theorems are *independent* of the fifth Euclidean postulate, the parallel postulate. That is, they can be proved from an axiom set that does not include the fifth postulate. Such results form the basis of what is called *absolute* or *neutral* geometry.

The first time that Euclid actually used the fifth postulate in a proof was for Proposition 29 of Book I. Recall that the fifth postulate states:

> If a straight line falling on two straight lines make the interior angles on the same side less than two right angles, the two straight lines, if produced indefinitely, meet on that side on which are the angles less than the two right angles.

As was noted in the last chapter, many mathematicians have attempted to prove this postulate or replace it with a more palatable alternative. One of these alternatives is Playfair's Postulate, which we restate here:

> Given a line and a point not on the line, it is possible to construct one and only one line through the given point parallel to the line.

Let's see how Playfair's Postulate can be used to prove Proposition 29, which is essentially the converse of Propositions 27 and 28.

prove

* **Theorem 2.9.** *(Prop. 29 of Book I) If a line n falls on two parallel lines l and m, then alternate interior angles are congruent, corresponding angles are congruent, and the sum of the measures of the interior angles on the same side of n is equal to two right angles (see Fig. 2.7).*

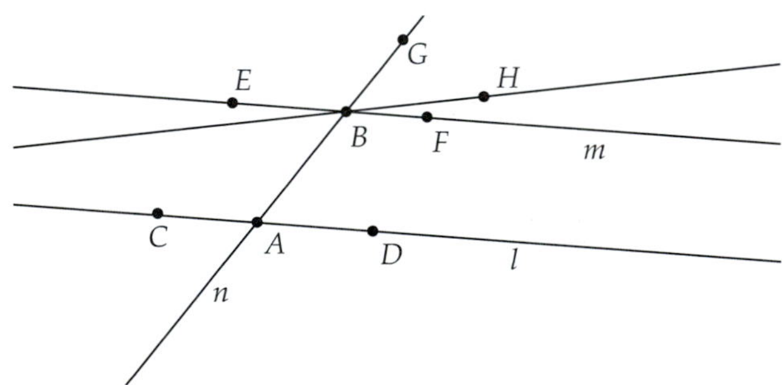

Fig. 2.7 Proposition 29

Proof: Let n be the line through A and B and let the two parallel lines be ones through A, D and B, F. We will prove the statement about the exterior angle and leave the other two parts of the proof as an exercise.

Assume $\angle FBG$ is not congruent to $\angle DAB$. We may assume that $\angle FBG$ is greater than $\angle DAB$. By Theorem 2.5 we can create a new angle $\angle HBG$ on $\overrightarrow{BG}$ so that $\angle HBG$ is congruent to $\angle DAB$. By Theorem 2.8 the line through H and B will be parallel to $\overleftrightarrow{AD}$. But this means that there are two different lines through B that are parallel to $\overleftrightarrow{AD}$, which contradicts Playfair's Postulate. □

In Euclid's original proof of Proposition 29, he bases the proof on the fifth Euclidean postulate, whereas we have used Playfair's Postulate. Actually, these two postulates are *logically equivalent*. Either postulate, when added to Euclid's first four postulates, produces an equivalent axiomatic system. How can we prove this?

The simplest way to show logical equivalence is to prove that each postulate is a *theorem* within the other postulate's axiomatic system. To show this, we need to prove two results: first, that Playfair's Postulate can be derived using the first five Euclidean postulates and second, that Euclid's fifth postulate can be derived using the first four Euclidean postulates plus Playfair's Postulate. These derivations are shown in the exercises.

Exercise 2.1.1. Finish the proof of Theorem 2.9. That is, show that $\angle DAB$ is congruent to $\angle EBA$ and that $m\angle DAB + m\angle ABF = 180$.

Exercise 2.1.2. Use the Exterior Angle Theorem to show that the sum of the measures of two interior angles of a triangle is always less than 180 degrees.

It is critical to have a clear understanding of the terms of an axiomatic system when working on a mathematical proof. One must use *only* the facts given in the definitions and not impose preconceived notions on the terms.

Exercise 2.1.3. Each of the following statements are about a specific term of this section. Determine, solely on the basis of the definition of that term, if the statement is true or false.

(a) A *right angle* is an angle whose measure is 90 degrees.

(b) An *angle* is the set of points lying between two rays that have a common vertex.

(c) An *exterior angle* results from a line crossing two other lines.

(d) A *line* is the union of two opposite rays.

Exercise 2.1.4. In this exercise we will practice defining terms. Be careful to use only previously defined terms, and take care not to use imprecise and colloquial language in your definitions.

(a) Define the term *midpoint* of a segment.

(b) Define the term *perpendicular bisector* of a segment.

(c) Define the term *triangle* defined by three non-collinear points A, B, C.

(d) Define the term *equilateral triangle.*

In the next set of exercises, we consider the logical equivalence of Euclid's fifth postulate with Playfair's Postulate. You may use any of the first 28 Propositions of Euclid (found in Appendix A) and/or any of the results from this section for these exercises.

Exercise 2.1.5. In this exercise you are to show that Euclid's fifth postulate implies Playfair's Postulate. Given a line l and a point A not on l, we can copy $\angle CBA$ to A to construct a parallel line n to l. (Which of Euclid's first 28 Propositions is this based on?) Suppose that there was another line t through A that was not identical to n. Use Euclid's fifth postulate to show that t cannot be parallel to l (Fig. 2.8).

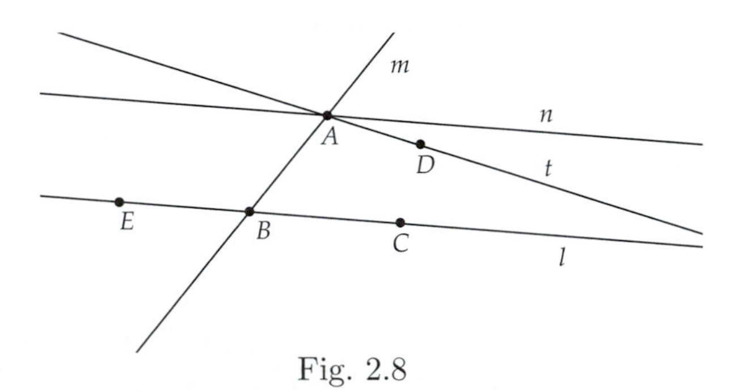

Fig. 2.8

Exercise 2.1.6. Now, we will prove the converse of the preceding exercise, that Playfair's Postulate implies Euclid's fifth postulate. Consider Fig. 2.8. Suppose line m intersects lines t and l such that the measures of angles $\angle CBA$ and $\angle BAD$ add up to less than 180 degrees. Copy $\angle CBA$ to A, creating line n, and then use Playfair's Postulate to argue that lines t and l must intersect. We now need to show that the lines intersect on the same side of m as D and C. We will prove this by contradiction. Assume that t and l intersect on the *other* side of m from point C, say at some point E. Use the exterior angle $\angle CBA$ to triangle ΔABE to produce a contradiction.

Exercise 2.1.7. Show that Playfair's Postulate is equivalent to the statement, Whenever a line is perpendicular to one of two parallel lines, it must be perpendicular to the other.

Exercise 2.1.8. Given triangle ΔABC, construct a parallel to $\overleftrightarrow{BC}$ at A. (How would we do this?) Use this construction to show that Playfair's Postulate (or Euclid's fifth) implies that the angle sum in a triangle is 180 degrees, namely, equal to two right angles. (The converse to this statement is also true: If the angle sum of a triangle is always 180 degrees, then Playfair's Postulate is true. For a proof see [41], [pages 21-23].)

Exercise 2.1.9. Show that Playfair's Postulate is equivalent to the statement, Two lines that are parallel to the same line are coincident (the same) or themselves parallel.

Exercise 2.1.10. Show that Playfair's Postulate is equivalent to the statement, If a line intersects but is not coincident with one of two parallel lines, it must intersect the other.

2.2 Congruent Triangles and Pasch's Axiom

One of the most useful tools in a geometer's toolbox is that of congruence, especially triangle congruence. In this section we will review the basic triangle congruence results found in Propositions 1–28 of Book I of *Elements.*

Definition 2.14. Two triangles are *congruent* if and only if there is some way to match vertices of one to the other such that corresponding sides are congruent and corresponding angles are congruent.

If ΔABC is congruent to ΔXYZ, we shall use the notation $\Delta ABC \cong \Delta XYZ$. We will use the symbol "$\cong$" to denote congruence in general for segments, angles, and triangles. Thus, $\Delta ABC \cong \Delta XYZ$ if and only if

$$\overline{AB} \cong \overline{XY}, \overline{AC} \cong \overline{XZ}, \overline{BC} \cong \overline{YZ}$$

and

$$\angle BAC \cong \angle YXZ, \angle CBA \cong \angle ZYX, \angle ACB \cong \angle XZY$$

Let's review a few triangle congruence theorems.

Theorem 2.10. *(***SAS: Side-Angle-Side***, Prop. 4 of Book I) If in two triangles there is a correspondence such that two sides and the included angle of one triangle are congruent to two sides and the included angle of another triangle, then the triangles are congruent.*

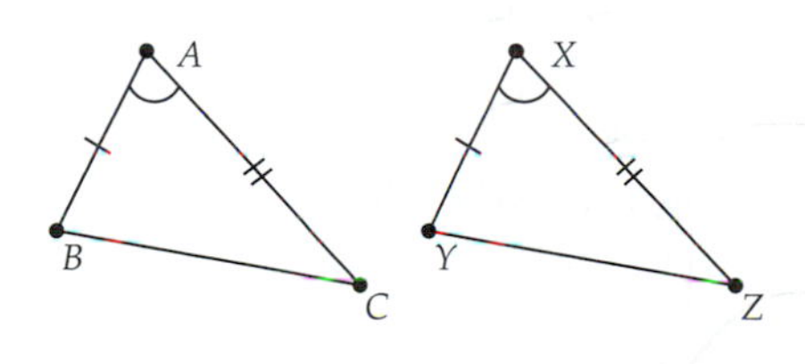

This proposition is one of the *axioms* in Hilbert's axiomatic basis for Euclidean geometry. Hilbert chose to make this result an axiom rather than a theorem to avoid the trap that Euclid fell into in his proof of the SAS result. In Euclid's proof, he *moves* points and segments so as to overlay one triangle on top of the other and thus prove the result. However, there is no axiomatic basis for such transformations in Euclid's original set of five postulates. Most modern treatments of Euclidean geometry assume SAS congruence as an axiom. Birkhoff chooses a slightly different triangle comparison result, the SAS condition for triangles to be *similar*, as an axiom in his development of Euclidean geometry.

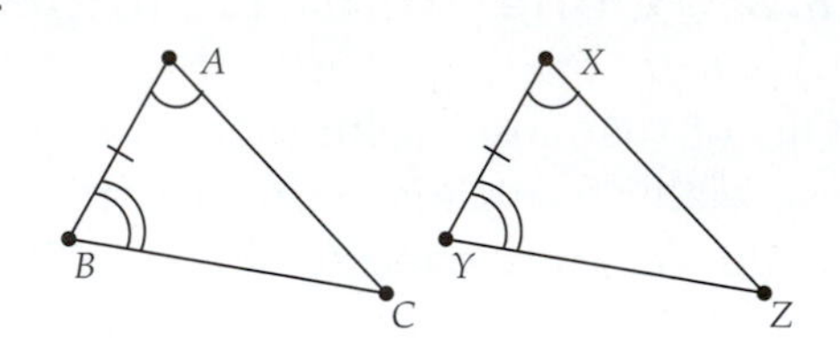

Theorem 2.11. *(***ASA: Angle-Side-Angle***, Prop. 26 of Book I) If in two triangles there is a correspondence in which two angles and the included side of one triangle are congruent to two angles and the included side of another triangle, then the triangles are congruent.*

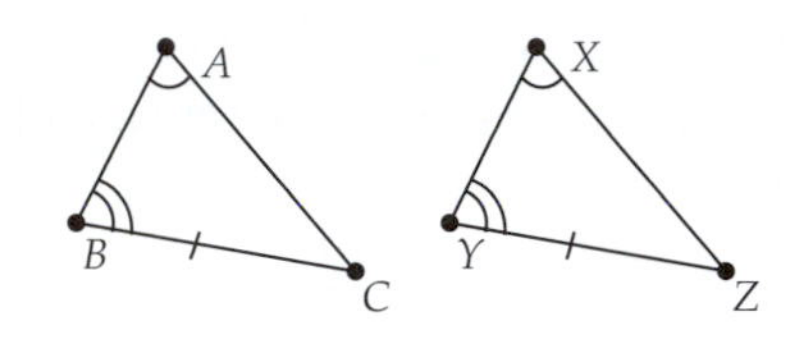

Theorem 2.12. *(***AAS: Angle-Angle-Side***, Prop. 26 of Book I) If in two triangles there is a correspondence in which two angles and the side subtending one of the angles are congruent to two angles and the side subtending the corresponding angle of another triangle, then the triangles are congruent.*

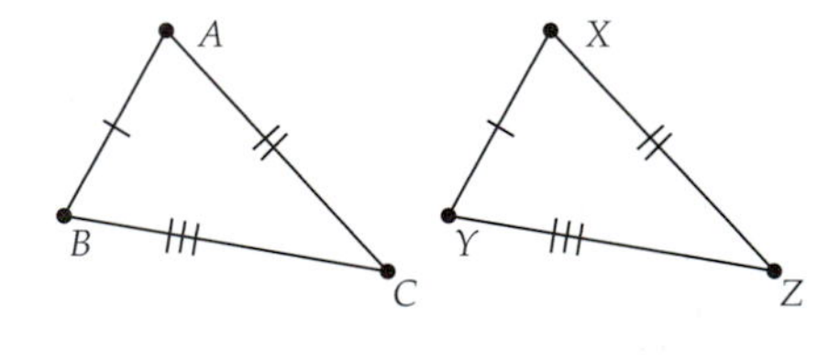

Theorem 2.13. *(***SSS: Side-Side-Side***, Prop. 8 of Book I) If in two triangles there is a correspondence in which the three sides of one triangle are congruent to the three sides of the other triangle, then the triangles are congruent.*

We note here for future reference that the four fundamental triangle congruence results are independent of the parallel postulate; that is, their proofs do not make reference to any result based on the parallel postulate.

Let's see how triangle congruence can be used to analyze isosceles triangles.

Definition 2.15. An *isosceles triangle* is a triangle that has two sides congruent. The two congruent sides are called the *legs* of the triangle and the third side is called the *base*. The *base angles* of the triangle are those angles sharing the base as a side.

Isosceles triangles were a critical tool for many of Euclid's proofs, and he introduced the next result very early in the *Elements*. It followed immediately after SAS congruence (Proposition 4).

Theorem 2.14. *(Prop. 5 of Book I) In an isosceles triangle, the two base angles are congruent.*

Proof: Let triangle ΔABC have sides $\overline{AB}$ and $\overline{BC}$ congruent, as shown in Fig. 2.9. Let $\overrightarrow{BD}$ be the bisector of $\angle ABC$, with D inside the triangle. Let $\overrightarrow{BD}$ intersect $\overline{AC}$ at E. Then, by SAS we have triangles ΔABE and ΔCBE congruent and thus $\angle EAB \cong \angle ECB$ and we're done. □

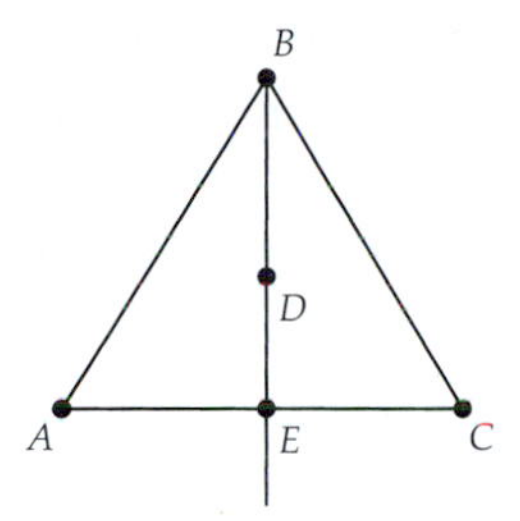

Fig. 2.9 Base Angles Congruence

This proof seems okay—it relies solely on definitions, on the existence of an angle bisector, and on the SAS congruence result. It is the proof one finds in many modern geometry texts, but is not the proof Euclid used. Why not?

If we were true to Euclid's development, we would have a major problem with this proof. First, the construction of angle bisectors comes in Proposition 9 of the *Elements*, which itself depends on Proposition 5, the result we are trying to prove. To use angle bisectors for this proof would be circular reasoning, if we assumed Euclid's development of axioms and theorems. Circular reasoning is an often subtle error that can creep into our attempts to prove a result.

Beyond the problem of circular reasoning, there is an even more fundamental axiomatic problem with this proof, as seen from within Euclid's system. The use of SAS congruence is fine, except for a hidden assumption that has slipped into the proof—that the bisector actually intersects the third side of the triangle. This seems intuitively obvious to us, as we *see* triangles having an *inside* and an *outside*; that is, we see the triangle separating the plane into two regions. In fact, Euclid assumes this *separation* property without proof and does not include it as one of his axioms.

Moritz Pasch (1843–1930) was the first to notice this hidden assumption of Euclid, and in 1882 he published *Vorlesungen uber neure Geometrie*, which for the first time introduced axioms on separation and ordering of points in the plane. One of these axioms specifically addresses the issue of triangle separation and has become known as *Pasch's Axiom*:

> Let A, B, C be three non-collinear points and let l be a line that does not pass through A, B, or C. If l passes through side $\overline{AB}$, it must pass through either a point on $\overline{AC}$ or a point on $\overline{BC}$, but not both.

Hilbert built on Pasch's work in his own axiomatic system. Hilbert's four axioms of "order" are essentially the axioms Pasch used in 1882. Using Pasch's Axiom, the hidden assumption in Euclid's proof of Theorem 2.14 can be justified.

In fact, the proof given above for the isosceles triangle theorem is perfectly correct within Hilbert's axiomatic system, as the triangle intersection property, as well as SAS congruence, are *axioms* in this system, and thus need no proof, and angle bisection properties are proved before this theorem appears.

We see, then, that the same proof can be valid in one system and not valid in another. What makes the difference is how the two systems progressively build up results that then are used for proving later results. Can we give a valid proof in Euclid's system? We know that we can safely use SAS as it is Proposition 4 of Book I. If we consider ΔABC in comparison to *itself* by the ordering ΔABC with ΔCBA, then we get the base angles congruent by a simple application of SAS. (Convince yourself of this.) This may seem like cheating, but if we look carefully at the statement of SAS, it never stipulated that the two triangles for comparison needed to be different.

This concludes our whirlwind review of the highlights of basic planar geometry (the first 28 Propositions of Book I of *Elements*). This basic material concerns points, lines, rays, segments and segment measure, angles and angle measure, parallels, and triangles as covered in many elementary geometry courses. We summarized these areas without taking the time to carefully derive the results from first principles, although we did take care to point out where more foundational work was needed.

Our brief treatment of formal issues is not an indication that careful axiomatic exposition of basic geometry is unimportant. Hilbert's axiomatic system is an outstanding achievement in the foundations of geometry. However, to better appreciate such foundational work, it is advisable to explore the power and complexity of more advanced geometric concepts and techniques, which we do in the rest of this chapter and in the following chapters.

Exercise 2.2.1. Is it possible for a line to intersect all three sides of a triangle? If so, does this contradict Pasch's Axiom?

Exercise 2.2.2. Use Pasch's Axiom to show that a line intersecting one side of a rectangle, at a point other than one of the vertices of the rectangle, must intersect another side of the rectangle. Will the same be true for a pentagon? A hexagon? An n-gon? If so, develop a proof for the case of a line intersecting a side of a regular n-gon (all sides and angles equal).

Exercise 2.2.3. If a line intersects three (or more) sides of a four-sided polygon (any shape made of four connected segments), must it contain a vertex of the polygon? If so, give a proof; if not, give a counter-example.

Let l be a line and A, B two points not on l. If $A = B$, or if segment $\overline{AB}$ contains no points on l, we will say that A and B are on the *same side* of l. Otherwise, they are on *opposite sides* (i.e., $\overline{AB}$ must intersect l). We say that the set of points all on one side of l are in the same *half-plane*.

Use Pasch's Axiom to prove the following pair of results, which Hilbert termed the *Plane Separation Property*. This property is sometimes used to replace Pasch's Axiom in modern axiomatic developments of geometry.

Exercise 2.2.4. For every line l and every triple of points A, B, C not on l, if A, B are on the same side of l and B, C are on the same side of l, then A, C are on the same side of l.

Exercise 2.2.5. For every line l and every triple of points A, B, C not on l, if A, B are on opposite sides of l and B, C are on opposite sides of l, then A, C are on the same side of l.

Definition 2.16. Define a point D to be in the *interior* of an angle $\angle ABC$ if D is on the same side of $\overleftrightarrow{AC}$ as B and also D is on the same side of $\overleftrightarrow{AB}$ as C. (The interior is the intersection of two half-planes.)

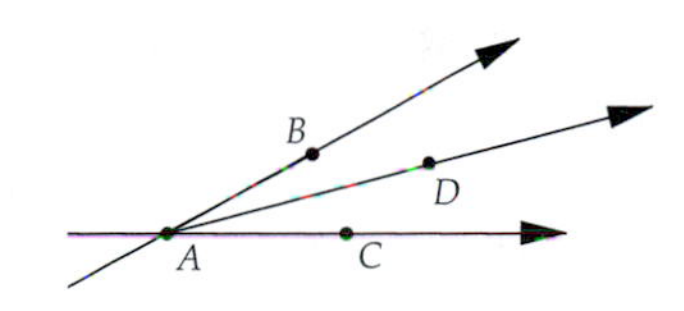

Exercise 2.2.6. Define precisely what is meant by the *interior* of a triangle.

The next set of exercises deal with triangle congruence.

Exercise 2.2.7. Prove the converse to the preceding isosceles triangle result; that is, show that if a triangle has two angles congruent, then it must be isosceles.

Exercise 2.2.8. Use congruent triangles to prove that the angle bisector construction discussed in section 2.1 is valid.

Exercise 2.2.9. Show that if two sides of a triangle are *not* congruent, then the angles opposite those sides are not congruent, and the larger angle is opposite the larger side of the triangle. [Hint: Use isosceles triangles.]

Exercise 2.2.10. Using SAS congruence, prove Angle-Side-Angle congruence.

Exercise 2.2.11. Show that for two right triangles, if the hypotenuse and leg of one are congruent to the hypotenuse and leg of the other, then the two triangles are congruent. [Hint: Try a proof by contradiction.]

Exercise 2.2.12. State and prove a SASAS (Side-Angle-Side-Angle-Side) congruence result for quadrilaterals.

2.3 Project 3 - Special Points of a Triangle

Triangles are the simplest two-dimensional shapes that we can construct with segments. The simplicity of their construction masks the richness of relationships exhibited by triangles. In this project we will explore several interesting points of intersection that can be found in triangles.

2.3.1 Circumcenter

Start *Geometry Explorer* and use connected segments to create ΔABC on the screen.

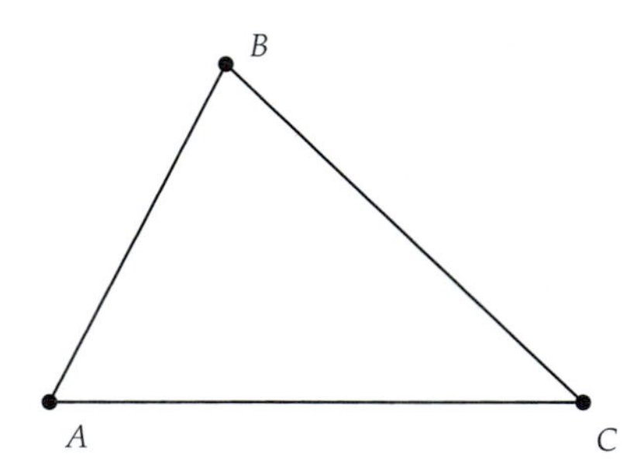

Multi-select all three sides (hold down the Shift key to do a multiple selection) and construct the midpoints D, E, F of the sides (click the Midpoint button in the Construct panel).

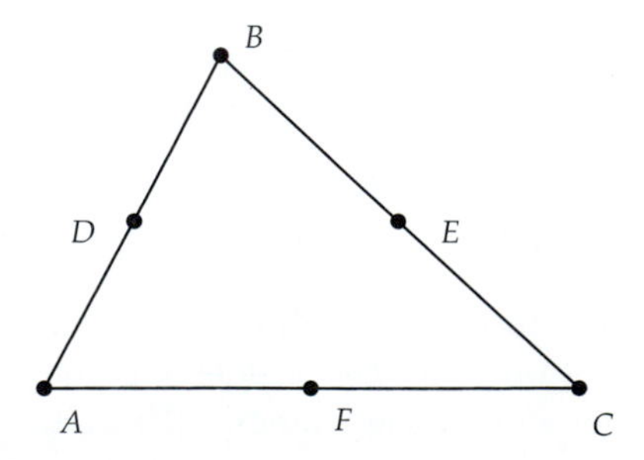

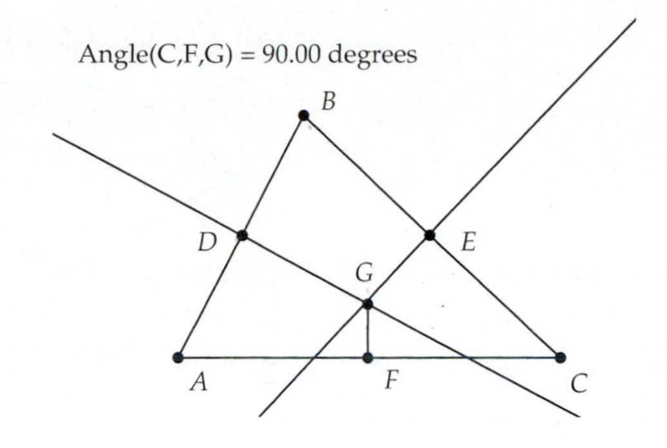

Now, construct the perpendicular bisectors of $\overline{AB}$ and $\overline{BC}$ by multi-selecting a side and the midpoint of that side and clicking on the Perpendicular tool. After constructing these two perpendiculars, multi-select the perpendiculars and click on the Intersect tool to construct the intersection point G. Create a segment from F to G and measure $\angle CFG$. (Multi-select C, F, G, in that order, and choose **Angle** (**Measure** menu). It appears that $\overleftrightarrow{GF}$ is perpendicular to $\overleftrightarrow{AC}$ and thus all three perpendicular bisectors meet at G. Drag the vertices of the triangle around. Does the common intersection property persist? This suggests a theorem.

Theorem 2.15. *The perpendicular bisectors of the sides of a triangle intersect at a common point called the* circumcenter *of the triangle.*

Let's see if we can prove this.

Consider the two perpendicular bisectors we constructed, the ones at D and E. Since $\overleftrightarrow{AB}$ and $\overleftrightarrow{BC}$ are not parallel, we can find their intersection at point G.

Exercise 2.3.1. Show that triangles ΔDGB and ΔDGA are congruent, as are triangles ΔEGB and ΔEGC. Use this to show that the line through F and G will be a perpendicular bisector for the third side, and thus all three perpendicular bisectors intersect at point G. (Make sure your labels match the ones shown above.)

Create a circle with center at the circumcenter (point G above) and radius point equal to one of the vertices of the triangle. (Be careful that you drag the radius point onto one of the triangle vertices, thus *attaching* the circle's radius point to a triangle vertex.) What do you notice about the circle in relation to the other vertices of the triangle?

Corollary 2.16. *The circle with center at the circumcenter of a triangle and radius out to one of the vertices will pass through the other vertices of the triangle. This is called the* circumscribed *circle of the triangle (Fig. 2.10).*

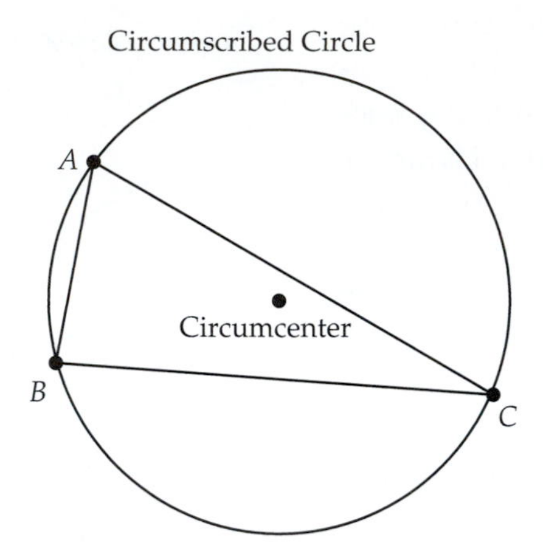

Fig. 2.10 Circumscribed Circle of a Triangle

Exercise 2.3.2. Prove Corollary 2.16.

2.3.2 Orthocenter

An *altitude* of a triangle will be a perpendicular to a side of the triangle that passes through the opposite vertex. In Fig. 2.11, line a is an altitude to ΔABC through vertex B.

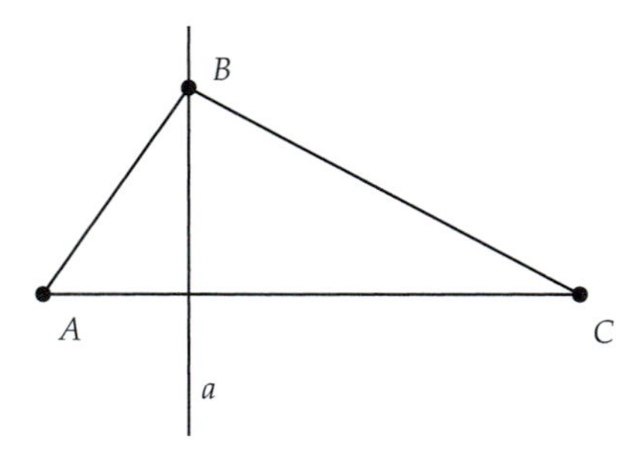

Fig. 2.11 Altitude

What can we say about the three altitudes of a given triangle? Do they intersect at a common point?

We will show that the altitudes of a triangle do intersect at a common point by showing that the altitudes are also the perpendicular bisectors of an associated triangle to the given triangle.

Clear the screen and create a triangle ΔABC. Multi-select a side and the vertex opposite and click on the Parallel tool. Repeat this construction for the other two sides to construct the three parallels to ΔABC as shown in the figure. Since the original sides of ΔABC were not parallel, each pair of the new parallels will intersect. These three intersection points will form a new triangle, ΔDEF. We will call this the *associated* triangle to ΔABC.

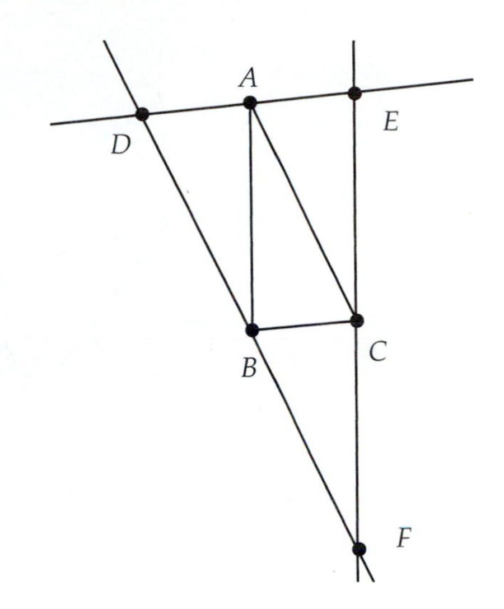

Drag the vertices around and notice the relationship among the four triangles ΔBAD, ΔABC, ΔFCB, and ΔCEA. What would your conjecture be about the relationship among these four triangles?

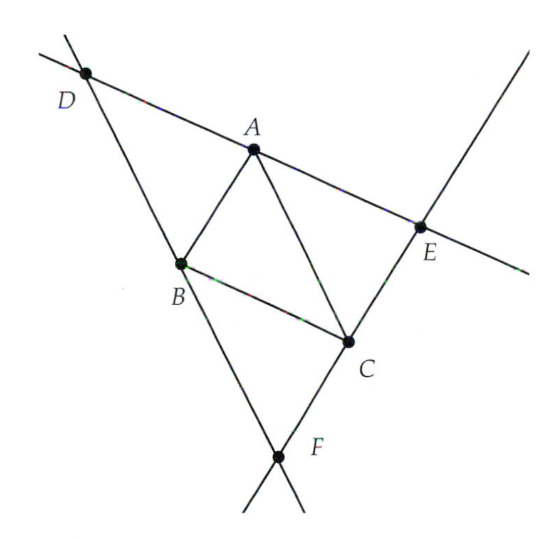

Exercise 2.3.3. Use Theorem 2.9 to verify the angle congruences illustrated in Fig. 2.12 for ΔABC and its associated triangle ΔDEF. Then, show that all four sub-triangles are congruent. That is,

$$\Delta ABC \cong \Delta BAD \cong \Delta FCB \cong \Delta CEA$$

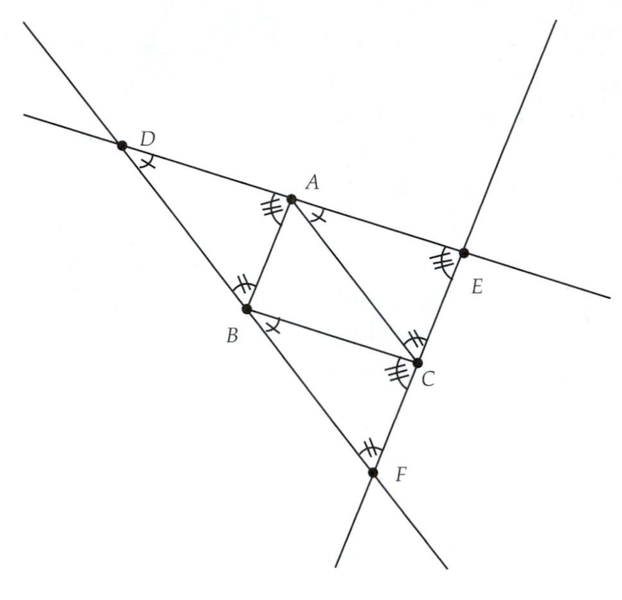

Fig. 2.12

This leads directly to the following result:

Theorem 2.17. *The altitudes of a triangle are the perpendicular bisectors of the associated triangle as we have described.*

Exercise 2.3.4. Prove Theorem 2.17. [Hint: Use the preceding exercise.]

Since the altitudes of a triangle are the perpendicular bisectors of the associated triangle, and since we have already shown that the three perpendicular bisectors of any triangle have a common intersection point, then we have the following.

Corollary 2.18. *The altitudes of a triangle intersect at a common point. This point is called the* orthocenter *of the triangle.*

Corollary 2.19. *The orthocenter of a triangle is the circumcenter of the associated triangle.*

Does the orthocenter have a nice circle like the circumscribed circle? Create a circle with center at the orthocenter and radius point equal to one of the vertices of triangle ΔABC. What do you notice that is different about this situation in comparison to the preceding circumscribed circle?

2.3.3 Incenter

Clear the screen and create ΔABC one more time. In this part of our project, we will find the angle bisectors of each of the three interior angles of the triangle. For example, to bisect $\angle CBA$ multi-select points C, B, A (in that order!) and click on the Angle Bisector construction tool. Note that angles are always *oriented* in *Geometry Explorer*, so you need to select the points in the correct order to define an angle. The selection order for the angles in the triangle shown would be C, B, A followed by B, A, C followed by A, C, B.

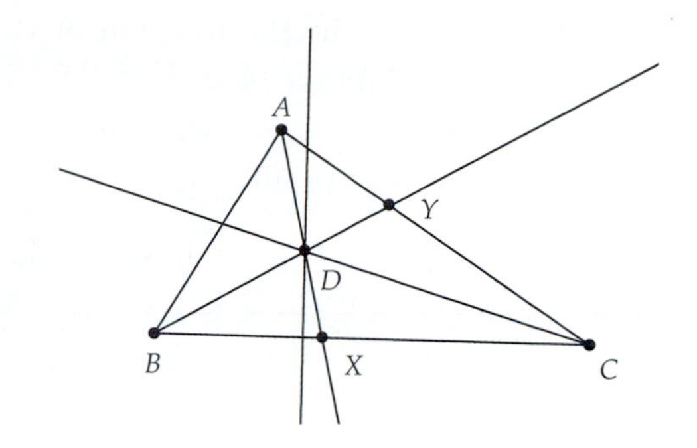

It appears that the three ray bisectors have a common intersection point, point D in the figure at right.

Theorem 2.20. *The angle bisectors of triangle ΔABC intersect at a common point. This point is called the* incenter *of the triangle.*

Proof: First note that the bisector at A will intersect side $\overline{BC}$ at some point X. The bisector at B will intersect $\overline{AC}$ at some point Y and $\overline{AX}$ at some point D that is interior to the triangle.

In the next exercise, we will show that the points on an angle bisector are equidistant from the sides of an angle and, conversely, if a point is interior to an angle and equidistant from the sides of an angle, then it is on the bisector. Thus, the distance from D to $\overleftrightarrow{AB}$ equals the distance from D to $\overleftrightarrow{AC}$. Also, the distance from D to $\overleftrightarrow{AB}$ equals the distance from D to $\overleftrightarrow{BC}$, as D is on the bisector at B. Thus, the distance from D to $\overleftrightarrow{AC}$ equals the distance from D to $\overleftrightarrow{BC}$, and D must be on the bisector at C. □

Exercise 2.3.5. Use congruent triangles to show that the points on an angle bisector are equidistant from the sides of the angle and, conversely, if a point is interior to an angle and equidistant from the sides of an angle, then it is on the bisector. [Hint: Distance from a point to a line is measured by dropping a perpendicular from the point to the line. You may also use the Pythagorean Theorem if you wish.]

Drop a perpendicular line from the incenter to one of the sides of the triangle and find the intersection of this perpendicular with the side. (For example, we get point Z in Fig. 2.13 by dropping a perpendicular from D to $\overleftrightarrow{BC}$.) Create a circle with center at the incenter and radius point equal to the intersection point.

Definition 2.17. The circle just constructed is called the *inscribed* circle to the triangle. A circle is inscribed in a triangle (or other polygon) if it intersects each side of the triangle (polygon) in a single point.

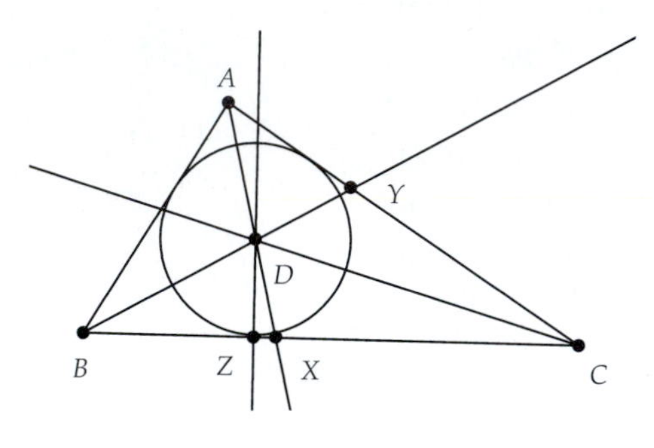

Fig. 2.13

Project Report

It is amazing what structure exists in a simple shape like a triangle! In your report discuss the similarities and differences you discovered in exploring these three special points of a triangle. Also, take care that all of your work on proofs and exercises is written in a logical and careful fashion.

2.4 Measurement and Area in Euclidean Geometry

Euclid, in his development of triangle congruence, had a very restrictive notion of what congruence meant for segments and angles. When we hear the word *congruence* our modern mathematical background prompts us to immediately think of *numbers* being equal. For example, we immediately associate a number with a segment and consider two segments congruent if their numerical lengths are equal.

This is very different from the way that Euclid viewed congruence. In fact, Euclid's notion of congruence was one of *coincidence* of figures. That is, two segments were congruent if one could be made to be coincident with the other, by movement of one onto the other. Euclid never used segments as substitutes for numerical values, or vice versa. If one looks at Book II of Euclid, where he develops what we would now call algebraic relationships, we see that these relationships are always based on geometric figures. When Euclid mentions the *square* on segment $\overline{AB}$, he literally means the geometric square constructed on $\overline{AB}$ and not the product of the length of $\overline{AB}$ with itself.

Modern axiomatic treatments of Euclidean geometry resolve this problem of dealing with numerical measure by *axiomatically* stipulating that there is a way to associate a real number with a segment or angle. Hilbert, staying true to Euclid, does not explicitly tie numbers to geometric figures, but instead provides two axioms of continuity that allow for the *development* of measurement through an ingenious but rather tedious set of theorems. Birkhoff, on the other hand, directly specifies a connection between real numbers and lengths in his first axiom and between numbers and angles in his third axiom (see Appendix C).

In the following project, we will continue to assume the standard connection between geometric figures and real numbers via lengths and angle degree measures. This connection will lead naturally to a definition of area based on segment length.

2.4.1 Mini-Project - Area in Euclidean Geometry

In this project we will create a definition for area, based on properties of four-sided figures. Whereas earlier projects utilized computer software, for this project we will primarily be using human "grayware," that is, the canvas of our minds.

Definition 2.18. A *quadrilateral* $ABCD$ is a figure comprised of segments $\overline{AB}$, $\overline{BC}$, $\overline{CD}$, and $\overline{DA}$ such that no three of the points of the quadrilateral are collinear and no pair of segments intersects, except at the endpoints. We call $\overline{AB}$, $\overline{BC}$, $\overline{CD}$, and $\overline{DA}$ the *sides* of the quadrilateral and $\overline{AC}$, $\overline{BD}$ the *diagonals* of the quadrilateral.

Definition 2.19. A *parallelogram* is a quadrilateral $ABCD$, where $\overleftrightarrow{AB}$ is parallel to $\overleftrightarrow{CD}$ and $\overleftrightarrow{BC}$ is parallel to $\overleftrightarrow{AD}$. That is, opposite sides are parallel.

Here's the first exercise to work through. The tools that will prove most helpful to you include the theorems covered so far on parallels and triangle congruence.

Exercise 2.4.1. Show that in a parallelogram opposite sides are *congruent.*

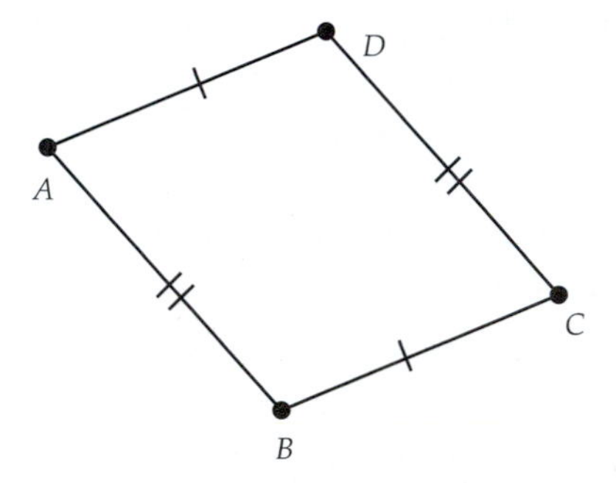

Exercise 2.4.2. (Prop. 34 of Book I) Show that in parallelogram $ABCD$, opposite angles are congruent and the diagonal bisects the parallelogram. (That is, the diagonal divides the parallelogram into two congruent triangles.)

Definition 2.20. We will call two geometric figures *equivalent* if each figure can be split into a finite number of polygonal pieces so that all pieces can be separated into congruent pairs.

Exercise 2.4.3. Why would it be acceptable to replace "polygonal" with "triangular" in Definition 2.20?

Exercise 2.4.4. (Prop. 35 of Book I) In this exercise we will show that two parallelograms having the same base and defined between the same parallels are *equivalent.*

To do this, let parallelograms $ABCD$ and $EBCF$ be two parallelograms sharing base $\overline{BC}$ with $\overleftrightarrow{AF}$ parallel to $\overleftrightarrow{BC}$. There are two possible configurations for the parallelograms: either D will be between E and F as shown in the top configuration, or D will not be on segment $\overline{EF}$ as shown in the second configuration. In either case use SAS to show that triangles ΔEAB and ΔFDC are congruent. Then, argue by adding and subtracting congruent figures that the two parallelograms are equivalent.

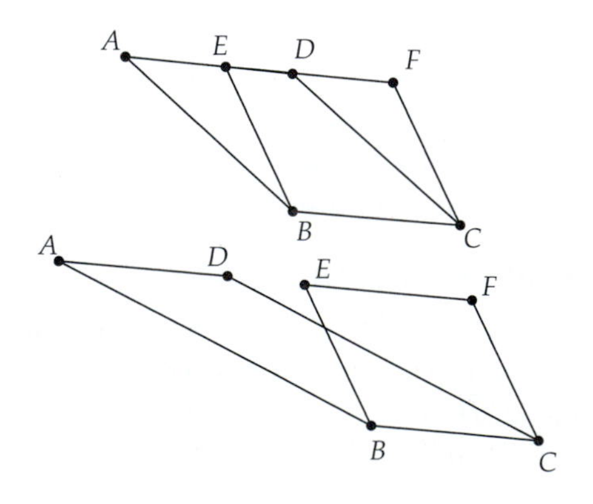

We now are in a position to define area in terms of rectangles!

Definition 2.21. A *rectangle* is a quadrilateral in which all the angles defined by the vertices are right angles. Given a rectangle $ABCD$, we can identify two adjacent sides and call the length of one of the sides a *base* of the rectangle and the length of the other a *height*. The *area* of the rectangle will be defined as the product of a base and height.

Exercise 2.4.5. Show that this definition of rectangular area is not ambiguous. That is, it does not depend on how we choose the sides for the base and height. [Be careful—this is not as trivial as it may appear. You need to base your argument only on known geometric theorems and not on analytic or numerical reasoning.]

Definition 2.22. If two figures can be made equivalent, we will say that they have the same area.

Definition 2.23. If we identify a side of a parallelogram to be a *base side*, then the *height* of the parallelogram, relative to that base side, is the perpendicular distance between the base side and the opposite side.

Exercise 2.4.6. Show that the area of a parallelogram $ABCD$ is the product of its base length and its height by showing the parallelogram is equivalent to a rectangle with those side lengths (Fig. 2.14). [Hint: Given parallelogram $ABCD$, with base AB, drop perpendiculars from A to $\overleftrightarrow{CD}$ at E and from B to $\overleftrightarrow{CD}$ at F. Show that $ABFE$ is a rectangle with the desired properties.]

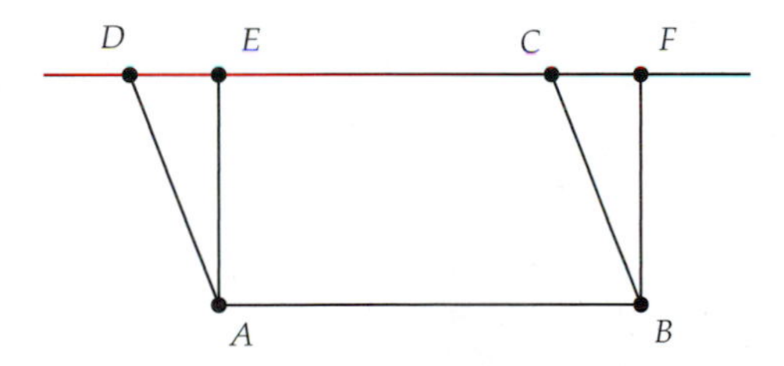

Fig. 2.14

The next set of theorems follows immediately from your work.

Theorem 2.21. *The area of a right triangle is one-half the product of the lengths of its legs, the legs being the sides of the right angle.*

Theorem 2.22. *(Essentially Prop. 41 of Book I) The area of any triangle is one-half the product of a base of the triangle with the height of the triangle.*

Project Report

In this project we have developed the notion of area from simple figures such as parallelograms and rectangles. From these simple beginnings, we could expand our idea of area to arbitrary polygonal figures, and from there find areas of curved shapes by using approximating rectilinear areas and limits, as is done in calculus when defining the integral.

Provide a summary of the proofs for each of the exercises in this project. If you finish early, discuss what aspects of area are still not totally defined. That is, were there hidden assumptions made in our proofs that would necessitate area *axioms*?

2.4.2 Cevians and Areas

We will now apply our results on areas to a very elegant development of the intersection properties of cevians. A full account of the theory of cevians can be found in the monograph *Geometry Revisited*, by Coxeter and Greitzer [11].

Definition 2.24. Given a triangle, a *cevian* is a segment from a vertex to a point on the opposite side.

Theorem 2.23. *Given ΔABC, and cevians $\overline{AX}$, $\overline{BY}$, and $\overline{CZ}$ (refer to Fig. 2.15), if the three cevians have a common intersection point P, then*

$$\frac{BX}{XC}\frac{CY}{YA}\frac{AZ}{ZB} = 1$$

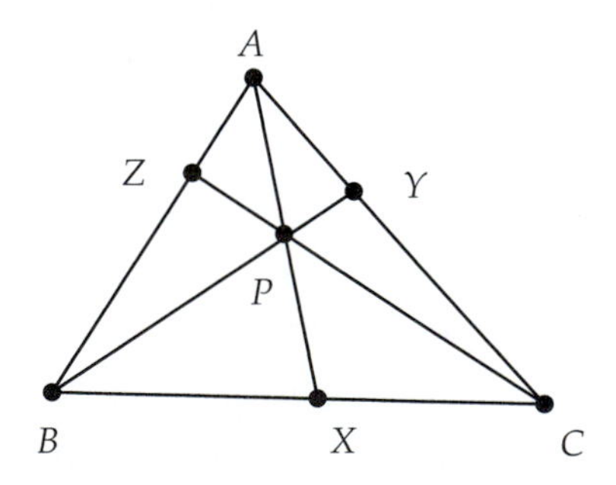

Fig. 2.15

Proof: Since two triangles with the same altitudes have areas that are proportional to the bases, then $\frac{area(\Delta ABX)}{area(\Delta AXC)} = \frac{BX}{XC}$, and $\frac{area(\Delta PBX)}{area(\Delta PXC)} = \frac{BX}{XC}$.

A little algebra shows

$$\frac{area(\Delta ABX) - area(\Delta PBX)}{area(\Delta AXC) - area(\Delta PXC)} = \frac{area(\Delta ABP)}{area(\Delta APC)} = \frac{BX}{XC}$$

Similarly, $\frac{area(\Delta BCP)}{area(\Delta APB)} = \frac{CY}{YA}$, and $\frac{area(\Delta CAP)}{area(\Delta BPC)} = \frac{AZ}{ZB}$.

Thus,

$$\frac{BX}{XC}\frac{CY}{YA}\frac{AZ}{ZB} = \frac{area(\Delta ABP)}{area(\Delta APC)}\frac{area(\Delta BCP)}{area(\Delta ABP)}\frac{area(\Delta APC)}{area(\Delta BCP)}$$

Cancellation of terms yields the desired result. □

This theorem has become known as *Ceva's Theorem*, in honor of Giovanni Ceva (1647–1734), an Italian geometer and engineer. The term *cevian* is derived from Ceva's name, in recognition of the significance of his theorem in the history of Euclidean geometry.

The converse to Ceva's theorem is also true.

Theorem 2.24. *If the three cevians satisfy*

$$\frac{BX}{XC}\frac{CY}{YA}\frac{AZ}{ZB} = 1$$

then they intersect at a common point (refer to Fig. 2.16).

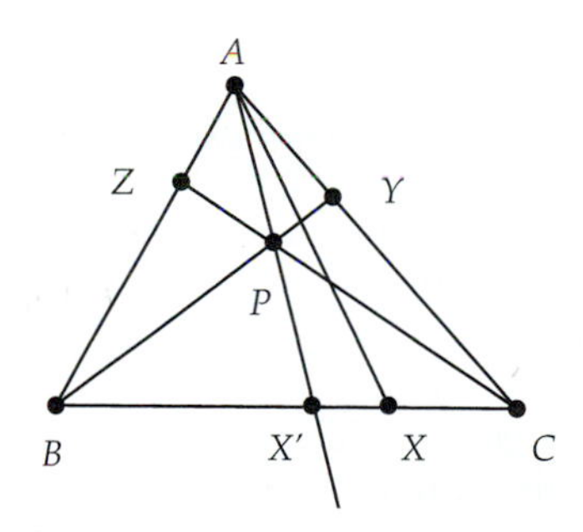

Fig. 2.16

Proof: $\overline{BY}$ will intersect $\overline{CZ}$ at some point P which is interior to the triangle. The ray $\overrightarrow{AP}$ will intersect $\overline{BC}$ at some point X' that is between B and C. By the previous theorem we have

$$\frac{BX'}{X'C}\frac{CY}{YA}\frac{AZ}{ZB} = 1$$

Thus, $\frac{BX'}{X'C} = \frac{BX}{XC}$.

Suppose that X' is between B and X. Then

$$\frac{BX}{XC} = \frac{BX' + X'X}{X'C - X'X}$$

Since $\frac{BX'}{X'C} = \frac{BX}{XC}$, we have

$$\frac{BX'}{X'C} = \frac{BX' + X'X}{X'C - X'X}$$

Cross-multiplying in this last equation gives

$$(BX')(X'C - X'X) = (X'C)(BX' + X'X)$$

Simplifying we get

$$-(BX')(X'X) = (X'C)(X'X)$$

Since X' is between B and C, the only way that this equation can be satisfied is for $X'X$ to be zero. Then $X' = X$.

A similar result would be obtained if we assumed that X' is between X and C. □

Definition 2.25. A *median* of a triangle is a segment from the midpoint of a side to the opposite vertex.

Exercise 2.4.7. Show that the medians of a triangle intersect at a common point. This point is called the *centroid* of the triangle.

Exercise 2.4.8. Use an area argument to show that the centroid must be the balance point for the triangle. [Hint: Consider each median as a "knife edge" and argue using balancing of areas.]

Exercise 2.4.9. The medians of a triangle split the triangle into six sub-triangles. Show that all six have the same area.

Exercise 2.4.10. Use Exercise 2.4.9 and an area argument to show that the medians divide one another in a 2:1 ratio.

2.5 Similar Triangles

One of the most useful tools for a more advanced study of geometric properties (e.g., the geometry used in surveying) is that of *similar* triangles.

Definition 2.26. Two triangles are *similar* if and only if there is some way to match vertices of one to the other such that corresponding sides are in the same ratio and corresponding angles are congruent.

If ΔABC is similar to ΔXYZ, we shall use the notation $\Delta ABC \sim \Delta XYZ$. Thus, $\Delta ABC \sim \Delta XYZ$ if and only if

$$\frac{AB}{XY} = \frac{AC}{XZ} = \frac{BC}{YZ}$$

and

$$\angle BAC \cong \angle YXZ, \angle CBA \cong \angle ZYX, \angle ACB \cong \angle XZY$$

Similar figures have the same *shape* but are of differing (or the same) *size*. One is just a scaled-up version of the other. The scaling factor is the constant of proportionality between corresponding lengths of the two figures. Thus, similarity has to do with the equality of ratios of corresponding measurements of two figures.

Let's review some basic similarity and proportionality results.

Theorem 2.25. *Suppose that a line is parallel to one side of a triangle and intersects the other two sides at two different points. Then, this line divides the intersected sides into proportional segments (refer to Fig. 2.17).*

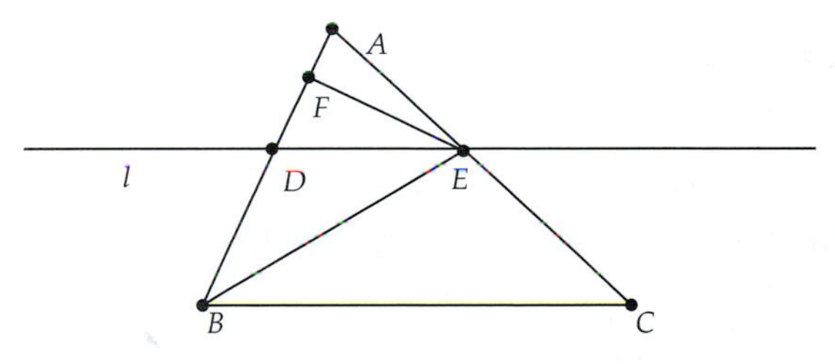

Fig. 2.17

Proof: Let line l be parallel to $\overleftrightarrow{BC}$ in ΔABC. Let l intersect sides $\overline{AB}$ and $\overline{AC}$ at points D and E. Construct a perpendicular from point E to $\overleftrightarrow{AB}$ intersecting at point F. Consider the ratio of the areas of triangles ΔBED and ΔAED:

$$\frac{Area(\Delta BED)}{Area(\Delta AED)} = \frac{\frac{1}{2}(BD)(EF)}{\frac{1}{2}(AD)(EF)} = \frac{BD}{AD}$$

Now, consider dropping a perpendicular from D to $\overleftrightarrow{AC}$, intersecting this line at point G (Fig. 2.18).

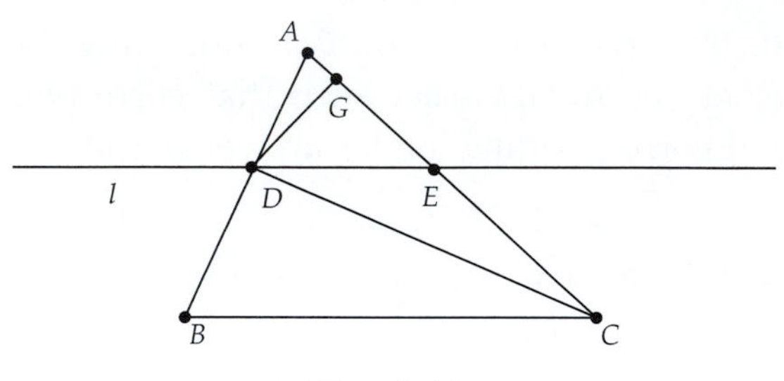

Fig. 2.18

Then,

$$\frac{Area(\Delta CED)}{Area(\Delta AED)} = \frac{\frac{1}{2}(CE)(DG)}{\frac{1}{2}(AE)(DG)} = \frac{CE}{AE}$$

Now, triangles BED and CED share base $\overline{DE}$ and so the heights of these two triangles will be the same, as both heights are the lengths of perpendiculars dropped to $\overleftrightarrow{DE}$ from points on another line parallel to $\overleftrightarrow{DE}$. Thus, these two triangles have the same area and

$$\frac{Area(\Delta BED)}{Area(\Delta AED)} = \frac{Area(\Delta CED)}{Area(\Delta AED)}$$

Thus,

$$\frac{BD}{AD} = \frac{CE}{AE}$$

This completes the proof. □

Corollary 2.26. *Given the assumptions of Theorem 2.25, it follows that*

$$\frac{AB}{AD} = \frac{AC}{AE}$$

Proof: Since $AB = AD + BD$ and $AC = AE + CE$, we have

$$\frac{AB}{AD} = \frac{AD + BD}{AD} = 1 + \frac{BD}{AD} = 1 + \frac{CE}{AE} = \frac{AE + CE}{AE} = \frac{AC}{AE}$$

□

The converse to this result is also true.

Theorem 2.27. *If a line intersects two sides of a triangle so that the segments cut off by the line are proportional to the original sides of the triangle, then the line must be parallel to the third side.*

Proof: Let line l pass through ΔABC at sides $\overline{AB}$ and $\overline{AC}$, intersecting at points D and E (Fig. 2.19).

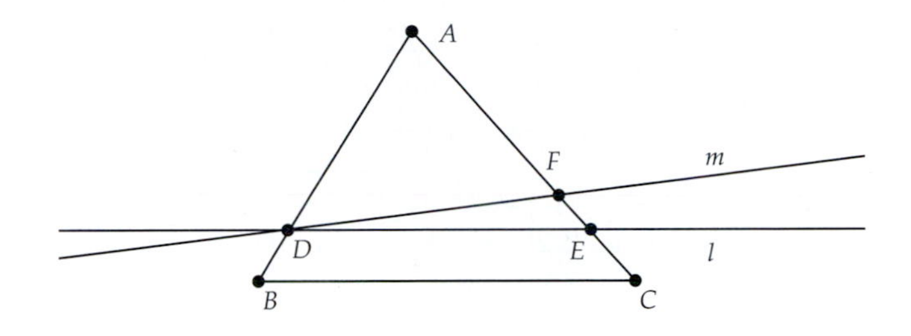

Fig. 2.19

It is given that

$$\frac{AB}{AD} = \frac{AC}{AE}$$

By Playfair's Axiom there is a unique parallel m to $\overleftrightarrow{BC}$ through D. Since m is parallel to $\overleftrightarrow{BC}$ and it intersects side $\overline{AB}$, then it must intersect side $\overline{AC}$ at some point F. But, Corollary 2.26 then implies that

$$\frac{AB}{AD} = \frac{AC}{AF}$$

Thus,

$$\frac{AC}{AE} = \frac{AC}{AF}$$

and $AE = AF$. This implies that points E and F are the same and lines l and m are the same. So, l is parallel to $\overleftrightarrow{BC}$. $\square$

We now have enough tools at our disposal to prove the following similarity condition.

Theorem 2.28. *(AAA Similarity Condition) If in two triangles there is a correspondence in which the three angles of one triangle are congruent to the three angles of the other triangle then the triangles are similar (Fig. 2.20).*

Proof: Let ΔABC and ΔDEF be two triangles with the angles at A, B, and C congruent to the angles at D, E, and F, respectively. If segments $\overline{AB}$ and $\overline{DE}$ are congruent, then the two triangles are congruent by the AAS congruence theorem and thus are also similar.

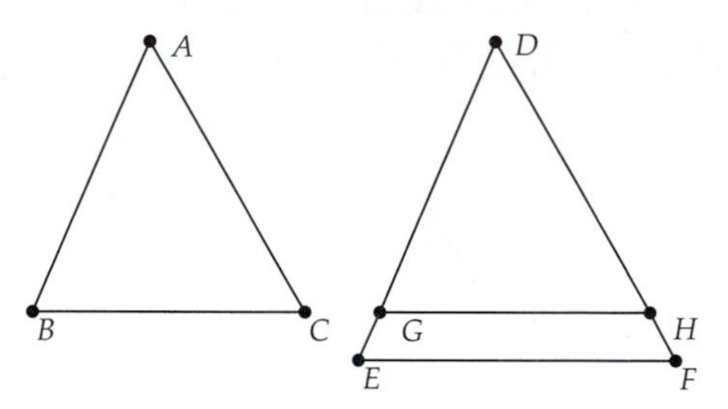

Fig. 2.20 AAA Similarity

Suppose that $\overline{AB}$ and $\overline{DE}$ are not congruent. We can assume that one is larger than the other, say $\overline{DE}$ is larger than $\overline{AB}$. Then, there is a point G between D and E such that $\overline{DG} \cong \overline{AB}$. Also, there must be a point H along the ray $\overrightarrow{DF}$ such that $\overline{DH} \cong \overline{AC}$. Then, since the angles at A and D are congruent, we have that $\Delta ABC \cong \Delta DGH$ (SAS) and $\angle DGH \cong \angle ABC$. Since $\angle ABC$ and $\angle DEF$ are assumed congruent, then $\angle DGH \cong \angle DEF$ and $\overline{GH} \parallel \overline{EF}$ (Theorem 2.7). This implies that H is between D and F. Applying Corollary 2.26, we get that

$$\frac{DG}{DE} = \frac{DH}{DF}$$

Since $\overline{DG} \cong \overline{AB}$ and $\overline{DH} \cong \overline{AC}$, we get

$$\frac{AB}{DE} = \frac{AC}{DF}$$

A similar argument can be used to show that

$$\frac{AB}{DE} = \frac{BC}{EF}$$

□

Another important similarity condition is the SAS condition.

Theorem 2.29. *(SAS Similarity Condition) If in two triangles there is a correspondence in which two sides of one triangle are proportional to two sides of the other triangle and the included angles are congruent, then the triangles are similar (Fig. 2.21).*

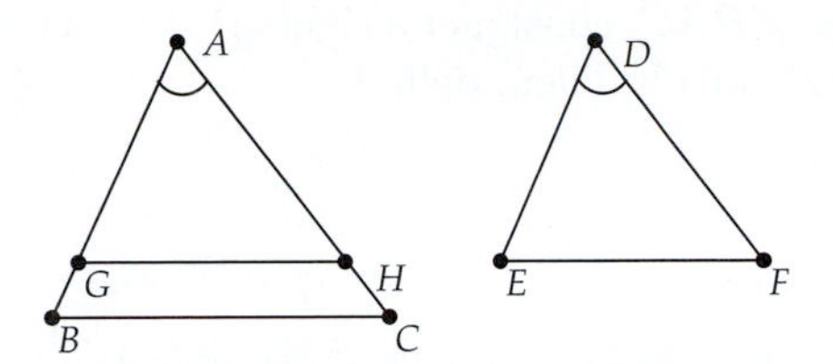

Fig. 2.21 SAS Similarity Condition

Proof: Let triangles ΔABC and ΔDEF be as specified in the theorem with

$$\frac{AB}{DE} = \frac{AC}{DF}$$

If $\overline{AB}$ and $\overline{AC}$ are congruent to their counterparts in ΔDEF, then the triangles are congruent by SAS and thus are similar as well. We may then assume that $\overline{AB}$ and $\overline{AC}$ are greater than their counterparts in ΔDEF. On $\overline{AB}$ and $\overline{AC}$ there must be points G and H such that $AG = DE$ and $AH = DF$. It is left as an exercise to show that $\overleftrightarrow{GH}$ and $\overleftrightarrow{BC}$ are parallel and thus the angles at B and C are congruent to the angles at E and F. □

Exercise 2.5.1. In ΔABC let points D and E be the midpoints of sides $\overline{AB}$ and $\overline{AC}$. Show that DE must be half the length of $\overline{BC}$.

Exercise 2.5.2. Fill in the gap in Theorem 2.29. That is, show in Fig. 2.21 that $\overleftrightarrow{GH}$ and $\overleftrightarrow{BC}$ are parallel.

Exercise 2.5.3. (SSS Similarity Condition) Suppose that two triangles have three sides that are correspondingly proportional. Show that the two triangles must be similar.

Exercise 2.5.4. Let ΔABC be a right triangle with the right angle at C. Let a, b, and c be the side lengths of this triangle. Let $\overline{CD}$ be the altitude to side $\overline{AB}$ (the hypotenuse). Use similarity to prove the Pythagorean Theorem; that is, show that

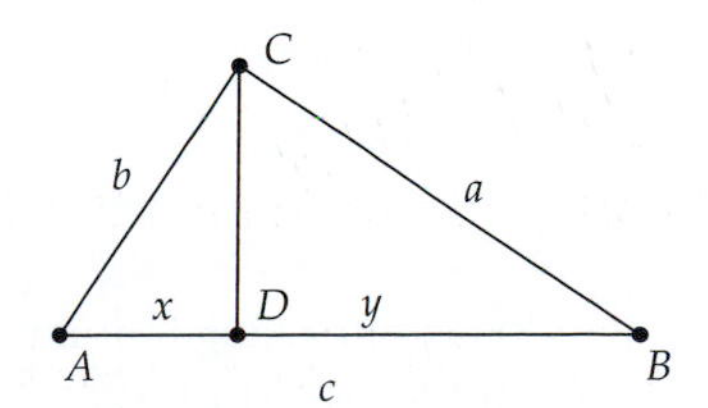

$$a^2 + b^2 = c^2$$

Exercise 2.5.5. The basic results on similar triangles allow us to define the standard set of trigonometric functions on angles as follows:

Given an acute angle $\angle BAC$, construct a right triangle with one angle congruent to $\angle BAC$ and right angle at C. Then, define

- $\sin(\angle A) = \frac{BC}{AB}$
- $\cos(\angle A) = \frac{AC}{AB}$

Show that these definitions are well defined; that is, they do not depend on the construction of the right triangle.

Exercise 2.5.6. We can extend the functions defined in the last exercise to *obtuse* angles (ones with angle measure greater than 90 degrees) by defining

- $\sin(\angle A) = \sin(\angle A')$
- $\cos(\angle A) = -\cos(\angle A')$

where $\angle A'$ is the supplementary angle to $\angle A$. We can extend the definitions to right angles by defining $\sin(\angle A) = 1$ and $\cos(\angle A) = 0$ if $\angle A$ is a right angle.

Using this extended definition of the trig functions, prove the *Law of Sines*, that in any triangle ΔABC, if a and b are the lengths of the sides opposite A and B, respectively, then

$$\frac{a}{b} = \frac{\sin(\angle A)}{\sin(\angle B)}$$

[Hint: Drop a perpendicular from C to AB at D and use the two right triangles ΔADC and ΔBDC.]

Exercise 2.5.7. In this exercise we will prove Menelaus's Theorem: Given triangle ABC, let R be a point outside of the triangle on the line through A and B. From R draw any line intersecting the other two sides of the triangle at points P, Q. Then

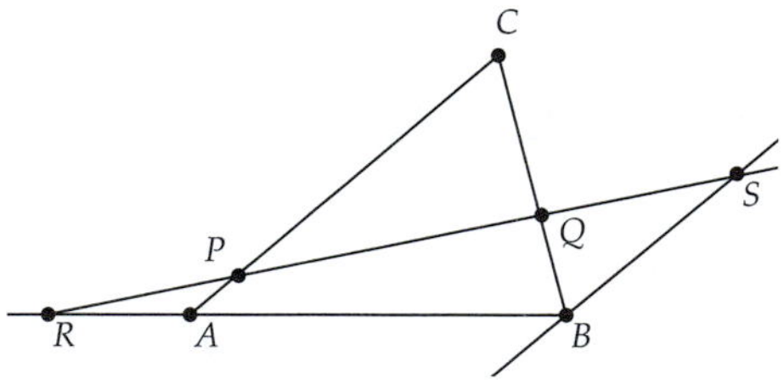

$$\frac{CP}{PA}\frac{AR}{RB}\frac{BQ}{QC} = 1$$

Prove this result using similar triangles. Specifically, construct a parallel to $\overleftrightarrow{AC}$ through B and let this intersect $\overleftrightarrow{RP}$ at point S. (Why must these lines intersect?) Then, show that ΔRBS and ΔRAP are similar, as are ΔPCQ and ΔSBQ, and use this to prove the result. (Menelaus was a Roman mathematician who lived in Alexandria from about 70–130 AD.)

Exercise 2.5.8. Blaise Pascal (1623–1662) did fundamental work in many areas of science and mathematics, including the physics of hydrostatics, the geometry of conic sections, and the foundations of probability and philosophy. He is also credited with inventing the first digital calculator. The following result is due to Pascal.

Given ΔABC, construct a line $\overleftrightarrow{DE}$, intersecting sides $\overline{AB}$ and $\overline{AC}$ at D, E, and parallel to $\overleftrightarrow{BC}$. Pick any point F on $\overline{BD}$ and construct $\overline{FE}$. Then, construct $\overleftrightarrow{BG}$ parallel to $\overleftrightarrow{FE}$, intersecting side $\overline{AC}$ at G. Then, $\overleftrightarrow{FC}$ must be parallel to $\overleftrightarrow{DG}$.

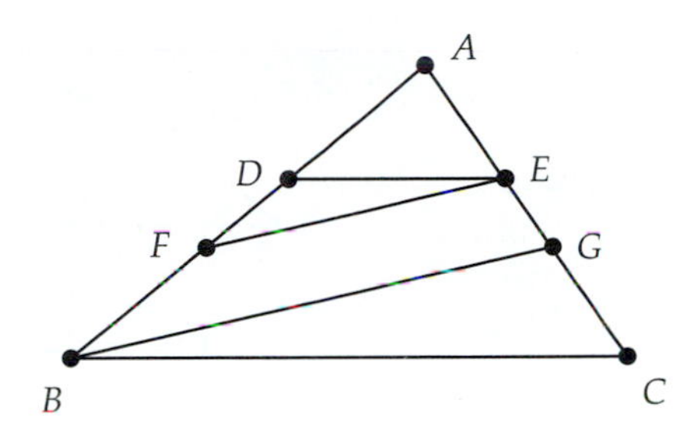

Prove Pascal's Theorem. [Hint: Use the theorems in this section on parallels and similar triangles to show that $\frac{AD}{AG} = \frac{AF}{AC}$.] Hilbert [21, page 46] used Pascal's Theorem extensively to develop his "arithmetic of segments," by which he connected segment length with ordinary real numbers.

2.5.1 Mini-Project - Finding Heights

A *surveyor* is one who measures things like distances and elevations. While modern surveyors use sophisticated equipment like laser range finders, the basic mathematics of surveying is that of triangle geometry and in particular similar triangles.

Consider the situation shown in Fig. 2.22. At C we have a water tower. At a particular time of the day, the sun will cast a shadow of the tower that hits the ground at A. Suppose that a person stands at exactly point B, where his or her shadow will match the shadow of the tower at A.

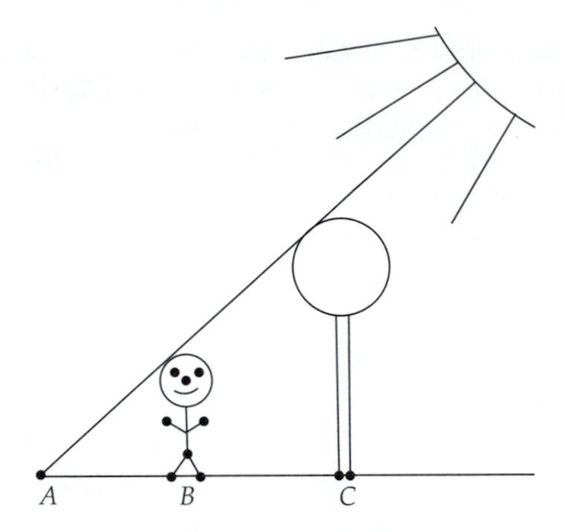

Fig. 2.22 Finding Heights

For this project, choose a partner and find a tall building (or tree) to measure in this fashion. You will need a long tape measure (or you can use a yard stick) and a pencil and paper for recording your measurements.

The report for this part of the project will include (1) a discussion of how to calculate the height using similar triangles, (2) a table of measurements needed to find the height of the building (or tree), and (3) the calculation of the height of your tall object.

Now, let's consider a second method of finding the height, as depicted in Fig. 2.23. For this method you will need a measuring device and a small mirror. Place the mirror flat on the ground some distance from the tall object whose height you wish to measure. Then, walk back from the mirror and stop when the top of the object is just visible in the mirror. Describe how you can find the height of the object using this mirror method. Include in your project report a table of measurements for this second method and a short discussion of which method is more accurate and/or practically useful.

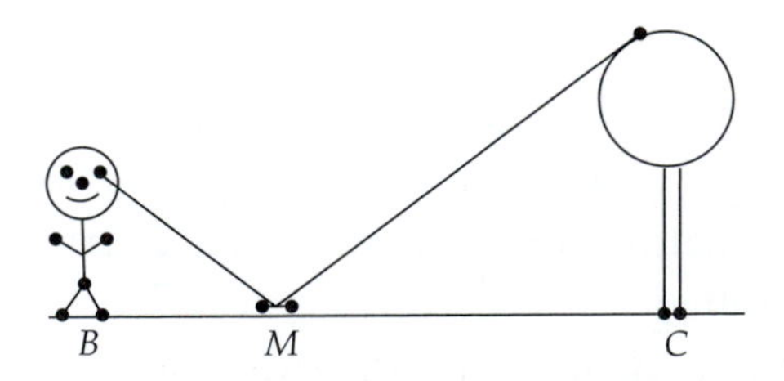

Fig. 2.23 Finding Heights by Mirror

2.6 Circle Geometry

One of the most beautiful of geometric shapes is the circle. Euclid devoted Book III of the *Elements* to a thorough study of the properties of circles.

To start our discussion of circles, we need a few definitions.

Definition 2.27. Let c be a circle with center O (Fig. 2.24). If A and B are distinct points on c, we call the segment $\overline{AB}$ a *chord* of the circle. If a chord passes through the center O, we say it is a *diameter* of the circle.

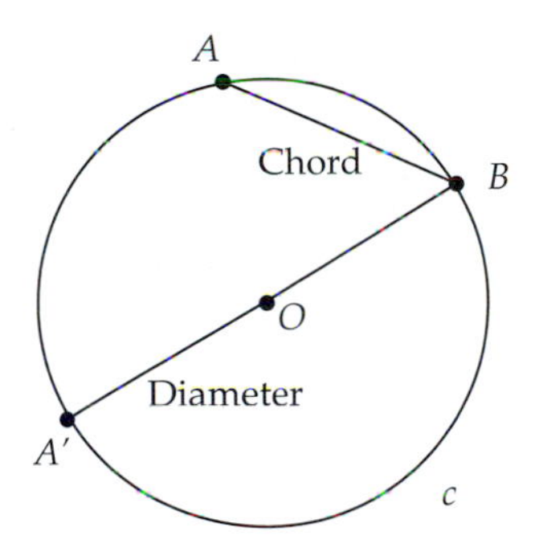

Fig. 2.24

Definition 2.28. A chord $\overline{AB}$ of a circle c with center O will divide the points of a circle c (other than A and B) into two parts, those points of c on one side of $\overleftrightarrow{AB}$ and those on the other side. Each of these two parts is called an *open arc* of the circle. An open arc determined by a diameter is called a *semi-circle.* If we include the endpoints A, B we would call the arc or semi-circle *closed.*

Definition 2.29. Note that one of the two open arcs defined by a (non-diameter) chord $\overline{AB}$ will be within the angle $\angle AOB$. This will be called a *minor* arc. This is arc $a1$ in Fig. 2.25. The arc that is exterior to this angle is called a *major* arc. This is arc $a2$ in the figure.

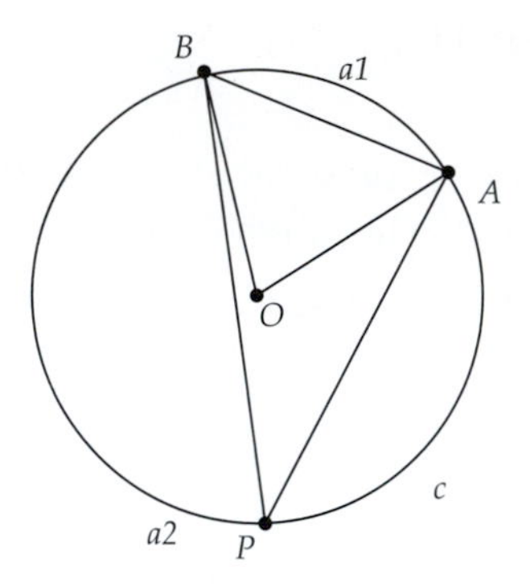

Fig. 2.25

Definition 2.30. A *central angle* of a circle is one that has its vertex at the center of the circle. For example, $\angle AOB$ in Fig. 2.25 is a central angle of the circle c. An *inscribed angle* of a circle within a major arc is one that has its vertex on the circle at a point on the major arc and has its sides intercepting the endpoints of the arc ($\angle APB$ in the figure).

Now, let's look at some basic properties of circles.

Theorem 2.30. *Given three distinct points, not all on the same line, there is a unique circle through these three points.*

Proof: Let the three points be A, B, C. Let l_1 and l_2 be the perpendicular bisectors of $\overline{AB}$ and $\overline{BC}$, with M_1, M_2 the midpoints of $\overline{AB}$ and $\overline{BC}$ (Fig. 2.26). We first show that l_1 and l_2 must intersect.

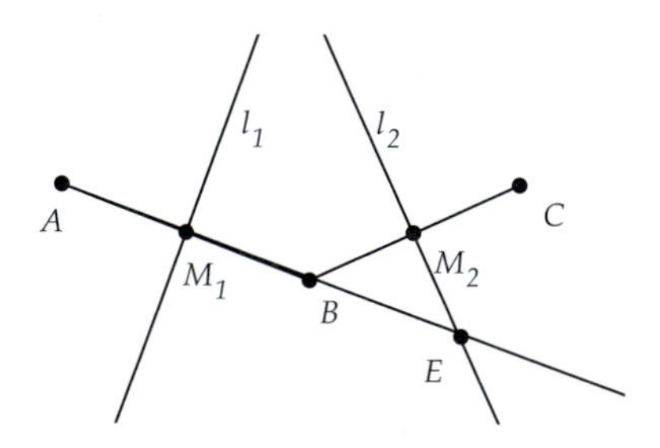

Fig. 2.26

Suppose l_1 and l_2 were parallel. Then, since $\overleftrightarrow{AB}$ intersects l_1, it must intersect l_2 at some point E (by Exercise 2.1.10). Since A, B, C are not collinear, then E cannot be M_2. Since l_1 and l_2 are parallel, then by Theorem

2.9 $\angle BEM_2$ must equal the right angle at M_1. But, this is impossible as then triangle ΔBEM_2 would have an angle sum greater than 180 degrees. This would contradict Exercise 2.1.8, which showed that the angle sum for a triangle is 180 degrees.

Thus l_1 and l_2 intersect at some point O (Fig. 2.27). By SAS we know that ΔAM_1O and ΔBM_1O are congruent, as are ΔBM_2O and ΔCM_2O, and thus $\overline{AO} \cong \overline{BO} \cong \overline{CO}$ and the circle with center O and radius $\overline{AO}$ passes through these three points.

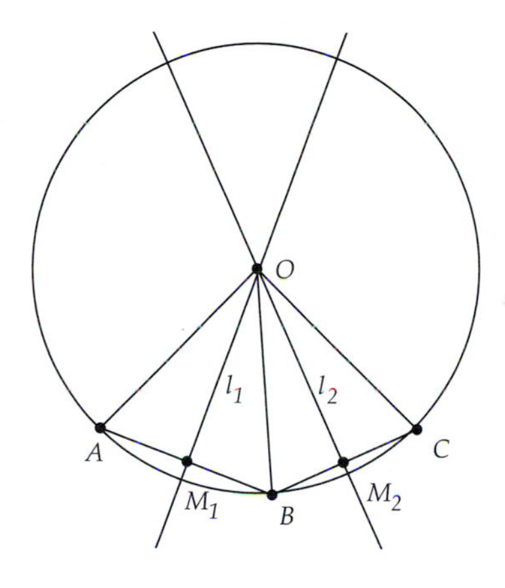

Fig. 2.27

Why is this circle unique? Suppose there is another circle c' with center O' passing through A, B, C. Then, O' is equidistant from A and B and thus must be on the perpendicular bisector l_1. (This can be seen from a simple SSS argument.) Likewise O' is on l_2. But, l_1 and l_2 already meet at O and thus $O = O'$. Then, $\overline{O'A} \cong \overline{OA}$ and c' must be equal to the original circle.
□

Theorem 2.31. *The measure of an angle inscribed in a circle is half that of its intercepted central angle.*

Proof: Let $\angle APB$ be inscribed in circle c having center O. Let $\overrightarrow{PO}$ intersect circle c at Q. Then, either A and B are on opposite sides of $\overleftrightarrow{PO}$ (as shown in Fig. 2.28), or A is on the diameter through $\overline{OP}$, or A and B are on the same side of $\overleftrightarrow{PO}$. We will prove the result in the first case and leave the remaining cases as an exercise.

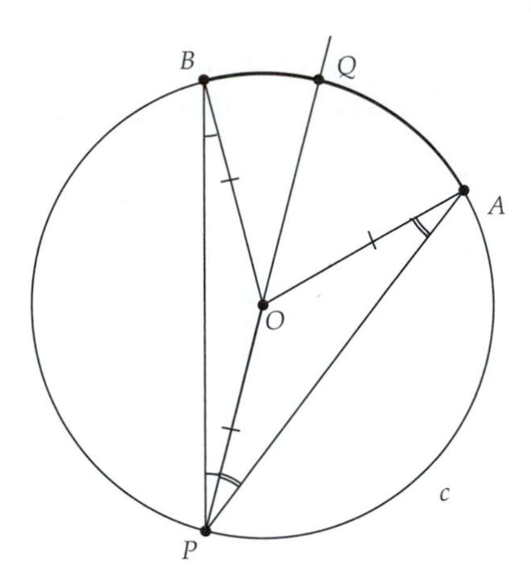

Fig. 2.28

Now, $\overline{OP} \cong \overline{OB} \cong \overline{OA}$ and ΔPOB and ΔPOA are isosceles. Thus, the base angles are congruent (Theorem 2.14). Let $\alpha = m\angle PBO$ and $\beta = m\angle PAO$. Then, $m\angle POB = 180 - 2\alpha$ and $m\angle POA = 180 - 2\beta$. Also, $m\angle QOB = 2\alpha$ and $m\angle QOA = 2\beta$.

Clearly, the measure of the angle at P, which is $\alpha+\beta$, is half the measure of the angle at O, which is $2\alpha + 2\beta$. □

The following two results are immediate consequences of this theorem.

Corollary 2.32. *If two angles are inscribed in a circle such that they share the same arc, then the angles are congruent.*

Corollary 2.33. *An inscribed angle in a semi-circle is always a right angle.*

Definition 2.31. A polygon is *inscribed* in a circle if its vertices lie on the circle. We also say that the circle *circumscribes* the polygon in this case.

Corollary 2.34. *If quadrilateral $ABCD$ is inscribed in a circle, then the opposite angles of the quadrilateral are supplementary.*

The proof is left as an exercise.

Theorem 2.35. *(Converse to Corollary 2.33) If an inscribed angle in a circle is a right angle, then the endpoints of the arc of the angle are on a diameter.*

Proof (refer to Fig. 2.29): Let $\angle BAC$ be a right angle inscribed in a circle c. We need to show that $\overline{BC}$ is a diameter of the circle.

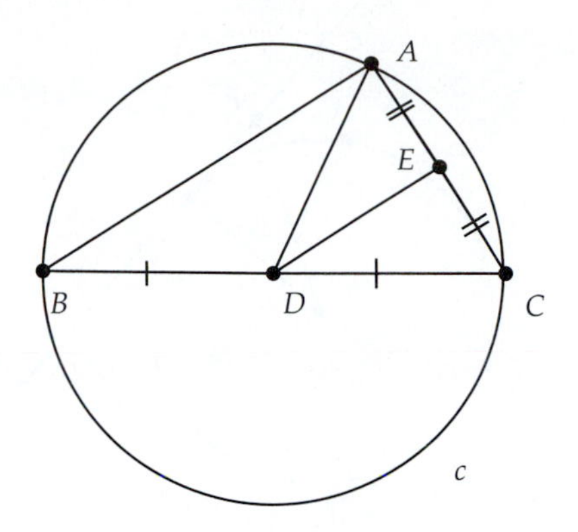

Fig. 2.29

Let D and E be the midpoints of segments $\overline{BC}$ and $\overline{AC}$. Then, since $\overline{BD}$ and $\overline{AE}$ are both in the same proportion to $\overline{BC}$ and $\overline{AC}$, we have by Theorem 2.27 that $\overleftrightarrow{DE}$ is parallel to $\overleftrightarrow{AB}$ and thus the angle at E is a right angle. Then, $\Delta AED \cong \Delta CED$ by SAS and $\overline{DC} \cong \overline{DA}$.

Similarly, $\overline{BD} \cong \overline{DA}$. Thus, the circle with center D and radius equal to $\overline{DA}$ passes through A, B, C. But, c is the unique circle through these points and thus $\overline{DA}$ is the radius for the circle and $\overline{BC}$ is a diameter. □

Among all lines that pass through a circle, we will single out for special consideration those lines that intersect the circle only once.

Definition 2.32. A line l is said to be *tangent* to a circle c if l intersects the circle at a single point T, called the *point of tangency.*

What properties do tangents have?

Theorem 2.36. *Given a circle c with center O and radius $\overline{OT}$, a line l is tangent to c at T if and only if l is perpendicular to $\overleftrightarrow{OT}$ at T.*

Proof (refer to Fig. 2.30): First, suppose that l is tangent to c at T. If l is not perpendicular to $\overleftrightarrow{OT}$ at T, then let P be the point on l where $\overleftrightarrow{OP}$ is perpendicular to l. On l we can construct a point Q opposite of T from P such that $\overline{PQ} \cong \overline{PT}$. Then, by SAS $\Delta OPT \cong \Delta OPQ$. Then, $\overline{OT} \cong \overline{OQ}$. But, this would imply that Q is on the circle, which contradicts T being the point of tangency.

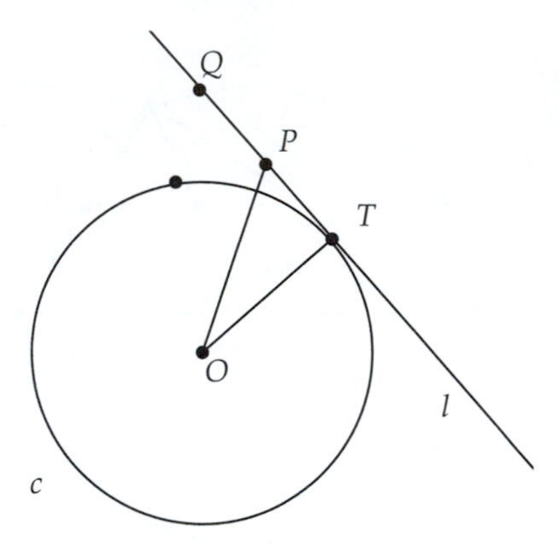

Fig. 2.30

On the other hand, suppose that l is perpendicular to $\overleftrightarrow{OT}$ at T. Let P be any other point on l. By the exterior angle theorem, the angle at P in ΔOPT is smaller than the right angle at T. Thus, since in any triangle the greater angle lies opposite the greater side and vice versa (Propositions 18 and 19 of Book I), we have that $OP > OT$ and P is not on the circle. Thus, l is tangent at T. □

Do tangents always exist? Theorem 2.36 shows us how to construct the tangent to a circle from a point on the circle. What if we want the tangent to a circle from a point outside the circle?

Theorem 2.37. *Given a circle c with center O and radius $\overline{OA}$ and given a point P outside of the circle, there are exactly two tangent lines to the circle passing through P.*

Proof (refer to Fig. 2.31): Let M be the midpoint of $\overline{OP}$ and let c' be the circle centered at M with radius $\overline{OM}$. Then, since P is outside circle c and O is inside, we know that c' and c will intersect at two points, T_1 and T_2. Since $\angle PT_1O$ is inscribed in a semi-circle of c', then it must be a right angle. Thus, $\overleftrightarrow{PT_1}$ is perpendicular to $\overline{OT_1}$ at T_1 and by the previous theorem T_1 is a point of tangency. Likewise, T_2 is a point of tangency.

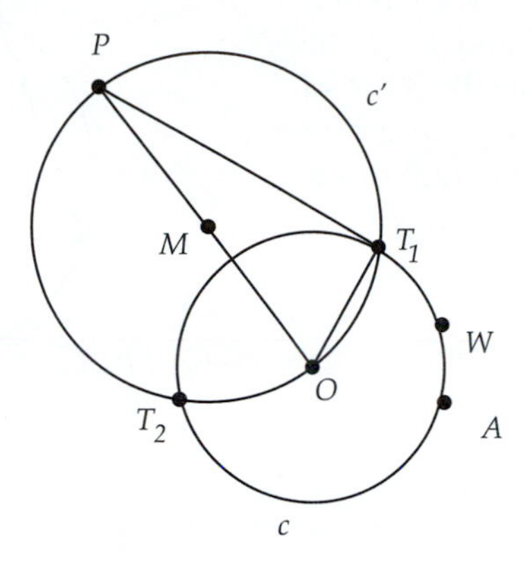

Fig. 2.31

Can there be any more points of tangency passing through P? Suppose that W is another point of tangency on c such that the tangent line at W passes through P. Then, $\angle PWO$ is a right angle. Let c'' be the circle through P, W, O. Then, by Corollary 2.33 we know that $\overline{OP}$ is a diameter of c''. But, then M would be the center of c'' and the radius of c'' would be $\overline{OM}$. This implies that c' and c'' are the same circle and that W must be one of T_1 or T_2. □

Exercise 2.6.1. Finish the proof for the remaining two cases in Theorem 2.31. That is, prove the result for the case where A is on the diameter through $\overline{OP}$ and for where A and B are on the same side of $\overleftrightarrow{PO}$.

Exercise 2.6.2. Prove Corollary 2.34.

Exercise 2.6.3. Two circles meet at points P and Q. Let $\overline{AP}$ and $\overline{BP}$ be diameters of the circles. Show that $\overline{AB}$ passes through the other intersection Q.

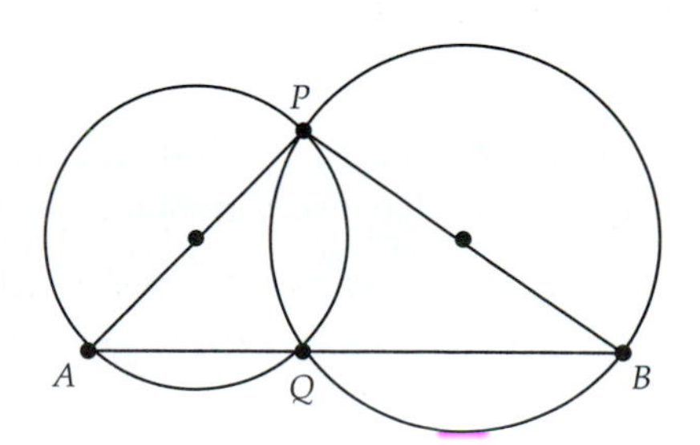

Exercise 2.6.4. Let c be the circumscribed circle of ΔABC and let P be the point on c where the bisector of $\angle ABC$ meets c. Let O be the center of c. Prove that the radius $\overline{OP}$ meets $\overline{AC}$ at right angles.

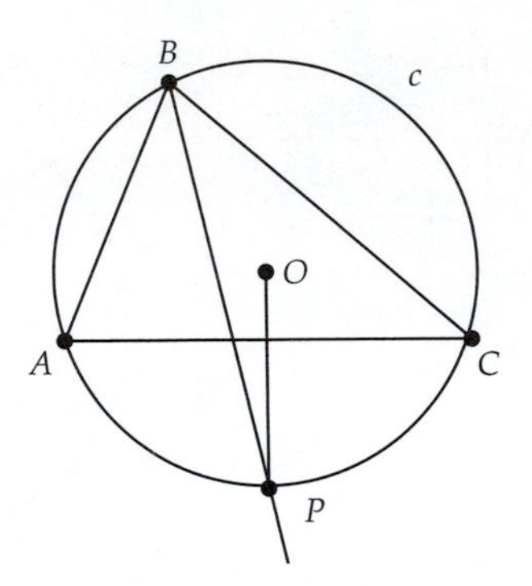

Exercise 2.6.5. Show that the line passing through the center of a circle and the midpoint of a chord is perpendicular to that chord, provided the chord is not a diameter.

Exercise 2.6.6. Let $\overline{AD}$ and $\overline{BC}$ be two chords of a circle that intersect at P. Show that $(AP)(PD) = (BP)(PC)$. [Hint: Use similar triangles.]

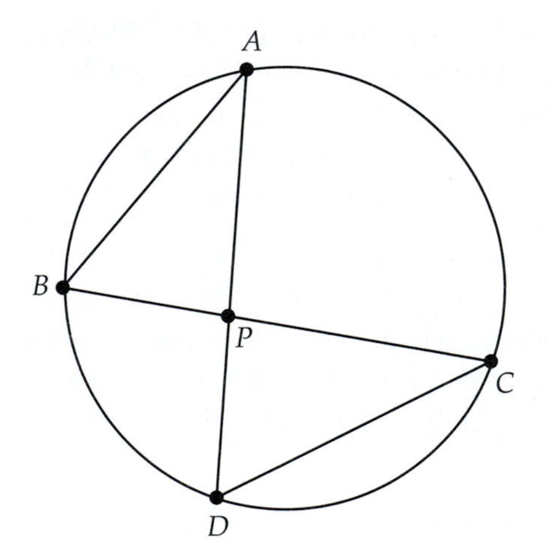

Exercise 2.6.7. In the last project we saw that every triangle can be circumscribed. Show that every rectangle can be circumscribed.

Definition 2.33. Two distinct circles $c_1 \neq c_2$ are *mutually tangent* at a point T if the same line through T is tangent to both circles.

Exercise 2.6.8. Show that two circles that are mutually tangent must have the line connecting their centers passing through the point of tangency.

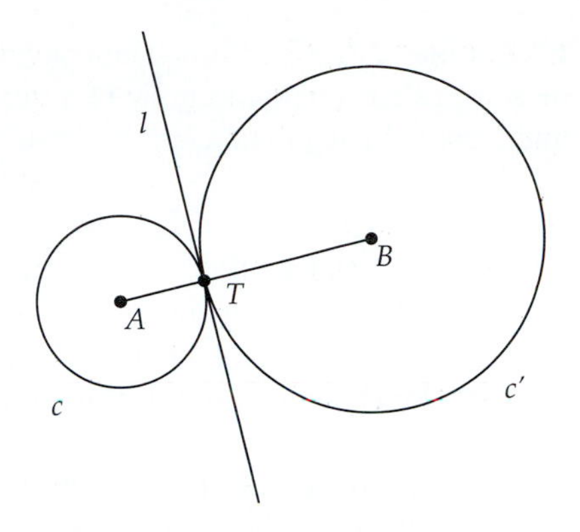

Exercise 2.6.9. Show that mutually tangent circles intersect in only one point. [Hint: Suppose they intersected at another point P. Use the previous exercise and isosceles triangles to yield a contradiction.]

Exercise 2.6.10. Show that two circles that are mutually tangent at T must either be

1. on *opposite* sides of the common tangent line at T, in which case we will call the circles *externally tangent*, or
2. on the *same* side of the tangent line, with one inside the other, in which case we will call the circles *internally tangent*.

Exercise 2.6.11. Given two circles externally tangent at a point T, let $\overline{AB}$ and $\overline{CD}$ be segments passing through T with A, C on c_1 and B, D on c_2. Show that $\overleftrightarrow{AC}$ and $\overleftrightarrow{BD}$ are parallel. [Hint: Show that the alternate interior angles at C and D are congruent.]

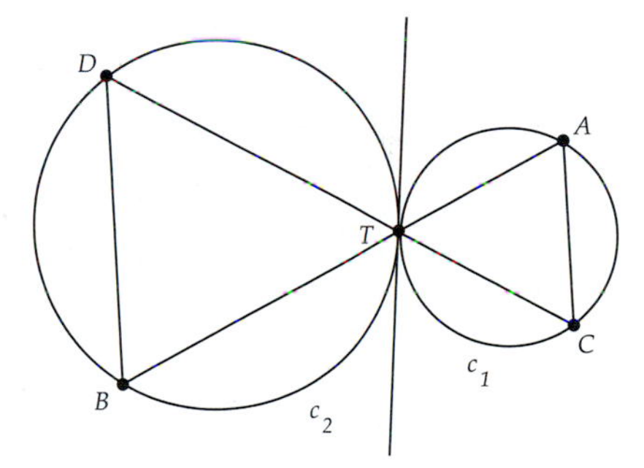

Exercise 2.6.12. Show that the line from the center of a circle to an outside point bisects the angle made by the two tangents from that outside point to the circle. [Hint: Use Exercise 2.2.11.]

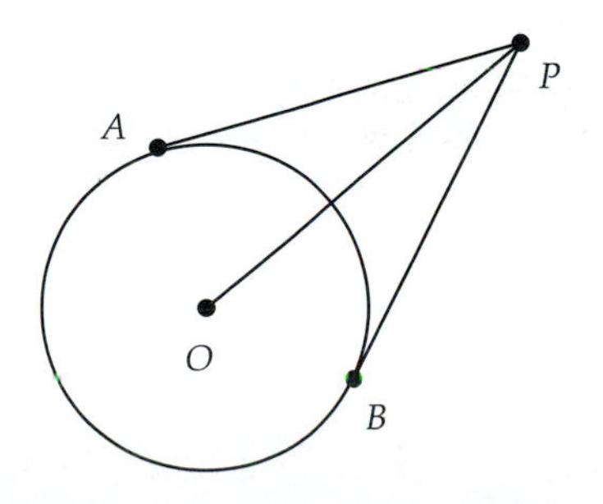

Exercise 2.6.13. Show the converse to the preceding exercise, that is, that the bisector of the angle made by two tangents from a point outside a circle to the circle must pass through the center of the circle. [Hint: Try a proof by contradiction.]

Exercise 2.6.14. Let c and c' be externally tangent at T. Show that there are two lines that are tangent to both circles (at points other than T). [Hint: Let m be the line through the centers. Consider the two radii that are perpendicular to m. Let l be the line through the endpoints of these radii on their respective circles. If l and m are parallel, show that l is a common line of tangency for both circles. If l and m intersect at P, let n be a tangent from P to one of the circles. Show n is tangent to the other circle.]

2.7 Project 4 - Circle Inversion and Orthogonality

In this project we will explore the idea of inversion through circles. Circle inversion will be a critical component of our construction of non-Euclidean geometry in Chapter 7.

We start out with the notion of the *power* of a point with respect to a given circle.

Start the *Geometry Explorer* program and create a circle c with center O and radius point A, and create a point P not on c.

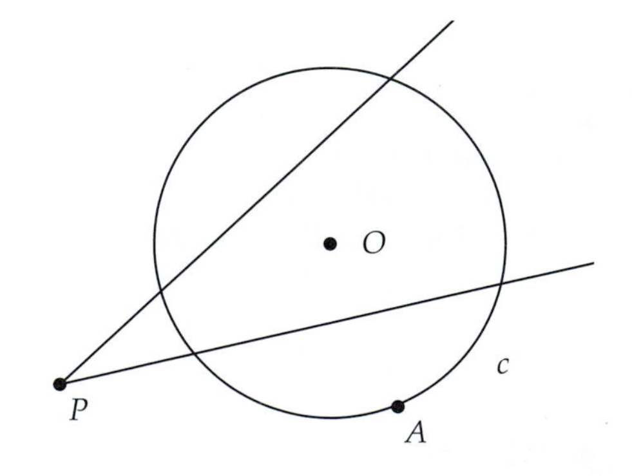

Now create two lines originating at P that pass through the circle. Find the two intersection points of the first line with the circle (call them P_1 and P_2) and the two intersection points of the second line with the circle (call them Q_1 and Q_2). Measure the four distances PP_1, PP_2, PQ_1, and PQ_2. (To measure distance, multi-select two points and choose **Distance** (**Measure** menu).)

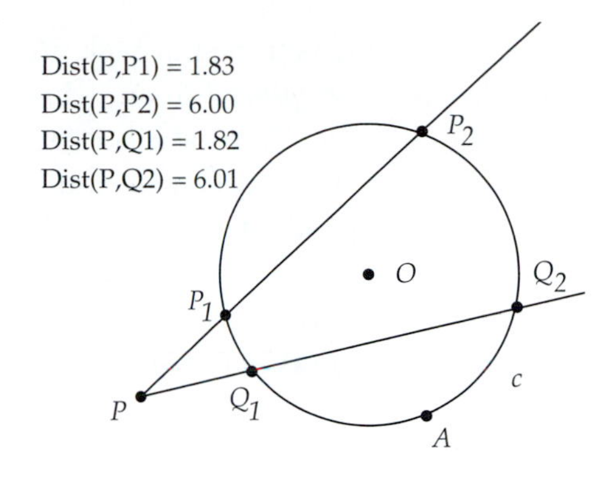

Now we will compare the product of PP_1 and PP_2 to the product of PQ_1 and PQ_2. To do this we will use the Calculator in *Geometry Explorer*. Go to the Help Web page (click on **Help** in the menu bar) and then go to the "View Menu" link and from there to the "Calculator" link. Read through this section to become familiarized with how to use the Calculator. Now, choose **Calculator** (**View** menu).

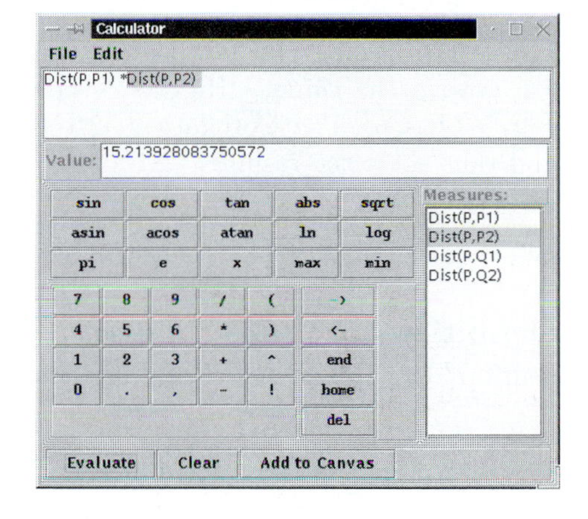

Notice that the four distance measurements are listed in the right half of the Calculator window. Double-click the first distance measure, then click on the Multiplication button (labeled "*"), and then double-click the second distance measurement. We have just created an expression for the product of PP_1 and PP_2.

To use this measurement back in the *Geometry Explorer* main window, we click the Evaluate button and then the Add to Canvas button. The new product measure will now be on the screen. Do the same for the product of PQ_1 and PQ_2. [Be sure to "Clear" the Calculator first.]

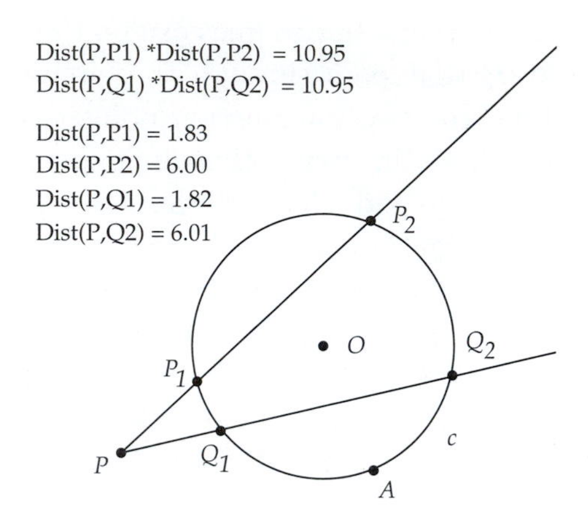

Interesting! It appears that these two products are the same. Drag point P around and see if this conjecture is supported.

Exercise 2.7.1. Our first task in this project is to *prove* that these two products are always the same. [Hint: Consider some of the inscribed angles formed by P_1, P_2, Q_1, Q_2. Use Corollary 2.32 to show that ΔPP_1Q_2 is similar to ΔPQ_1P_2 and thus show the result.]

Exercise 2.7.2. Show that the product of PP_1 and PP_2 (or PQ_1 and PQ_2) can be expressed as $PO^2 - r^2$, where r is the radius of the circle.

Definition 2.34. Given a circle c with center O and radius r and given a point P, we define the *Power of P with respect to c* as:

$$Power\ of\ P = PO^2 - r^2.$$

Note that by Exercise 2.7.2 the Power of P is also equal to the product of PP_1 and PP_2 for any line l from P, with P_1 and P_2 the intersections of l with the circle c.

Also note that the Power of P can be used to classify whether P is inside ($Power < 0$), on ($Power = 0$), or outside ($Power > 0$) the circle.

Now we are ready to define circle inversion.

Definition 2.35. The *inverse of P with respect to c* is the unique point P' on ray $\overrightarrow{OP}$ such that $OP' = \frac{r^2}{OP}$ (or $(OP')(OP) = r^2$).

Note that if the circle had unit radius ($r = 1$), and if we considered O as the origin in Cartesian coordinates with $OP = x$, then the inverse P' of P can be interpreted as the usual multiplicative inverse; that is, we would have $OP' = \frac{1}{x}$.

How do we construct the inverse point?

Clear the screen and create a circle c with center O and radius point A and then create a point P inside c. Create the ray $\overrightarrow{OP}$. At P construct the perpendicular to $\overrightarrow{OP}$ and find the intersection points (T and U) of this perpendicular with the circle. Create segment $\overline{OT}$ and find the perpendicular to $\overline{OT}$ at T. Let P' be the point where this second perpendicular intersects $\overrightarrow{OP}$.

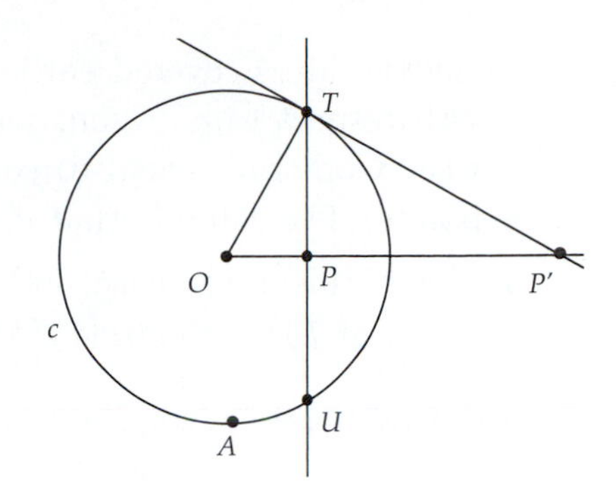

Measure the distances for segments $\overline{OP}$ and $\overline{OP'}$ and measure the radius of the circle. Use the Calculator to compute the product of OP and OP' and the square of the radius as shown in the figure.

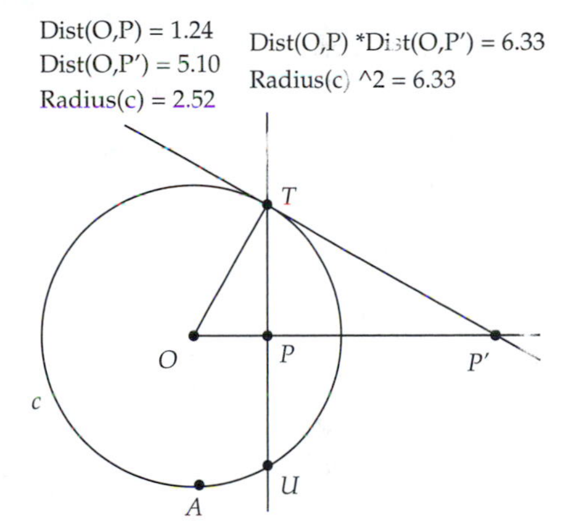

It appears that we have constructed the inverse!

Exercise 2.7.3. Prove that this construction actually gives the inverse of P. That is, show that $(OP)(OP') = r^2$.

In the last part of this lab, we will use the notion of circle inversion to construct a circle that meets a given circle at right angles.

Definition 2.36. Two circles c and c' that intersect at distinct points A and B are called *orthogonal* if the tangents to the circles at each of these points are perpendicular.

Suppose we have a circle c and two points P and Q inside c, with P not equal to Q and neither point equal to the center O of the circle. The goal is to construct a circle through P and Q that meets c at right angles.

Using the ideas covered earlier in this project, construct the inverse P' of P with respect to c. Then, select P, P', and Q and click on the Circle tool in the Construct panel to construct the unique circle c' through these three points. The claim is that c' is orthogonal to c.

To see if this is the case, let's first find the center of c'. Let R be the intersection of $\overleftrightarrow{TP}$ with circle c' (Fig 2.32).

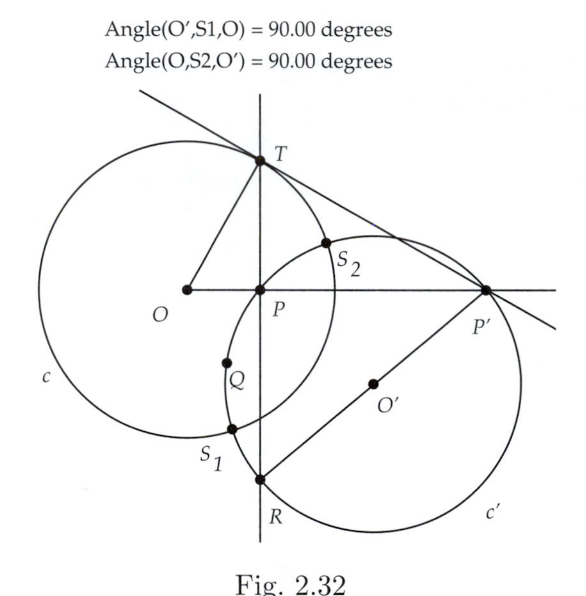

Fig. 2.32

Then $\angle RPP'$ is a right angle in circle c' as $\angle OPT$ is a right angle. Thus, by Theorem 2.33 $\overline{RP'}$ is a diameter of c'. The midpoint O' of $\overline{RP'}$ will be the center of c'. Let S_1 and S_2 be the intersection points of c with c'. Measure $\angle O'S_1O$ and $\angle OS_2O'$ and check that they are right angles. Since the tangents to c and c' are orthogonal to $\overline{OS_1}$, $\overline{OS_2}$, $\overline{O'S_1}$, and $\overline{O'S_2}$, then the tangents to the circles at S_1 and S_2 must also be orthogonal and the circles are orthogonal. Note that this *evidence* of the orthogonality of c and c' is not a rigorous proof. The proof will be covered when we get to Theorem 2.38.

Exercise 2.7.4. What do you think will happen to circle c' as one of the points P or Q approaches the center O of circle c? Try this out and then explain why this happens.

Project Report

The ability to construct orthogonal pairs of circles is crucial to developing a model of hyperbolic geometry, where parallels to a line through a point are "abundant." We will look at this model in detail in Chapter 7.

For the project report, provide detailed analysis of the constructions used in this project and complete answers to the exercises.

2.7.1 Orthogonal Circles Redux

Here is a proof of orthogonality of the circles constructed in the text preceding Exercise 2.7.4.

Theorem 2.38. *Given a circle c with center O and radius $\overline{OA}$ and given two points P and Q inside c, with P not equal to Q and neither point equal to O, there exists a unique circle c' (or line) that passes through P and Q that is orthogonal to the given circle (Fig. 2.33).*

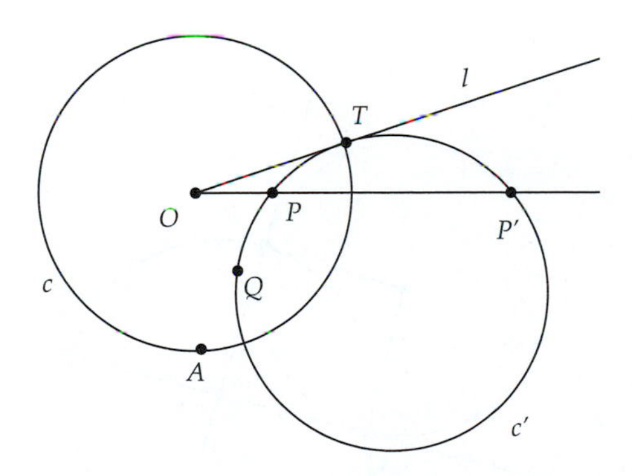

Fig. 2.33

Proof: It is clear that if P and Q lie on a diameter of c, then there is a unique line (coincident with the diameter) that is orthogonal to c. So, in the rest of this proof we assume that P and Q are not on a diameter of c.

Suppose that one or the other of P or Q, say P, is strictly inside c. As above construct the inverse P' to P and let c' be the unique circle passing through Q, P, P'. Construct a tangent l to circle c' that passes through O. Let l be tangent to c' at T. (To construct l, use the construction discussed in Theorem 2.37.) We claim that T is also on circle c. To see this, consider the power of O with respect to circle c':

$$Power\ of\ O = (OP)(OP') = (OT)^2$$

But, $(OP)(OP') = r^2$ (r being the radius of c) since P' is the inverse point to P with respect to c. Thus, $(OT)^2 = r^2$ and T is on circle c, and the circles are orthogonal at T.

To see that this circle is unique, suppose there was another circle c'' through P and Q that was orthogonal to c. Let P'' be the intersection of $\overrightarrow{OP}$ with c''. Let T'' be a point where c and c'' intersect. Then $(OP)(OP'') = (OT'')^2$. But, $(OT'')^2 = r^2$ and thus, P'' must be the inverse P' to P, and c'' must then pass through Q, P, P' and must be the circle c.

The final case to consider is when both P and Q are on the boundary of c (Fig. 2.34). Then, any circle through P and Q that is orthogonal to c must have its tangents at P and Q lying along $\overline{OP}$ and $\overline{OQ}$. Thus, the diameters of this circle must lie along tangent lines to c at P and Q. Thus, the center of the orthogonal circle must lie at the intersection of these tangents, which is a unique point.

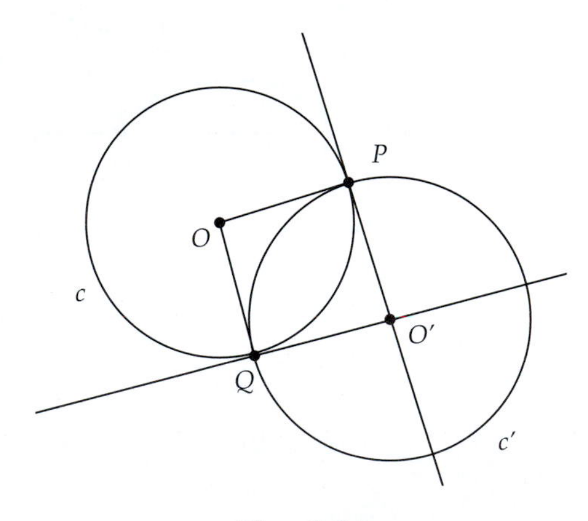

Fig. 2.34

□

We conclude this section on orthogonal circles with two results that will prove useful when we study non-Euclidean geometry in Chapter 7.

Theorem 2.39. *Let c and c' be two circles and let P be a point that is not on c and is not the center O of c. Suppose that c' passes through P. Then, the two circles are orthogonal if and only if c' passes through the inverse point P' to P with respect to c.*

Proof: First, suppose that c' passes through the inverse point P' (refer to Fig. 2.35). We know from the proof of Theorem 2.30 that the center O' of c' lies on the perpendicular bisector of $\overline{PP'}$. Since P and P' are inverses with respect to c, then they both lie on the same side of ray $\overrightarrow{OP}$. Thus, O is not between P and P' and we have that $O'O > O'P$. Thus, O is outside of c'. We then can construct two tangents from O to c' at points T_1 and T_2 on c'. Using the idea of the power of points with respect to c', we have $(OT_1)^2 = (OP)(OP')$. But, $(OP)(OP') = r^2$ by assumption, where r is the radius of c. Therefore, $(OT_1)^2 = r^2$, and T_1 is on c. A similar argument shows that T_2 is also on c. This implies that the two circles are orthogonal.

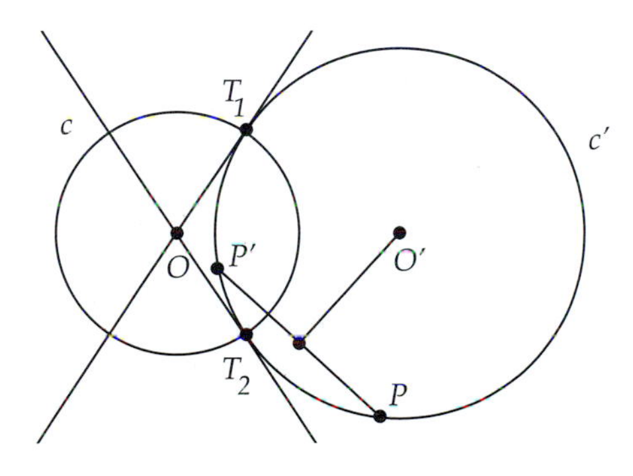

Fig. 2.35

Conversely, suppose that c and c' are orthogonal at points T_1 and T_2. The tangent lines to c' at these points then pass through O, which implies that O is outside c'. Thus, $\overrightarrow{OP}$ must intersect c' at another point P'. Using the power of points, we have $r^2 = (OT_1)^2 = (OP)(OP')$, and thus P' is the inverse point to P with respect to circle c. $\square$

Corollary 2.40. *Suppose circles c and c' intersect. Then c' is orthogonal to c if and only if the circle c' is mapped to itself by inversion in the circle c.*

Proof: Suppose the circles are orthogonal, and let P be a point on c'. If P is also on c, then it is fixed by inversion through c. If P is not on c, then by the proof of Theorem 2.39, we know that P is also not the center O of c. Thus, Theorem 2.39 implies that the inverse P' of P with respect to c is on c'. Thus, for all points P on c', we have that the inverse point to P is again on c'.

Conversely, suppose c' is mapped to itself by inversion in the circle c. Let P be a point on c' that is not on c and which is not the center O of c. Then, the inverse point P' with respect to c is again on c'. By Theorem 2.39, the circles are orthogonal. □

Chapter 3

Analytic Geometry

> There once was a very brilliant horse who mastered arithmetic, plane geometry, and trigonometry. When presented with problems in analytic geometry, however, the horse would kick, neigh, and struggle desperately. One just couldn't put Descartes before the horse.
>
> —Anonymous

In 1637 Réne Descartes (1596–1650) published the work *La Geometrie*, in which he laid out the foundations for one of the most important inventions of modern mathematics—the Cartesian coordinate system and analytic geometry.

In classical Euclidean geometry, points, lines, and circles exist as idealized objects independent of any concrete context. In this strict *synthetic* geometry, algebraic relations can be discussed, but *only* in relation to underlying geometric figures.

For example, the Pythagorean Theorem, in Euclid's geometry, reads as follows: "The square on the hypotenuse equals the squares on the two sides." This statement literally means that the square constructed on the hypotenuse equals the other two squares, and Euclid's proof of the Pythagorean Theorem amounts to showing how one can rearrange the square figures to make this equivalence possible.

From the time of Euclid until Descartes, there was always this insistence on tying algebraic expressions precisely to geometric figures. The Arabs were the first to introduce symbols, such as x^2, for algebraic quantities, but they too insisted on strict geometric interpretations of algebraic variables—x^2 literally meant the *square* constructed on a segment of length x.

Descartes was the first person to assume that algebraic relationships need not be tied to geometric figures. For Descartes, expressions like x^2, x^3, xy, and the like, were all *numbers*, or lengths of segments, and algebraic expressions could be thought of as either arithmetic or geometric expressions.

This was a great leap forward in the level of mathematical abstraction, in that it opened up mathematical avenues of study that were artificially closed. For example, equations involving arbitrary powers of x were now possible, since one was no longer restricted to segments (x), squares (x^2), and cubes (x^3).

Descartes' geometry was groundbreaking, although by no means what we think of today as *Cartesian* geometry. There were no coordinate axes, and Descartes did not have algebraic expressions for such simple figures as straight lines.

A much more modern-looking attempt to merge algebra with geometry was that of Pierre de Fermat (1601–1655), a contemporary of Descartes working in Toulouse, who had a system of perpendicular axes and coordinate equations describing lines, quadratics, cubics, and the conic sections. Fermat also developed a general method for finding tangents and areas enclosed by such algebraic expressions. In this regard, his work foreshadowed the development of calculus by Newton and Leibniz.

The great insight of Descartes and Fermat was to embed the study of geometric figures in a grid system, where a point is precisely located by its distances from two fixed lines that are perpendicular to one another. These two distances are called the *coordinates* of a point and are customarily labeled "x" and "y."

By studying the set of coordinates for a geometric figure, one can identify patterns in these coordinates. For example, a line is a set of points (x, y) where x and y have a relationship of the form $ax + by + c = 0$, with a, b, and c constants. Similarly, the points making up a circle have their own x-y relationship.

The coordinate geometry of Descartes and Fermat ultimately led to the notion of *functions* and to the creation of calculus by Newton and Leibniz toward the end of the seventeenth century. The great achievement of analytic geometry is that it allows one to enrich the traditional synthetic geometry of figures by the study of the equations of the x-y relationships for those figures.

In this chapter we will construct analytic geometry from first principles. Initially, we will develop this geometry from a synthetic geometric base, similar to the development of Descartes and Fermat. Later in the chapter, we will develop analytic geometry from a modern *axiomatic* basis. In the chapters following, we will make optimum use of both synthetic and analytic geometry, using whichever approach to Euclidean geometry is most transparent in devising proofs.

3.1 The Cartesian Coordinate System

In analytic geometry we create a *coordinate system* in the plane so that we can have a way of uniquely referencing points by numerical values. The simplest such coordinate system is the *rectangular* coordinate system (Fig. 3.1).

Let l be a line in the plane and let O be a point on that line. Let m be a perpendicular to l through O. We will call these two lines the *axes* of our coordinate system, with O being the *origin*.

By the continuity properties of the line, we know that for every real number x there is a unique point X on l such that the length of $\overline{OX}$ is x. We pick one side (ray) of l from O to be the positive side where the length of $\overline{OX}$ is $x \geq 0$ and let -x be the coordinate of $\overline{OX'}$ on the opposite ray such that the length of $\overline{OX'}$ is also x. Likewise, there is a similar correspondence on m. We call x and y the *coordinates* of the points on the axes.

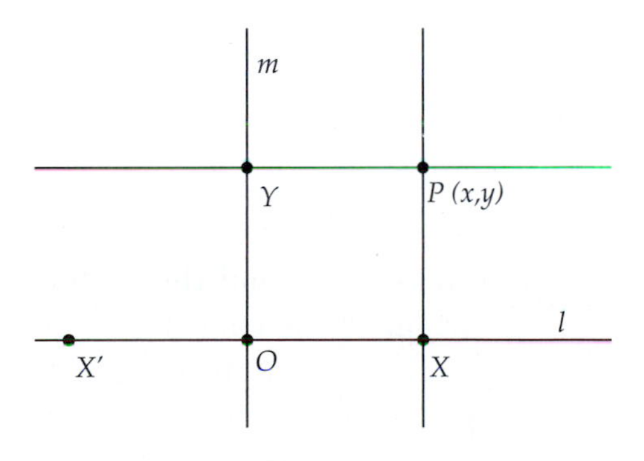

Fig. 3.1

Now, let P be any other point in the plane, not on l or m. We can drop (unique) perpendiculars down to l and m at points X and Y, represented by coordinates x and y. The lines will form a rectangle $OXPY$. Conversely, given coordinates x and y, we can construct perpendiculars that meet at P. Thus, there is a one-to-one correspondence between pairs of coordinates (x, y) and points in the plane. This system of identification of points in the plane is called the *rectangular* (or *Cartesian*) coordinate system.

To see the power of this method, let's look at the equation of a circle, that is, the algebraic relationship between the x and y coordinates of points on a circle. The circle is defined by starting with two points O, R and consists of points P such that $\overline{OP} \cong \overline{OR}$.

Let's create a coordinate system with origin at O and x-axis along $\overline{OR}$. Let r be the length of $\overline{OR}$. Given any point P on the circle, let (x, y) be its coordinates (Fig. 3.2). The angle $\angle OXP$ is a right angle. By the Pythagorean Theorem the length of $\overline{OP}$ is $\sqrt{x^2 + y^2}$, and this must equal the length of $\overline{OR}$, which is r. Thus, the circle is the set of points (x, y) such that $x^2 + y^2 = r^2$, a familiar equation from basic analysis.

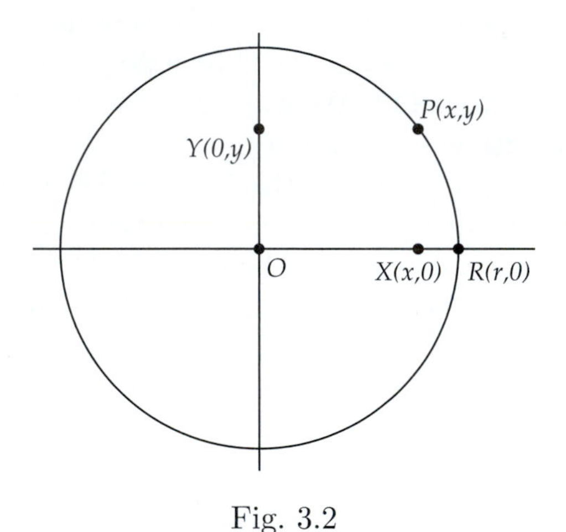

Fig. 3.2

In the preceding discussion on the equation of a circle, we saw that the length between a point (x, y) on a circle and the center of the circle is given by $\sqrt{x^2 + y^2}$, assuming the origin of the coordinate system is at the center of the circle. If the coordinate system is instead constructed so that the center of the circle is at (x_0, y_0), then clearly the new length formula will be given by

$$\sqrt{(x - x_0)^2 + (y - y_0)^2}$$

In general this will give the distance between the points (x, y) and (x_0, y_0) and is called the *distance formula in the plane.*

Here are some other useful algebraic-geometric facts.

Theorem 3.1. *Let $A = (x, y)$ and $B = (\text{-}x, \text{-}y)$. Then, the line through A, B passes through the origin $(0, 0)$ with A and B on opposite sides of the origin on this line.*

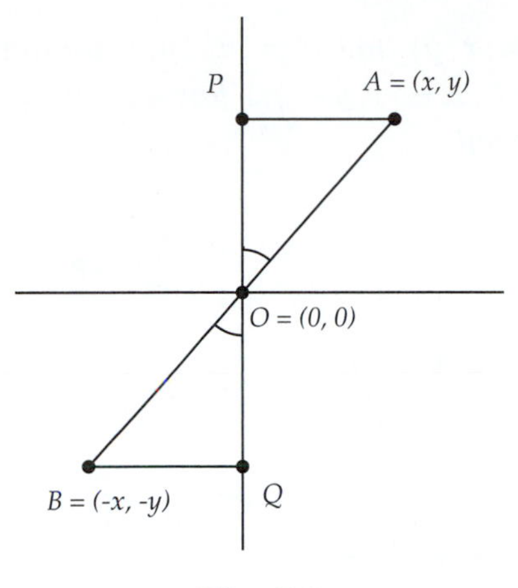

Fig. 3.3

Proof (refer to Fig. 3.3): If A is on either of the two axes, then so is B and the result is true by the definition of positive and negative coordinates on the axes.

Otherwise, we can assume that A (and thus B) is not on either axis. Drop perpendiculars from A and B to the y-axis at P and Q. Then, the distance from P to the origin will be y as is the distance from Q to the origin. By the distance formula $\overline{PA} \cong \overline{QB}$ and $\overline{AO} \cong \overline{BO}$. Thus, by SSS $\Delta AOP \cong \Delta BOQ$ and the angles at O in these triangles are congruent. Since $\angle BOQ$ and $\angle BOP$ are supplementary, then so are $\angle AOP$ and $\angle BOP$. Thus, A, O, and B lie on a line.

If A and B were on the same side of O on the line through A,B, then since this line intersects the x-axis only at O, they must be on the same side of the x-axis. But, then P and Q would have to be on the same side of the x-axis, which is not the case. □

Theorem 3.2. *Let $A = (x, y)$ and k be a number. Then, $B = (kx, ky)$ is on the line through A and the origin, and the distance from B to the origin is equal to k times the distance from A to the origin.*

Proof: This can be proved using similar triangles and is left as an exercise. □

Theorem 3.3. *If $A = (x, y) \neq (0, 0)$ and $B = (x_1, y_1)$ are points on the same line through the origin, then $(x_1, y_1) = k(x, y)$ for some number k.*

The proof of this theorem is left as an exercise.

Theorem 3.4. *If $A = (x, y)$ and $B = (x_1, y_1)$ are not collinear with the origin, then the point $C = (x + x_1, y + y_1)$ forms, with O, A, B, a parallelogram where $\overline{OC}$ is the diagonal.*

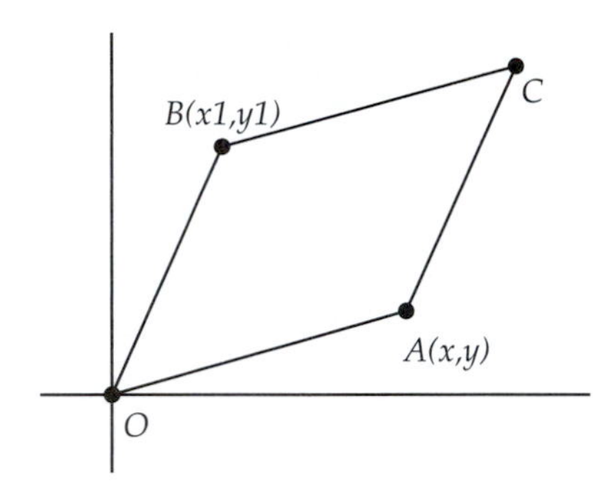

Fig. 3.4

Proof: First we have that $AC = \sqrt{(x + x_1 - x)^2 + (y + y_1 - y)^2}$ (refer to Fig. 3.4). Thus, $AC = \sqrt{(x_1)^2 + (y_1)^2} = BO$. Similarly, $AO = BC$. A simple triangle congruence argument shows that $OACB$ is a parallelogram and $\overline{OC}$ is the diagonal. □

The choice of a coordinate system will divide the set of non-axes points into four groups, called *quadrants*. Quadrant I consists of points with both coordinates positive. Quadrant II is the set of points where $x < 0$ and $y > 0$. Quandrant III has $x < 0$ and $y < 0$, and quadrant IV has $x > 0$ and $y < 0$. Thus, all points are either on an axis or in one of the four quadrants.

3.2 Vector Geometry

Analytic geometry has been incredibly useful in modeling natural systems such as gravity and temperature flow. One reason for analytic geometry's effectiveness is the ease of representing physical properties using coordinates. Many physical systems are governed by variables such as force, velocity, electric charge, and so forth, that are uniquely determined by their size and direction of action. Such variables are called *vectors*.

The concept of a *vector* can be traced back to the work of William Rowan Hamilton (1805–1865) and his efforts to treat complex numbers both as ordered pairs and as algebraic quantities. His investigations into the algebraic properties of vector systems led to one of the greatest discoveries of the nineteenth century, that of *quaternions*. While trying to create an algebra of multiplication for three-dimensional vectors (a task at which he repeatedly failed), he realized that by extending his vector space by one dimension, into four dimensions, he could define an operation of multiplication on vectors.

In 1843, as he was walking along the Royal Canal in Dublin, Hamilton came to a realization:

> And here there dawned on me the notion that we must admit, in some sense, a fourth dimension of space for the purpose of calculating with triples . . . An electric circuit seemed to close, and a spark flashed forth. [39]

This realization had such a profound effect on Hamilton that he stopped by a bridge on the canal and carved into the stone this famous formula:

$$i^2 = j^2 = k^2 = ijk = -1$$

Here i, j, and k are unit length vectors along the y, z, and w axes in four dimensions. Other important figures in the development of vector algebra include Arthur Cayley (1821–1895) and Hermann Grassman (1809–1877). Vector methods have come to dominate fields such as mathematical physics. We will cover the basics of vector methods and then use vectors to study geometric properties.

Definition 3.1. A *vector* is a quantity having a length and a direction. Geometrically, we represent a vector in the plane by a directed line segment or arrow. The starting point of the vector is called the *tail* and the point of direction is called the *head*. Two parallel directed segments with the same length and direction will represent the same vector. Thus, a vector is really a *set* of equivalent directed segments in the plane.

What is the coordinate representation of a vector? Let the vector be represented by the segment from (h, k) to (x, y) as shown in Fig. 3.5.

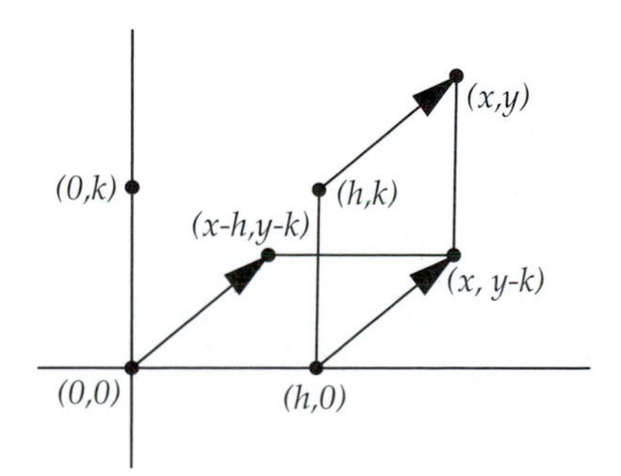

Fig. 3.5

If we consider the vector from $(h, 0)$ to $(x, y-k)$, which has its tail on the x-axis, then, the quadrilateral formed by the heads and tails of these two vectors forms a parallelogram, since by the distance formula the sides will be congruent in pairs. These two directed segments are equivalent as vectors since they have the same length and direction. Similarly, the directed segment from $(0, 0)$ to $(x-h, y-k)$ represents the same vector.

Thus, given any vector in the plane, there is a unique way of identifying that vector as a directed segment from the origin. The vector from $A = (h, k)$ to $B = (x, y)$ can be represented as the ordered pair $(x-h, y-k)$ if we assume that the vector's tail is at the origin. This is the standard way to represent vectors.

Definition 3.2. Let v be a vector represented by the ordered pair (x, y). The *norm* of v will be defined as the length of the directed segment v and will be denoted by $\|v\|$. Thus,

$$\|v\| = \sqrt{x^2 + y^2}$$

It will be convenient to define certain algebraic operations with vectors.

Definition 3.3. Let u, v be vectors and k a positive number. Then

- The vector $u+v$ is the vector representing the diagonal of the parallelogram determined by u and v, if u, v are not on the same line through the origin. If u, v are on the same line through the origin, then $u+v$ is the vector representing the sum of the x and y coordinates of the vectors.

- The vector $-v$ is the vector having the same length but opposite direction from v.

- The vector ku is the vector in the same direction as u whose length is k times the length of u.

It follows from the definitions that the vector $u - v$, which is $u + (-v)$, is the vector from v to u (Fig. 3.6).

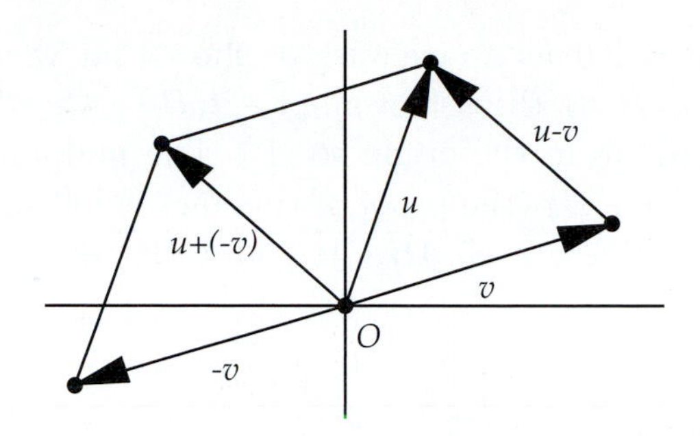

Fig. 3.6

Theorem 3.5. *Let $u = (a, b)$, $v = (s, t)$ be two vectors in a coordinate system and k be a number. Then*

- *The vector $u + v$ has coordinates $(a + s, b + t)$.*
- *The vector -v has coordinates (-s, -t).*
- *The vector ku has coordinates (ka, kb).*

Proof: Since vectors can be represented by segments from the origin to points in the plane, then this theorem is basically a restatement of the algebraic-geometric properties discussed earlier in our discussion of coordinate systems. □

Vectors can be used to give some very elegant geometric proofs. For example, consider the following theorem about the medians of a triangle.

Theorem 3.6. *The medians of a triangle intersect at a common point that lies two-thirds of the way along each median.*

Let's see how we can use vectors to prove this theorem (Fig. 3.7).

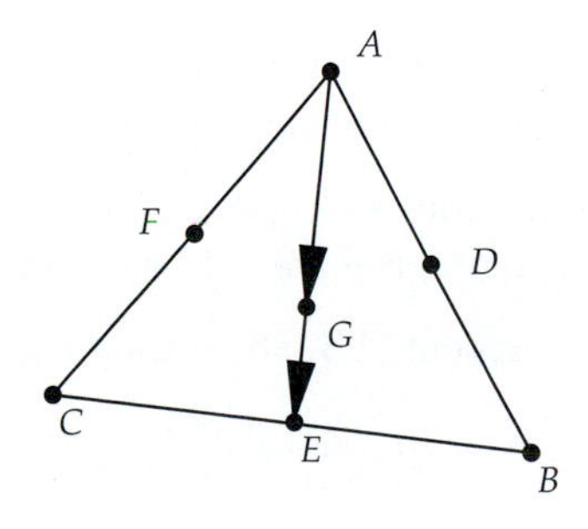

Fig. 3.7

Proof: First, it is left as an exercise to show that we can represent the line through A and B by the set of all $\vec{A} + t(\vec{B} - \vec{A})$, where t is real. ($\vec{A}$ stands for the vector from the origin to A.) The midpoint of the segment $\overline{AB}$ will then be $\frac{1}{2}(\vec{A} + \vec{B})$ (the proof of this fact is left as an exercise).

Similarly, the midpoints of $\overline{AB}$, $\overline{AC}$, and $\overline{BC}$ are given by the three vectors

$$\begin{aligned}\vec{D} &= \frac{1}{2}(\vec{A} + \vec{B}) \\ \vec{E} &= \frac{1}{2}(\vec{B} + \vec{C}) \\ \vec{F} &= \frac{1}{2}(\vec{A} + \vec{C})\end{aligned} \tag{3.1}$$

Consider the median $\overline{AE}$. Let G be the point two-thirds along this segment from A to E. Then

$$\vec{G} - \vec{A} = \vec{AG} = \frac{2}{3}\vec{AE} = \frac{2}{3}(\vec{E} - \vec{A}) = \frac{2}{3}(\frac{1}{2}(\vec{B} + \vec{C}) - \vec{A}) = (\frac{1}{3}\vec{B} + \frac{1}{3}\vec{C} - \frac{2}{3}\vec{A})$$

Adding $\vec{A}$ to both sides yields

$$\vec{G} = \frac{1}{3}(\vec{A} + \vec{B} + \vec{C})$$

A similar argument shows that the point that is two-thirds along the other two medians is also $\frac{1}{3}(\vec{A} + \vec{B} + \vec{C})$. Thus, these three points can be represented by the same vector, and thus the three medians meet at the point represented by this vector. □

Exercise 3.2.1. Prove Theorem 3.2.

Exercise 3.2.2. Prove Theorem 3.3.

Exercise 3.2.3. Let l be the line through a point $P = (a, b)$ and parallel to vector $v = (v_1, v_2)$. Show that if $Q = (x, y)$ is another point on l, then $(x, y) = (a, b) + t(v_1, v_2)$ for some t. [Hint: Consider the vector from P to Q.]

Exercise 3.2.4. Use the previous exercise to show that given a line l, there are constants A, B, and C such that $Ax + By + C = 0$, for any point (x, y) on l.

Exercise 3.2.5. Given a segment $\overline{AB}$, show that the midpoint is represented as the vector $\frac{1}{2}(\vec{A} + \vec{B})$.

Exercise 3.2.6. Given a quadrilateral $ABCD$, let W, X, Y, Z be the midpoints of sides $\overline{AB}, \overline{BC}, \overline{CD}, \overline{DA}$, respectively. Use vectors to prove that $WXYZ$ is a parallelogram.

3.3 Project 5 - Bézier Curves

So far we have considered fairly simple geometric figures such as lines and circles, for which we have correspondingly simple equations in x and y. While these have incredibly useful features for the design of many everyday objects, they are not so useful for describing complex shapes such as the curves one finds in automobile designs or in the outlines of the letters of a font.

For such non-linear and non-circular curves it is very difficult to derive simple polynomial functions of x and y that completely represent the curve. In general, the more a curve oscillates, the higher the degree of polynomial we need to represent the curve. For a designer, having to calculate high-order polynomials in order to sketch curves such as the fender of a car is not a pleasant or efficient prospect.

In the 1950s and 1960s the problem of mathematically describing curves of arbitrary shape was a primary area of research for engineers and mathematicians in the aircraft and automobile industries. As computational tools for design became more and more widely used, the need for efficient mathematical algorithms for modeling such curves became one of the highest priorities for researchers.

To image a curve on a computer screen, one has to plot each individual screen pixel that makes up the curve. If one pre-computes all the (x, y) points needed for accurately representing the curve to the resolution of the screen, then this will consume significant portions of the computer's system memory. A better solution would be to have an algorithm for computing the curve that stored just a few special points and then computed the rest of the points "on the fly."

In this project we will look at a clever method of computing smooth curves by using a simple set of "control" points in the plane. This method is due to two automobile designers: Pierre Bézier (1910–1999), who worked for the French automaker Rénault, and Paul de Casteljau, who also worked for a French automaker—Citroën. The curves which these engineers discovered have become known as *Bézier curves*, whereas the algorithm we will consider for computing them has become known as *de Casteljau's algorithm*. A complete review of Bézier curves, and other curves used in computer-aided design, can be found in textbooks on computer graphics such as [23] or [16].

To start our discussion, let's consider how we would create a curved path in the plane by using a small number of defining points. Clearly, one point cannot define a two-dimensional curve and two points uniquely define a line. Thus, to have any hope of defining a truly curved path in the plane, we need at least three points.

Start *Geometry Explorer* and create three (non-collinear) points A, B, and C joined by two segments as shown.

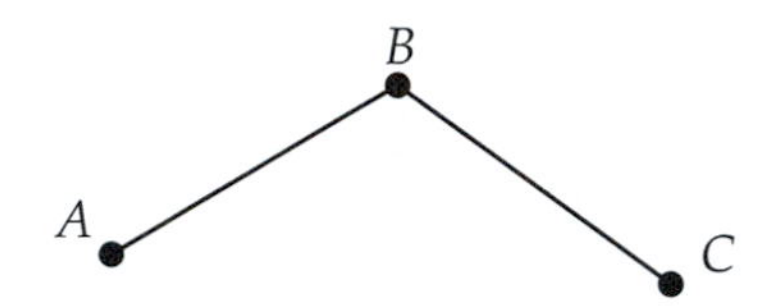

Let's imagine a curve that passes through points A and C and that is pulled toward B like a magnet. We will define such a curve *parametrically*. That is, the curve will be given as a vector function $\vec{c}(t)$ where t will be a parameter running from 0 to 1. We want $\vec{c}(0) = A$ and $\vec{c}(1) = C$, and for each $0 < t < 1$ we want $\vec{c}(t)$ to be a point on a smooth curve bending toward B. By *smooth* we mean a curve that has no sharp corners, that is, one whose derivative is everywhere defined and continuous.

So, our task is to find a way to associate the parameter t with points $\vec{c}(t)$ on the curve we want to create.

Bézier's solution to this problem was to successively *linearly interpolate* points on the segments to define $\vec{c}(t)$. Attach a point A' on the segment $\overline{AB}$ as shown and measure the ratio of AA' to AB by multi-selecting A', A, and B (in that order) and choosing **Ratio** (**Measure** menu).

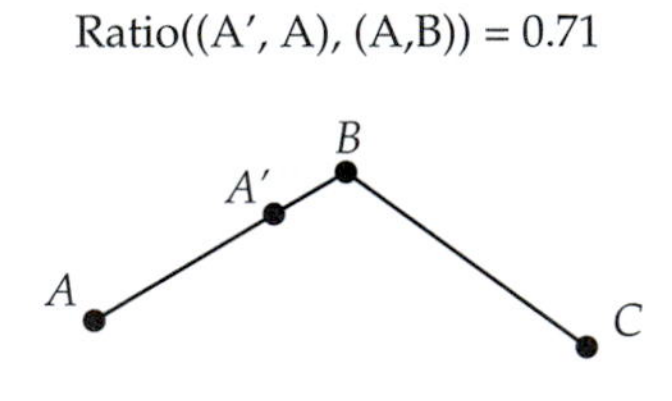

The value of this ratio is the value of t that would appear as the parameter value for the point A' in a parametric equation for $\overline{AB}$. This parametric equation is given by $\vec{l_1}(t) = \vec{A} + t(\vec{B} - \vec{A})$, with $0 \leq t \leq 1$. (Refer to Exercise 3.2.3.) We will use this ratio parameter as the defining parameter for the curve $\vec{c}(t)$ we are trying to create. Note that when $t = 0$ we are at A, which is what we want. But when $t = 1$, we are only at B and not at C.

We will now interpolate a point between B and C that has the same ratio as the point A' did along the segment $\overline{AB}$. To do this, we will use the dilating (or scaling) capability of *Geometry Explorer*. A dilation is defined by a center point and a numerical ratio. To define these two values we will use the **Mark** pop-up menu in the Transform panel. This is a button that, when clicked, will pop up a menu of items that can be selected.

To define the dilation that will create the desired point between B and C, we first select B and choose **Center** under the **Mark** pop-up menu. We will use the already created ratio measurement as our dilation ratio. To do this, click the mouse somewhere on the text of the measurement on the screen. A red box will appear surrounding the measurement to show it has been selected.

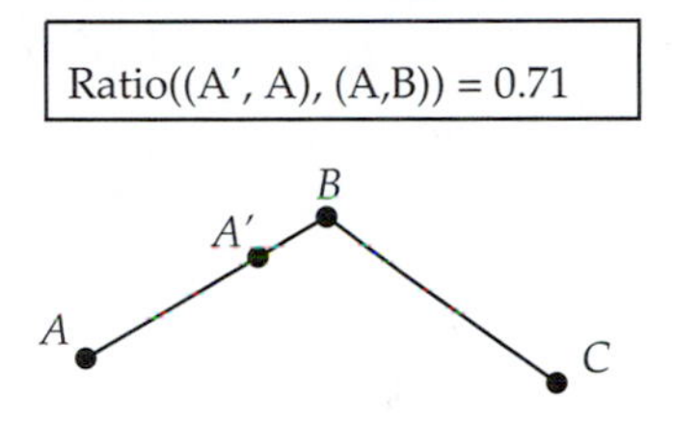

Next, choose **Ratio** under the **Mark** pop-up menu to define this measurement as the dilation ratio we will use to scale C toward B. Select C and click on the Dilate button in the Transform panel (first button in second row). Point B' will be created on $\overline{BC}$ with the same relative parameter value as A' has on $\overline{AB}$. To convince yourself of this, measure the ratio of $B'B$ to BC as shown. (Multi-select B', B, and C (in that order) and choose **Ratio** (**Measure** menu)).

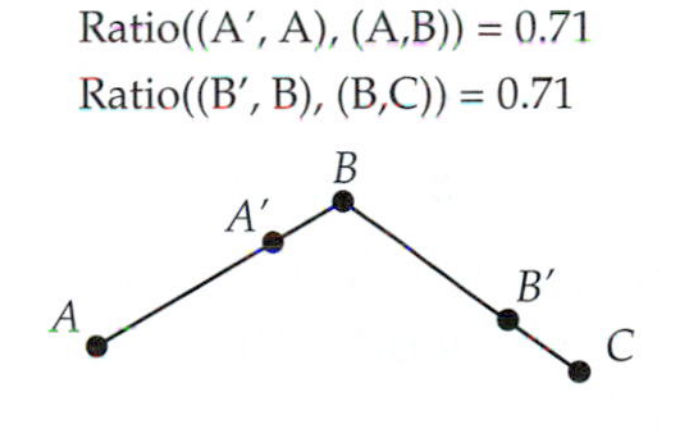

We now have parameterized both segments with two vector functions $\vec{l_1}(t) = \vec{A} + t(\vec{B} - \vec{A})$ and $\vec{l_2}(t) = \vec{B} + t(\vec{C} - \vec{B})$, where the parameter t represents the relative distance points are along their respective segments.

Since linear interpolation has worked nicely so far, we carry it one step further on the new segment $\overline{A'B'}$. Create this segment and then select A' and choose **Center** under the **Mark** pop-up menu. We will scale B' toward A' by the same factor t as for the other segments. To do this, select B' and click on the Dilate button.

Ratio((A′, A), (A,B)) = 0.71
Ratio((B′, B), (B,C)) = 0.71

This new point is the point we will use for our definition of $\vec{c}(t)$. Let's calculate the precise form for $\vec{c}(t)$.

Since $\vec{c}(t)$ is the interpolated point along $\overline{A'B'}$, then it will have equation

$$\vec{c}(t) = \vec{A'} + t(\vec{B'} - \vec{A'})$$

We also know that $\vec{A'} = \vec{l_1}(t) = \vec{A} + t(\vec{B} - \vec{A})$ and $\vec{B'} = \vec{l_2}(t) = \vec{B} + t(\vec{C} - \vec{B})$. Substituting these values into $\vec{c}(t)$ we get

$$\begin{aligned} \vec{c}(t) &= \vec{A} + t(\vec{B} - \vec{A}) + t((\vec{B} + t(\vec{C} - \vec{B})) - (\vec{A} + t(\vec{B} - \vec{A}))) \\ &= \vec{A} + t(2\vec{B} - 2\vec{A}) + t^2(\vec{C} - 2\vec{B} + \vec{A}) \end{aligned}$$

We note that $\vec{c}(0) = \vec{A} = A$ and $\vec{c}(1) = \vec{C} = C$, which is exactly what we wanted! Also, this function is a simple quadratic function in t and thus is perfectly smooth. But, what does $\vec{c}(t)$ really look like?

To answer this question, hide the two measurements and hide the label for $\vec{c}(t)$, to unclutter our figure. (To hide a label, Shift-click on it while the cursor is in text mode.) Then, select $\vec{c}(t)$ and choose **Trace On** (**Edit** menu). This makes *Geometry Explorer* trace a point as it moves. Drag point A' back and forth along $\overline{AB}$ and see how $\vec{c}(t)$ traces out the curve we have just calculated.

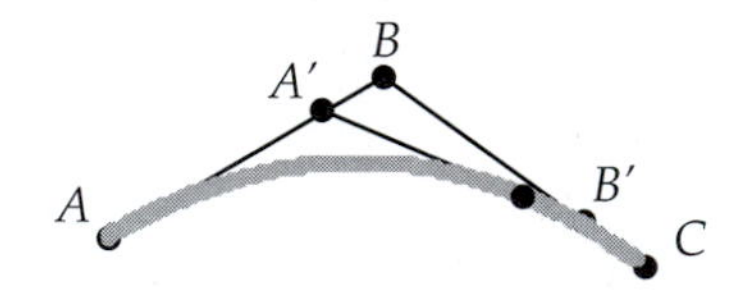

Exercise 3.3.1. It appears from our picture that the curve $\vec{c}(t)$ is actually *tangent* to the vector $\vec{B}-\vec{A}$ at $\vec{c}(0)$ and to the vector $\vec{C}-\vec{B}$ at $\vec{c}(1)$. Calculate the derivative of $\vec{c}(t)$ and prove that this is actually the case.

By controlling the positions of A, B, and C we can create curves that are of varying shapes, but that have only one "bump." As we discussed earlier, for more oscillatory behavior, we will need higher order functions. A simple way to do this is to have more control points.

Clear the screen and draw four points connected with segments as shown. Attach a point A' to $\overline{AB}$ and then measure the ratio of $A'A$ to AB.

Now, review our methods of the previous construction, and construct B' and C' on $\overline{BC}$ and $\overline{CD}$ with the same parameter as A' has on $\overline{AB}$. (In your project write-up, describe the steps you take to accomplish this step and the next set of steps.)

Next, construct $\overline{A'B'}$ and $\overline{B'C'}$. Now, we have just three points, as we did at the start of the previous construction. Construct points A'' and B'' on $\overline{A'B'}$ and $\overline{B'C'}$ using the same parameter.

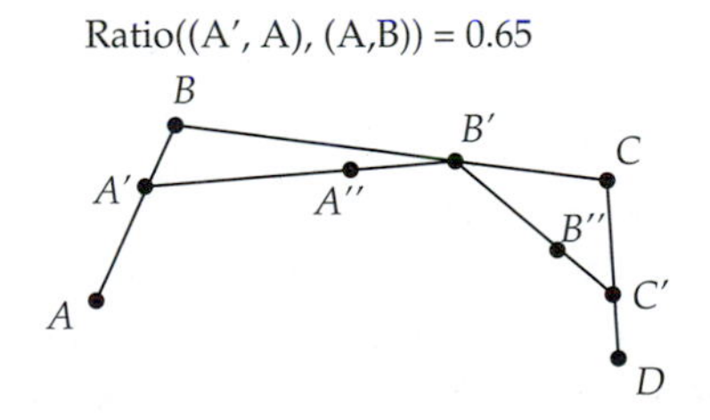

Finally, carry out the interpolation process one more time to get $\vec{c}(t)$.

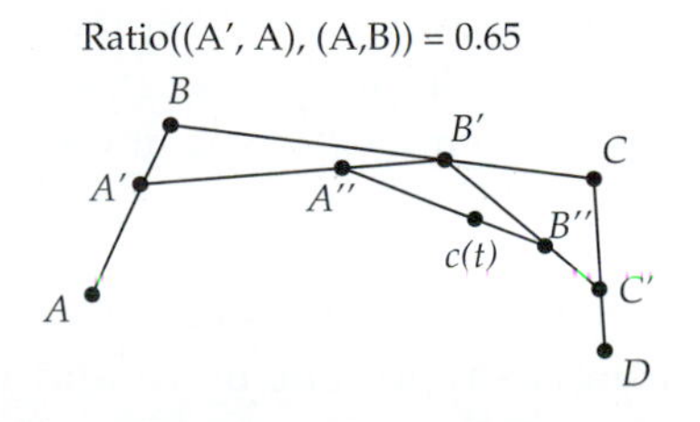

If we now trace $\vec{c}(t)$ we again get a nice smooth curve that passes through the endpoints A and D.

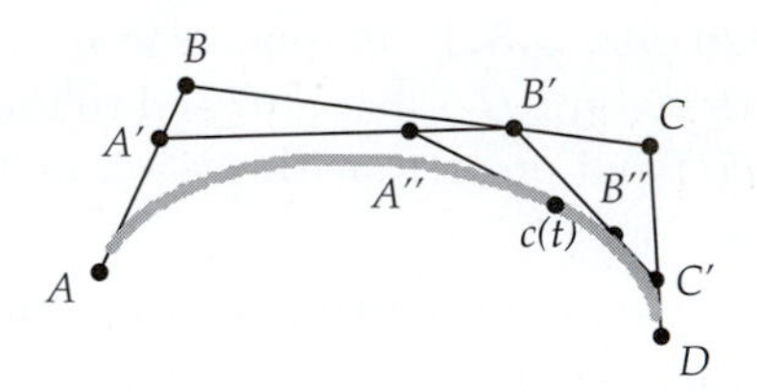

If we move C to the other side of $\overleftrightarrow{AD}$ we get an oscillation in the curve. (Make sure to clear the tracing (**Clear All Traces** (**Edit** menu)) and then re-start the tracing after you move C.)

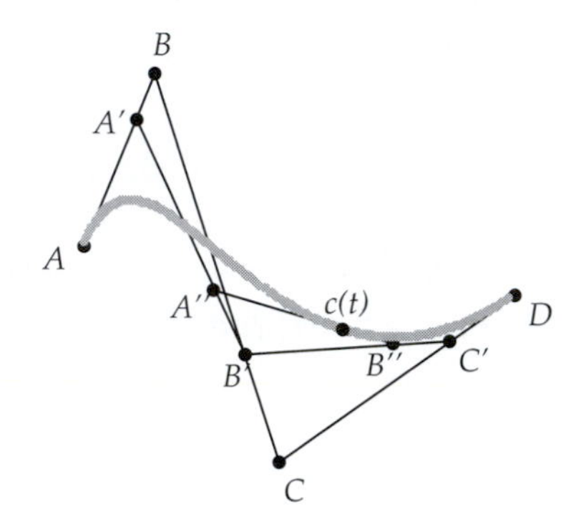

What is the equation for this new four-point Bézier curve? Clearly, the parametric form for $\vec{c}(t)$ will be

$$\vec{c}(t) = \vec{A''} + t(\vec{B''} - \vec{A''})$$

From our previous work we know that $\vec{A''} = \vec{A} + t(2\vec{B} - 2\vec{A}) + t^2(\vec{C} - 2\vec{B} + \vec{A})$ and $\vec{B''} = \vec{B} + t(2\vec{C} - 2\vec{B}) + t^2(\vec{D} - 2\vec{C} + \vec{B})$.

Exercise 3.3.2. Substitute these values of $\vec{A''}$ and $\vec{B''}$ into the equation of $\vec{c}(t)$ and show that

$$\vec{c}(t) = \vec{A} + t(3\vec{B} - 3\vec{A}) + t^2(3\vec{C} - 6\vec{B} + 3\vec{A}) + t^3(D - 3\vec{C} + 3\vec{B} - \vec{A})$$

Bézier curves generated from four points are called *cubic* Bézier curves due to the t^3 term in $\vec{c}(t)$. Bézier curves generated from three points are called *quadratic* Bézier curves.

Exercise 3.3.3. Show that $\vec{c}(t)$ has the same tangent properties as the quadratic Bézier curve. That is, show that the tangent at $t = 0$ is in the direction of $\vec{B} - \vec{A}$ and the tangent at $t = 1$ is in the direction of $\vec{D} - \vec{C}$.

We could continue this type of construction of Bézier curves for five control points, or six control points, or as many control points as we wish, in order to represent more and more complex curves. However, as we add more points, the degree of our parameteric Bézier curve increases as well. As the number of operations to compute a point increases, so too does the numerical instability inherent in finite precision computer algebra. Also, the oscillatory behavior of higher order curves is not always easy to control. Altering one control point to achieve a bump in one part of the curve can have unwanted ripple effects on other parts of the curve.

In practice, quadratic and cubic Bézier curves are the most widely used curves for computer graphics. They are easy to calculate, relatively easy to control as to shape, and can model highly complex figures by joining together separately defined curves to form one continuous curve.

As an example, let's look at one of the most important uses of Bézier curves in computer graphics—the representation of fonts. Suppose we want to represent the letter "S".

Here we have defined two four-point (cubic) Bézier curves to approximate the two bumps in the "S" shape. One curve is controlled by A, B, C, and D, with A_1 the initial parameter point and c_1 the curve point. The other curve is controlled by D, E, F, and G with A_2 the parameter point and c_2 the curve point. By "doubling up" on the point D we make the two curve pieces join together, and by making C, D, and E lie along a line, we make the curve look smooth across the joining region (the curves' tangents line up).

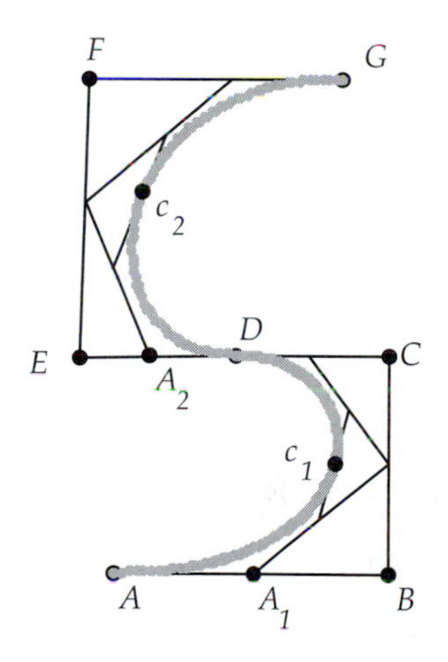

Exercise 3.3.4. Use cubic (or quadratic) Bézier curves to design your own version of the letter "b." Describe your construction in detail and provide a printout of the curves used, similar to that in the preceding example.

In the actual design of a font, we would need to define more than a simple curve representing the shape. Fonts are not made of one-dimensional curves, but are made of filled-in areas that are bounded by curves. For example, look closely at the bold letter **G** of this sentence. Notice how the vertical bulge on the left side of the shape is much thicker than the top and bottom parts. When designing a font, what we actually design is the *outline* of each letter shape; that is, the curve that surrounds the shape. Once this is defined, we fill in the outline to form the completed shape.

Project Report

The ability to model complex geometric figures is a critical component of modern computer graphics systems. Computational geometry methods, like the ones explored in this project, are at the heart of computer-aided design and also appear in the computer-generated images we see on television and in film.

For your project report, provide detailed answers to each of the questions, along with an explanation of the steps you took to construct the four-point cubic Bézier curve.

3.4 Angles in Coordinate Geometry

As mentioned in the previous chapter, angles have a very precise definition in classical Euclidean geometry, a definition that is independent of questions of orientation. In this section we will expand the notion of angle in coordinate geometry to include the idea of orientation. We start off by defining two familiar functions.

Definition 3.4. Let $\theta = \angle CAB$ be an acute angle in the plane. Let c be the circle centered at A of radius equal to AB. Let l be the perpendicular to $\overleftrightarrow{AC}$ through B, intersecting $\overleftrightarrow{AC}$ at D.

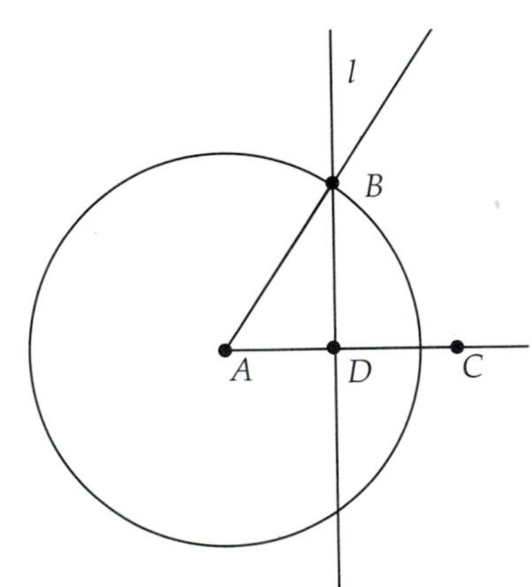

Then

$$\cos(\theta) = \frac{AD}{AB}$$
$$\sin(\theta) = \frac{BD}{AB}$$

These are the standard trig formulas everyone learns to love in high school trigonometry. Note that this definition is consistent with the definition presented in Exercise 2.5.5 of Chapter 2.

Theorem 3.7. *Let (x, y) be a point in the plane making an acute angle of θ with the x-axis. Let r be the distance from (x, y) to the origin. Then*

$$x = r\ \cos(\theta),\ y = r\ \sin(\theta)$$

Proof: This is an immediate consequence of the definition of sine and cosine. □

Corollary 3.8. *If (x, y) is a point on the unit circle (in the first quadrant) making an angle of θ with the x-axis, then $x = \cos(\theta)$ and $y = \sin(\theta)$.*

We now want to extend the definition of sine and cosine to *obtuse* angles. If $90 < \theta \leq 180$, we define

$$\begin{aligned} \cos(\theta) &= -\cos(180 - \theta) \\ \sin(\theta) &= \sin(180 - \theta) \end{aligned}$$

It is a simple exercise to show that this new definition still has $x = \cos(\theta)$ and $y = \sin(\theta)$, where (x, y) is a point on the unit circle making an obtuse angle of θ with the x-axis.

To extend this correspondence between angles and points on the unit circle into the third and fourth quadrants, we need to expand our definition of angle measure. Note that for angles whose initial side is on the x-axis, and whose measures are between 0 and 180 degrees, there is a one-to-one correspondence between the set of such angles and the lengths of arcs on the unit circle. This correspondence is given by the length of the arc subtended by the angle. Thus, it makes sense to *identify* an angle by the arclength swept out by the angle on the unit circle. We call this arclength the *radian measure* of the angle.

Definition 3.5. An angle has *radian measure* θ if the angle subtends an arc of length θ on the unit circle. If θ is positive, then the arclength is swept out in a *counterclockwise* fashion. If θ is negative, the arc is swept out in a *clockwise* fashion.

Thus, an angle of π radians corresponds to an angle of 180 degrees swept out in the counterclockwise direction from the x-axis along the unit circle. Likewise, an angle of 2π would be the entire circle, or 360 degrees, while an angle of -$\frac{\pi}{2}$ would correspond to sweeping out an angle of 90 degrees *clockwise* from the x-axis. In this definition of angle, the *direction* or *orientation* of the angle is measured by the positive or negative nature of the radian measure.

Consistent with this extension of the definition of angle measure, we extend the definitions of sine and cosine in the usual fashion consistent with the x, y values on the unit circle.

The next result is a generalization of the Pythagorean Theorem.

Theorem 3.9. *(The Law of Cosines) Let ABC be a triangle with sides of lengths a, b, c, with a opposite A, b opposite B, and c opposite C. Then*

$$c^2 = a^2 + b^2 - 2ab\ \cos(\angle ACB).$$

Proof: Let $\overline{AD}$ be the altitude to side $\overline{BC}$ at D. There are four cases to consider.

First, suppose that $D = B$ or $D = C$. If $D = B$, then $\cos(\angle ACB) = \frac{a}{b}$ and we would have $c^2 = a^2 + b^2 - 2a^2 = b^2 - a^2$, which is just a restatement of the Pythagorean Theorem. If $D = C$, we again have a restatement of the Pythagorean Theorem.

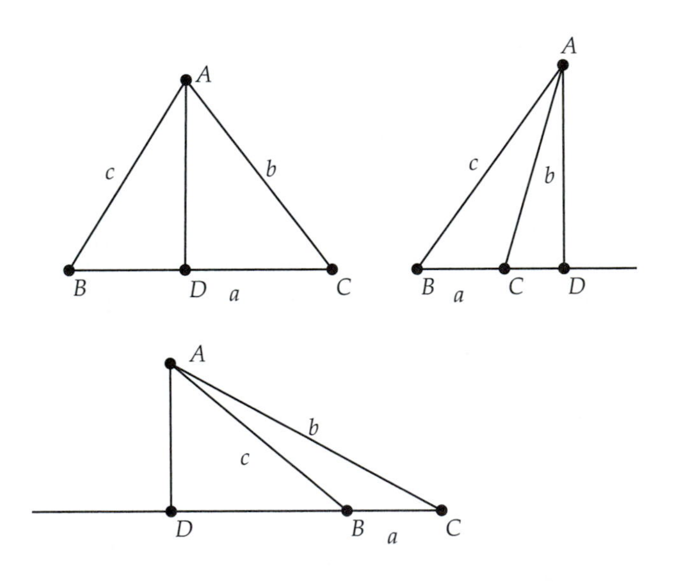

Fig. 3.8

If D is not B or C, then either D is between B and C, or D is on one or the other of the sides of $\overline{BC}$ on $\overleftrightarrow{BC}$. In all three cases (Fig. 3.8) we have

$$DB^2 + AD^2 = c^2$$

and

$$DC^2 + AD^2 = b^2$$

Solving for AD^2, we get

$$b^2 - DC^2 = c^2 - DB^2$$

Suppose that D is between B and C. Then

$$a^2 = DB^2 + 2(DB)(DC) + DC^2$$

and solving for DB^2 and substituting into the equation above, we get

$$c^2 - a^2 + 2(DB)(DC) + DC^2 = b^2 - DC^2$$

or

$$c^2 = a^2 + b^2 - 2(DB + DC)DC = a^2 + b^2 - 2a(DC)$$

Since $\cos(\angle ACB) = \frac{DC}{b}$, we have

$$c^2 = a^2 + b^2 - 2ab\,\cos(\angle ACB)$$

A similar argument can be used to show the result in the other two cases.
□

Another very useful result connecting angles and triangles is the Law of Sines.

Theorem 3.10. *(Law of Sines) Let ABC be a triangle with sides of lengths a, b, c, with a opposite A, b opposite B, and c opposite C. Then*

$$\frac{a}{\sin(\angle A)} = \frac{b}{\sin(\angle B)} = \frac{c}{\sin(\angle C)} = d \tag{3.2}$$

where d is the diameter of the circle that circumscribes the triangle.

Proof: Let σ be the circumscribing circle, which can be constructed using the techniques from the triangle project in section 2.3 of Chapter 2. Let O be the center of σ (Fig. 3.9).

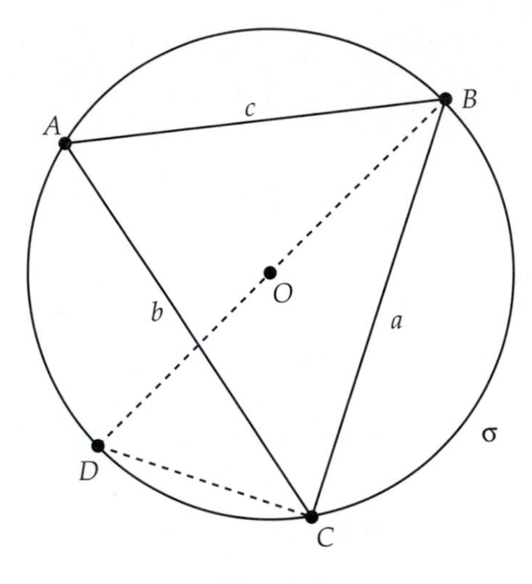

Fig. 3.9

Let D be the intersection of $\overrightarrow{BO}$ with σ. If A and D are on the same side of $\overleftrightarrow{BC}$, then $\angle BAC$ and $\angle BDC$ are congruent, as they share the same arc (refer to Corollary 2.32 in Chapter 2). Thus, $\sin(\angle A) = \sin(\angle D)$, and since $\triangle DCB$ is a right triangle (refer to Corollary 2.33), we have $a = \sin(\angle D)(BD) = \sin(\angle A)d$.

If A and D are on opposite sides of $\overleftrightarrow{BC}$, as illustrated in Fig. 3.10, it is left as an exercise to show that $\sin(\angle A) = \sin(\angle D)$ and thus $a = \sin(\angle A)d$. Similar arguments can be used to show $b = \sin(\angle B)d$ and $c = \sin(\angle C)d$.□

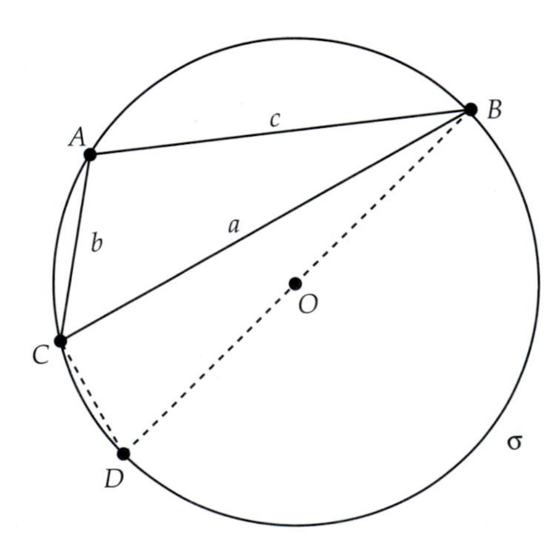

Fig. 3.10

We can connect vectors and angles via the following results.

Definition 3.6. The *dot product* of two vectors $v = (v_1, v_2)$ and $w = (w_1, w_2)$ is defined as

$$v \bullet w = v_1 w_1 + v_2 w_2$$

Theorem 3.11. *Let v, w be two vectors with tails at the origin in a coordinate system, as illustrated in Fig. 3.11. Let θ be the angle between these two vectors. Then*

$$v \bullet w = \|v\| \|w\| \cos(\theta)$$

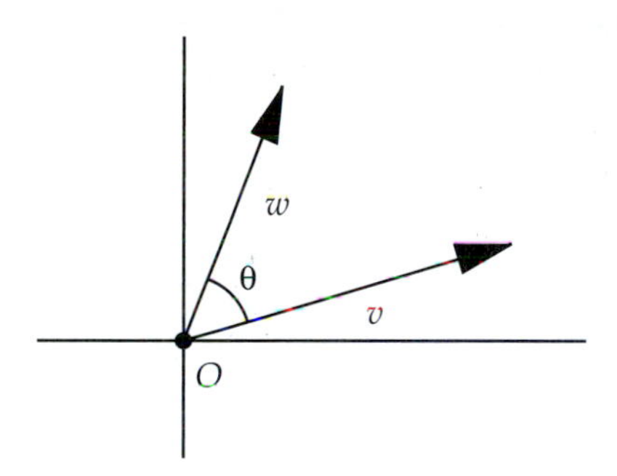

Fig. 3.11

Proof: Let $v = (v_1, v_2)$ and $w = (w_1, w_2)$. The vector from v to w will be the vector $w - v$. Let θ be the angle between v and w. By the Law of Cosines, we have that

$$\|v - w\|^2 = \|v\|^2 + \|w\|^2 - 2\|v\| \|w\| \cos(\theta)$$

Also, since $\|v\|^2 = v \bullet v$, we have that

$$\|v - w\|^2 = (v - w) \bullet (v - w) = v \bullet v - 2v \bullet w + w \bullet w$$

and, thus,

$$\|v - w\|^2 = \|v\|^2 + \|w\|^2 - 2v \bullet w$$

Clearly, $v \bullet w = \|v\| \|w\| \cos(\theta)$. □

We list here some of the useful identities from trigonometry. The proof of the first identity is clear from the definition of cosine and sine and the fact that the point $(\cos(\alpha), \sin(\alpha))$ lies on the unit circle. The proofs of the other two identities are left as exercises.

Theorem 3.12. *For any angles α, β,*

- $\sin^2(\alpha) + \cos^2(\alpha) = 1$
- $\sin(\alpha + \beta) = \sin(\alpha)\cos(\beta) + \sin(\beta)\cos(\alpha)$
- $\cos(\alpha + \beta) = \cos(\alpha)\cos(\beta) - \sin(\beta)\sin(\alpha)$

Exercise 3.4.1. Let $A = (\cos(\alpha), \sin(\alpha))$ and $B = (\cos(\beta), \sin(\beta))$. Use the definition of the dot product, and the preceding result connecting the dot product to the angle between vectors, to prove that $\cos(\alpha - \beta) = \cos(\alpha)\cos(\beta) + \sin(\beta)\sin(\alpha)$.

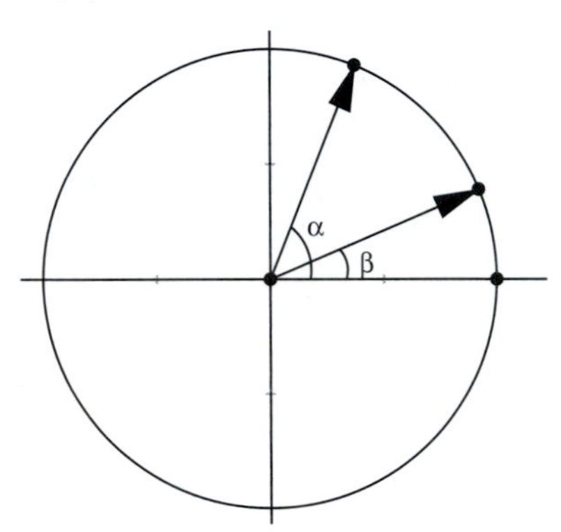

Exercise 3.4.2. Use the previous exercise to show that

$$\cos(\alpha + \beta) = \cos(\alpha)\cos(\beta) - \sin(\beta)\sin(\alpha)$$

Exercise 3.4.3. Show that

$$\sin(\alpha + \beta) = \cos(\frac{\pi}{2} - (\alpha + \beta))$$

and then use this to prove that

$$\sin(\alpha + \beta) = \sin(\alpha)\cos(\beta) + \sin(\beta)\cos(\alpha)$$

Exercise 3.4.4. Finish the proof of the Law of Sines in the case where the points A and D are on opposite sides of $\overleftrightarrow{BC}$.

3.5 The Complex Plane

Complex numbers were created to extend the real numbers to a system where the equation $x^2 + 1$ has a solution. This requires the definition of a new number (symbol), $\sqrt{-1}$. The Swiss mathematician Leonhard Euler (1707–1783) introduced the notation i for this new number in the late 1700s.

By adding this number to the reals, we get a new number system consisting of real numbers and products of real numbers with powers of i. Since $i^2 = -1$, these combinations can be simplified to numbers of the form

$$\alpha = a + ib$$

where a and b are real. The set of all possible numbers of this form will be called the set of *complex numbers* and the plane containing these numbers will be called the *complex plane*.

Historically, complex numbers were created strictly for solving algebraic equations, such as $x^2 + 1 = 0$. In the early part of the nineteenth century, Carl Friedrich Gauss (1777–1855) and Augustin Louis Cauchy (1789–1857) began working with complex numbers in a *geometric* fashion, treating them as points in the plane.

Thus, a complex number has an interesting dual nature. It can be thought of geometrically as a vector (i.e., an ordered pair of numbers), or it can be thought of algebraically as a single (complex) number having real components.

Definition 3.7. If $z = x + iy$ is a complex number, then x is called the *real part* of z, denoted $Re(z)$, and y is called the *imaginary part*, denoted $Im(z)$.

Given a complex number $z = x + iy$, or two complex numbers $z_1 = x_1 + iy_1$ and $z_2 = x_2 + iy_2$, we define basic algebraic operations as follows:

Addition-Subtraction $\quad z_1 \pm z_2 = (x_1 \pm x_2) + i(y_1 \pm y_2)$

Multiplication by Real Scalar $\quad kz = kx + iky$, for k a real number

Multiplication $\quad z_1 z_2 = (x_1 x_2 - y_1 y_2) + i(x_1 y_2 + x_2 y_1)$

Complex Conjugate $\quad \overline{z} = x - iy$

Modulus $\quad |z| = \sqrt{z\overline{z}} = \sqrt{x^2 + y^2}$

Note that complex addition (subtraction) is defined so that this operation satisfies the definition of vector addition (subtraction). In particular, complex addition satisfies the parallelogram property described earlier in this chapter.

The modulus of z is the same as the norm (length) of the vector that z represents. The conjugate of z yields a number that is the reflection of z (considered as a vector) across the x-axis.

3.5.1 Polar Form

The x and y coordinates of a complex number z can be written as

$$x = r\ \cos(\theta),\ y = r\ \sin(\theta)$$

where r is the length of $v = (x, y)$ and θ is the angle that v makes with the x-axis. Since $r = |z|$, then

$$z = x + iy = |z|(\cos(\theta) + i\ \sin(\theta))$$

The term $(\cos(\theta) + i\ \sin(\theta))$ can be written in a simpler form using the following definition.

Definition 3.8. The complex exponential function e^z is defined as

$$e^z = e^{x+iy} = e^x(\cos(y) + i\ \sin(y))$$

From this definition we can derive *Euler's Formula*:

$$e^{i\theta} = \cos(\theta) + i\ \sin(\theta)$$

Thus, the polar form for a complex number z can be written as

$$z = |z|e^{i\theta}$$

All of the usual power properties of the real exponential hold for e^z, for example, $e^{z_1+z_2} = e^{z_1}e^{z_2}$. Thus,

$$e^{i\theta}e^{i\phi} = e^{i(\theta+\phi)}$$

We also note that if $z = |z|e^{i\theta}$, then

$$\overline{z} = |z|e^{-i\theta}$$

The angle coordinate for z will be identified as follows:

Definition 3.9. Given $z = |z|e^{i\theta}$, the *argument* or *arg* of z is a value between 0 and 2π defined by

$$arg(z) = \theta\,(\text{mod } 2\pi)$$

We use here the modular arithmetic definition that a mod n represents the remainder (in $[0, n)$) left when a is divided by n. For example, 24 mod 10 is 4.

From the definition of arg, and using the properties of the complex exponential, we see that if z and w are complex numbers, then

$$arg(zw) = (arg(z) + arg(w))(\text{mod } 2\pi)$$

The proof is left as an exercise. Also, if $z = |z|e^{i\theta}$ and $w = |w|e^{i\phi}$, then $wz = |w||z|e^{i(\phi+\theta)}$.

3.5.2 Complex Functions

A complex function f in a region R of the plane is a rule that assigns to every $z \in R$ a complex number w. The relationship between z and w is designated by $w = f(z)$. In the last section, $f(z) = e^z$ defined a complex function on the entire complex plane.

Every complex function is comprised of two real-valued functions. By taking the real and imaginary parts of $w = f(z)$, we get that

$$f(x + iy) = u(x, y) + iv(x, y)$$

For example, if $f(z) = z^2$, then $u(x, y) = x^2 - y^2$ and $v(x, y) = 2xy$.

One of the simplest classes of complex functions is the set of polynomials with complex coefficients. One of the most significant results in the area of complex numbers is that every complex polynomial has at least one root, and therefore has a complete set of roots (see [28] for a proof).

Theorem 3.13. *(Fundamental Theorem of Algebra) Let $p(z)$ be a non-constant polynomial. Then, there is a complex number a with $p(a) = 0$.*

The Point at Infinity and The Extended Complex Plane

The set of points for which the function $w = \frac{1}{z}$ is defined will include all complex numbers, except $z = 0$. As z approaches 0, the modulus of w will increase without bound.

Also, for all $w \neq 0$ there is a point z for which $w = \frac{1}{z}$. Thus, $f(z) = \frac{1}{z}$ defines a one-to-one function from the complex plane (minus $z = 0$) to the complex plane (minus $w = 0$).

We call a function f that maps a set S to a set S' *one-to-one* (1-1) if it has the property that whenever $f(s) = f(t)$, then $s = t$, for s and t in S. We will call f*onto* if for all elements s' in S', there is an s in S such that $f(s) = s'$.

In order to make f a function defined on all points of the complex plane, we extend the complex plane by adding a new element, the *point at infinity*, denoted by ∞. To be more precise, we define the point at infinity as follows:

Definition 3.10. The *point at infinity* is the limit point of every sequence $\{z_n\}$ of complex numbers that is increasing without bound. A sequence is increasing without bound if for all $L > 0$ we can find N such that $|z_n| > L$ for all $n > N$.

What properties does the point at infinity have? If $\{z_n\}$ increases without bound, then $\{\frac{1}{z_n}\}$ must converge to zero. So, if $\infty = \lim_{n\to\infty} z_n$, then $0 = \lim_{n\to\infty} \frac{1}{z_n}$.

Thus, it makes sense to define $\frac{1}{\infty} = 0$ and $\frac{1}{0} = \infty$. Then, $f(z) = \frac{1}{z}$ will be a one-to-one map of the *extended* complex plane (the complex plane plus the point at infinity) onto itself.

Whereas we can conceptualize the set of complex numbers as the Euclidean (x, y) plane, the extended complex plane, with an ideal point at infinity attached, is harder to conceptualize. It turns out that the extended complex plane can be identified with a sphere through a process called *stereographic projection.*

Stereographic Projection

In a three-dimensional Euclidean space with coordinates (X, Y, Z), let S be the unit sphere, as shown in Fig. 3.12.

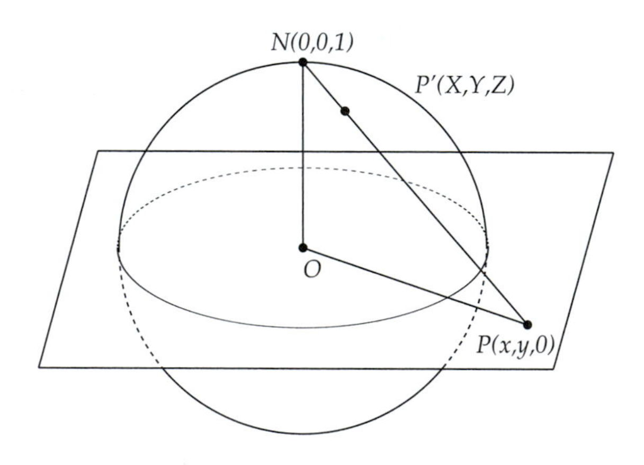

Fig. 3.12

Let N be the north pole of the sphere, the point at $(0, 0, 1)$. The sphere is cut into two equal hemispheres by the X-Y plane, which we will identify with the complex plane. Given a point $P = z = (x + iy)$ in the complex plane, we map P onto the sphere by joining N to P by a line and finding the intersection point P' of this line with the sphere.

Clearly, points for which $|z| < 1$ will map to the lower hemisphere and points for which $|z| > 1$ will map to the upper hemisphere. Also, all points in the complex plane will map to a point of the sphere, covering the sphere entirely, except for N. If we identify the point at infinity with N, we get a one-to-one correspondence between the extended complex plane and the sphere S. The coordinate equations for this map are

$$X = \frac{2x}{|z|^2 + 1}, \; Y = \frac{2y}{|z|^2 + 1}, \; Z = \frac{|z|^2 - 1}{|z|^2 + 1} \tag{3.3}$$

The derivation of these coordinate equations is left as an exercise.

The map π given by $\pi(P') = P$ identifies points on the sphere with points in the complex plane. This map is called the *stereographic projection* of S onto the complex plane.

An important property of stereographic projection is that it maps circles or lines to circles or lines.

Theorem 3.14. *Let c be a circle or line on the unit sphere. Then, the image of c under π is again a circle or line.*

Proof: We note that c is the intersection of some plane with the sphere. Planes have the general equation $AX + BY + CZ = D$, where A, B, C, and D are constants. Then, using equation 3.3 we have

$$A\frac{2x}{|z|^2 + 1} + B\frac{2y}{|z|^2 + 1} + C\frac{|z|^2 - 1}{|z|^2 + 1} = D$$

Simplifying, we get

$$2Ax + 2Bx + C(x^2 + y^2 - 1) = D(x^2 + y^2 + 1)$$

or

$$(C - D)x^2 + (C - D)y^2 + 2AX + 2By = C + D$$

If $C - D = 0$ we get the equation of a line. Otherwise, this is the equation of a circle. □

Stereographic projection also has the property that it preserves *angles*. This is true of any map of the extended complex plane that takes circles and lines to circles and lines (see [9, page 90] or [22, pages 248–254]).

3.5.3 Analytic Functions and Conformal Maps (Optional)

Two very important properties of a complex function f are its differentiability and its geometric effect on regions in the plane.

Definition 3.11. A complex function $f(z)$ is *differentiable* at z_0 if

$$\lim_{z \to z_0} \frac{f(z) - f(z_0)}{z - z_0}$$

exists. The value of the limit will be denoted as $f'(z_0)$.

The complex derivative of a function satisfies the same rules as for a real derivative: the power rule, product and quotient rules, and the chain rule. However, the fact that the limit defining the derivative is complex yields some interesting differences that one would not expect from comparison with real functions.

For example, the function $f(z) = \overline{z}$ (complex conjugation) is *not* differentiable. To see this, let $z = z_0 + h$. Then

$$\begin{aligned} \lim_{z \to z_0} \frac{f(z) - f(z_0)}{z - z_0} &= \lim_{h \to 0} \frac{f(z+h) - f(z)}{h} \\ &= \lim_{h \to 0} \frac{\overline{z+h} - \overline{z}}{h} \\ &= \lim_{h \to 0} \frac{\overline{h}}{h} \end{aligned}$$

If h is real, this limit is 1 and if h is pure imaginary, this limit is -1. Thus, the complex conjugate function is not differentiable.

The functions of most interest to us are those differentiable not only at a point, but in a region about a point.

Definition 3.12. A function $f(z)$ is *analytic* at a point z_0 if it is differentiable at z_0 and at all points in some small open disk centered at z_0.

An amazing difference between complex variables and real variables is the fact that an analytic function is not just one-times differentiable, but is in fact infinitely differentiable and has a power series expansion about any point in its domain. The proof of these results would take us far afield of our main focus of study. For a complete derivation of these results on analytic functions, see [28] or [24].

We will make extensive use of the *geometric* properties of analytic functions in Chapter 8. Analytic functions have the geometric property that angles and lengths will *conform* or be in harmony as they are transformed by the action of the function.

Definition 3.13. Let $f(z)$ be a function defined on an open subset D of the complex plane. Then, we say that

- f *preserves angles* if given two differentiable curves c_1 and c_2 intersecting at z_0 with an angle of θ between their tangents (measured from c_1 to c_2), the composite curves $f \circ c_1$ and $f \circ c_2$ have well-defined tangents that intersect at the same angle θ (measured from $f \circ c_1$ to $f \circ c_2$).

- f *preserves local scale* if for z near z_0, we have $|f(z) - f(z_0)| \approx k|z - z_0|$, with k a positive real constant, and

$$\lim_{z \to z_0} \frac{|f(z) - f(z_0)|}{|z - z_0|} = k$$

Definition 3.14. A continuous function $f(z)$ defined on an open set D is said to be *conformal* at a point z_0 in D if f preserves angles and preserves local scale.

It turns out that an analytic function is conformal wherever its derivative is non-zero.

Theorem 3.15. *If a function $f(z)$ is analytic at z_0, and if $f'(z_0) \neq 0$, then f is conformal at z_0.*

Proof: Let $z(t) = c(t)$ be a curve with $c(0) = z_0$. The tangent vector to this curve at $t = 0$ is $c'(0)$, and we can assume this tangent vector is non-zero. (To check conformality, we need to have well-defined angles.) Also, $arg(c'(0))$ measures the angle this tangent makes with the horizontal.

The image of c under f is given by $w(t) = f(c(t))$, and the tangent vector to this curve at $t = 0$ will be $\frac{dw}{dt}$ at $t = 0$. Since

$$\frac{dw}{dt} = f'(c(t))c'(t)$$

the tangent to $w(t)$ at $t = 0$ is $w'(0) = f'(c(0))c'(0) = f'(z_0)c'(0)$. Thus, $w'(0) \neq 0$ and $arg(w'(0)) = arg(f'(z_0)) + arg(c'(0))$. We see that the change in angle between the original tangent to c and the tangent to the image curve w is always $arg(f'(z_0))$. Thus, any two curves meeting at a point will be mapped to a new pair of intersecting curves in such a way that their tangents will both be changed by this constant angle, and thus the angle between the original tangents will be preserved.

For showing preservation of scale, we note that

$$|f'(z_0)| = \lim_{z \to z_0} \frac{|f(z) - f(z_0)|}{|z - z_0|}$$

Thus, for z_0 close to z, $|f(z) - f(z_0)| \approx |f'(z_0)||z - z_0|$ and f preserves local scale. □

We note here that if f is analytic, then it preserves not only the size of angles, but also their *orientation*, since the angle between two curves is modified by adding $arg(f'(z_0))$ to both tangents to get the new angle between the images of these curves.

The converse to Theorem 3.15 also holds.

Theorem 3.16. *If a function $f(z)$ is conformal in a region D, then f is analytic at $z_0 \in D$, and $f'(z_0) \neq 0$.*

For a proof of this theorem see [1].

In Chapter 8 we will be particularly interested in conformal maps defined on the complex plane or on the extended complex plane.

Theorem 3.17. *A conformal map f that is one-to-one and onto the complex plane must be of the form $f(z) = az + b$, where $a \neq 0$ and b are complex constants.*

Proof: By the previous theorem we know that f is analytic and thus must have a Taylor series expansion about $z = 0$, $f(z) = \sum_{k=0}^{\infty} a_k z^k$. If the series has only a finite number of terms, then f is a polynomial of some degree n. Then, f' is a polynomial of degree $n-1$, and if it is non-constant, then by the Fundamental Theorem of Algebra, f' must have a zero. But, $f' \neq 0$ anywhere, and thus $f(z) = az + b, a \neq 0$.

Suppose the series for f has an infinite number of terms. Then there are points α in the plane for which $f(z) = \alpha$ has an infinite number of solutions, which contradicts f being one-to-one. (The point α exists by the *Casorati-Weierstrass Theorem* and the fact that $f(\frac{1}{z})$ has an *essential singularity* at $z = 0$ (see [8, page 105] for more details)). □

Theorem 3.18. *A conformal map f that is one-to-one and onto the extended complex plane must be of the form $f(z) = \frac{az+b}{cz+d}$, where $ad - bc \neq 0$.*

Proof: If the point at infinity gets mapped back to itself by f, then f is a conformal map that is one-to-one and onto the regular complex plane; thus

$f(z) = az + b, a \neq 0$ by the previous theorem. Then $f(z) = \frac{az+b}{0z+d}$, where $d = 1$ and $ad - bc = a \neq 0$.

Otherwise, suppose that $z = \alpha$ is the point that gets mapped to infinity. Let $\zeta = \frac{1}{z-\alpha}$. Consider $w = f(\zeta)$. At $z = \alpha$ the value of ζ becomes infinite. Thus, for $w = f(\zeta)$, the point at infinity gets mapped to itself, and so $w = f(\zeta) = a\zeta + b, a \neq 0$. Then

$$w = a\left(\frac{1}{z-\alpha}\right) + b = \frac{a + b(z-\alpha)}{z-\alpha} = \frac{bz - (a + b\alpha)}{z-\alpha}$$

Since $-b\alpha - (-(a + b\alpha)(1)) = a \neq 0$, we have proved the result.

□

Definition 3.15. Functions f of the form $f(z) = \frac{az+b}{cz+d}$ are called *bilinear* transformations, or *linear fractional* transformations. If $ad - bc \neq 0$, then f is called a *Möbius transformation.*

Möbius transformations will play a critical role in our study of non-Euclidean geometry in Chapter 8. We note here that an equivalent definition of Möbius transformations would be the set of $f(z) = \frac{az+b}{cz+d}$ with $ad - bc = 1$. (The proof is left as an exercise.)

Exercise 3.5.1. Prove that $e^{i\theta}e^{i\phi} = e^{i(\theta+\phi)}$ using Euler's Formula and the trigonometric properties from section 3.4.

Exercise 3.5.2. Use Euler's Formula to prove the remarkable identity $e^{i\pi} + 1 = 0$, relating five of the most important constants in mathematics.

Exercise 3.5.3. Show that $arg(zw) = (arg(z) + arg(w))(\text{mod } 2\pi)$, where z and w are complex.

Exercise 3.5.4. Use the polar form of a complex number to show that every non-zero complex number has a multiplicative inverse.

Exercise 3.5.5. Express these fractions as complex numbers by *rationalizing the denominator*, namely, by multiplying the numerator and denominator by the conjugate of the denominator.

$$\frac{1}{2i}, \frac{1+i}{1-i}, \frac{1}{2+4i}$$

Exercise 3.5.6. Derive the stereographic equations (Equation 3.3). [Hint: If N, P', and P are on a line, then show that $(X, Y, Z - 1) = t(x, y, -1)$. Solve this for X, Y, Z and use the fact that $X^2 + Y^2 + Z^2 = 1$ to find t.]

Exercise 3.5.7. Show that stereographic projection (the map π described above) has the equation $\pi(P') = \frac{1}{1-Z}(X, Y)$. [Hint: The line through N, P', and P in Fig. 3.12 will have the form $N + t(P' - N)$. The third coordinate of points on this line will be given by $1 + t(Z - 1)$. Use this to find t.]

Exercise 3.5.8. Show that stereographic projection is a one-to-one map.

Exercise 3.5.9. Show that stereographic projection is onto the complex plane.

Exercise 3.5.10. Show that the set of Möbius transformations can be defined as the set of $f(z) = \frac{az+b}{cz+d}$ with $ad - bc = 1$.

Exercise 3.5.11. Let $f(z) = \overline{z}$. Show that f has the local scale preserving property, but has the angle-preserving property only up to a switch in the sign of the angle between tangent vectors. Such a map is called *indirectly conformal.*

3.6 Birkhoff's Axiomatic System for Analytic Geometry

In this last section of the chapter, we will consider Birkhoff's axiomatic system, a system that is quite different from Euclid's original axiomatic system. Euclid's set of axioms is somewhat cumbersome in developing analytic geometry. One has to *construct* the set of real numbers and also to develop the machinery necessary for computing distance and working with angle measure. Birkhoff's system, on the other hand, *assumes* the existence of the reals and gives four axioms for angles and distances directly.

Birkhoff's system starts with two undefined terms (*point* and *line*) and assumes the existence of two real-valued functions—a *distance* function, $d(A, B)$, which takes two points and returns a non-negative real number, and an *angle* function, $m(\angle A, O, B)$, which takes an ordered triple of points ($\{A, O, B\}$ with $A \neq O$ and $B \neq O$) and returns a number between 0 and 2π.

Birkhoff's Axioms are as follows:

The Ruler Postulate The points of any line can be put into one-to-one correspondence with the real numbers x so that if x_A corresponds to A and x_B corresponds to B, then $|x_A - x_B| = d(A, B)$ for all points A, B on the line.

The Euclidean Postulate One and only one line contains any two distinct points.

The Protractor Postulate Given any point O, the rays emanating from O can be put into one-to-one correspondence with the set of real numbers (mod 2π) so that if a_m corresponds to ray m and a_n corresponds to ray n and if A, B are points (other than O) on m, n, respectively, then $m(\angle AOB) = a_m - a_n$ (mod 2π). Furthermore, if the point B varies continuously along a line n not containing O, then a_n varies continuously also.

The SAS Similarity Postulate If in two triangles ABC and $A'B'C'$, and for some real number $k > 0$, we have $d(A', B') = k\, d(A, B)$, $d(A', C') = k\, d(A, C)$, and $m(\angle BAC) = m(\angle B'A'C')$, then the remaining angles are pair-wise equal in measure and $d(B', C') = k\, d(B, C)$.

With these four axioms and the assumed structure of the real numbers, Birkhoff was able to derive the standard set of results found in planar Euclidean geometry [6].

For example, given three distinct points A, B, and C, Birkhoff defines *betweenness* as follows: B is *between* A and C if $d(A, B)+d(B, C) = d(A, C)$. From this definition, Birkhoff defines a *segment* $\overline{AB}$ as the points A and B along with all points between them. Birkhoff then defines rays, triangles, and so on, and shows the standard set of results concerning betweenness in the plane.

The protractor postulate allows us to define *right angles* as follows: two rays m,n from a point O form a right angle if $m(\angle AOB) = \pm\frac{\pi}{2}$, where A, B are points on m, n. In this case, we say the rays are *perpendicular*. Parallel lines are defined in the usual way—as lines that never meet.

Let's see how analytic geometry can be treated as a *model* for Birkhoff's system. That is, we will see that Birkhoff's postulates can be satisfied by the properties of analytic geometry.

Define a *point* as an ordered pair (x, y) of numbers and define the distance function as

$$d((x_1, y_1), (x_2, y_2)) = \sqrt{(x_1 - x_2)^2 + (y_1 - y_2)^2}$$

Note that this definition does not suppose any *geometric* properties of points. All we are assuming is that points are ordered pairs of numbers.

Define a *line* as the set of points (x, y) satisfying an equation of the form $ax + by + c = 0$, where a, b, c are real constants, uniquely given up to a common scale factor.

We next define vectors as ordered pairs and define the angle determined by two vectors by

$$\cos(\theta) = \frac{v \bullet w}{\|v\|\|w\|}$$

where the dot product and norm of vectors is defined algebraically as before, and the cosine function can be defined as a Taylor series, thus avoiding any geometric interpretation. The angle determined by two rays emanating from a point will be defined as the angle determined by two vectors along these rays.

Does this definition of angles satisfy Birkhoff's Protractor Postulate? First of all, given any point O we can identify a ray from O with a direction vector $w = (x, y)$, with $w \neq (0, 0)$. To create a correspondence between angles and numbers, we will fix a particular direction given by $v = (1, 0)$ and measure the angle determined by w and v:

$$\cos(\theta) = \frac{v \bullet w}{\|v\|\|w\|} = \frac{x}{\|w\|}$$

We then define the sine function by

$$\sin(\theta) = \frac{y}{\|w\|}$$

If we divide w by its length, we still have a vector in the same direction as w and thus we can assume $w = (x, \sqrt{1-x^2})$, or $w = (x, -\sqrt{1-x^2})$ with $-1 \leq x \leq 1$. If $w = (x, \sqrt{1-x^2})$ we get from the above equation that $-1 \leq \cos(\theta) \leq 1$, and thus $0 \leq \theta \leq \pi$. We can extend this correspondence to angles between π and 2π in the usual way by identifying those angles between π and 2π with values of w, where $w = (x, -\sqrt{1-x^2})$.

Thus, we have created a correspondence between rays from O (and thus angles) and real numbers between 0 and 2π. Now, to check whether the second part of the Protractor Postulate holds, we suppose m, n are two rays from O, with direction vectors $w_1 = (x_1, y_1)$ and $w_2 = (x_2, y_2)$. If θ_1, θ_2 are the angles made by w_1, w_2 under the correspondence described, then

$$\begin{aligned} \cos(\theta_1) &= \frac{x_1}{\|w_1\|} \\ \cos(\theta_2) &= \frac{x_2}{\|w_2\|} \end{aligned}$$

Also, using the definitions of sine and cosine, we can show algebraically that

$$\cos(\theta_1 - \theta_2) = \cos(\theta_1)\cos(\theta_2) + \sin(\theta_1)\sin(\theta_2)$$

(Review the exercises at the end of section 3.3 to convince yourself that this formula can be proved using only the dot product and cosine expression above and the algebraic properties of sine and cosine.)

Now, by definition, if θ is the angle formed by m,n, then

$$\begin{aligned}
\cos(\theta) &= \frac{w_1 \bullet w_2}{\|w_1\|\|w_2\|} = \frac{x_1x_2 + y_1y_2}{\|w_1\|\|w_2\|} \\
&= \frac{x_1}{\|w_1\|}\frac{x_2}{\|w_2\|} + \frac{y_1}{\|w_1\|}\frac{y_2}{\|w_2\|} \\
&= \cos(\theta_1)\cos(\theta_2) + \sin(\theta_1)\sin(\theta_2) \\
&= \cos(\theta_1 - \theta_2)
\end{aligned}$$

Thus, under the correspondence described above between rays and numbers, $a_m = \theta_1$ and $a_n = \theta_2$ are the "angles" assigned to m and n, and then $m(\angle AOB) = a_n - a_m (\mathrm{mod}\ 2\pi)$, where A, B are points on n and m.

The analytic geometry model also satisfies Birkhoff's Euclidean Postulate. Let (x_1, y_1) and (x_2, y_2) be two distinct points. To find the "line" these points are on, we need to find the constants a, b, c in the line equation $ax + by + c = 0$.

Since (x_1, y_1) and (x_2, y_2) are on this line, we have

$$ax_1 + by_1 + c = 0$$

$$ax_2 + by_2 + c = 0$$

Subtracting, we get $a(x_1 - x_2) + b(y_1 - y_2) = 0$. Since (x_1, y_1) and (x_2, y_2) are distinct, we know that one of $x_1 - x_2$ or $y_1 - y_2$ are non-zero. Suppose $y_1 - y_2 \neq 0$. Then

$$b = -a\frac{x_1 - x_2}{y_1 - y_2}$$

and

$$c = -ax_1 - by_1$$

Since the constants need only be defined up to a constant multiple, we can choose $a = 1$, and the line through the two points is now well defined and, in fact, uniquely determined (up to the constant scale factor).

The Ruler Postulate follows from first choosing a point $O = (x_0, y_0)$ on a given line l to serve as the origin. Let (x_1, y_1) be a second point on l. Let $\Delta x = x_1 - x_0$ and $\Delta y = y_1 - y_0$. Then

$$a(\Delta x) = -b(\Delta y)$$

If (x, y) is any other point on l, then

$$a(x - x_0) = -b(y - y_0)$$

If $x - x_0 = t(\Delta x)$, then clearly, $y - y_0 = t(\Delta y)$, in order for the preceding equations to hold. Thus, we can write the line's equation in vector form as

$$(x, y) = (x_0, y_0) + t(\Delta x, \Delta y)$$

For the correspondence required in the Ruler Postulate, simply associate the point (x, y) with the value of $t\sqrt{(\Delta x)^2 + (\Delta y)^2}$. It is left as an exercise to check that $|x_A - x_B| = d(A, B)$ for all points A, B on l.

Finally, does the analytic geometry model satisfy the last postulate, the SAS condition for similar triangles?

Consider the Law of Cosines, described earlier in this chapter. Given a triangle ABC with side lengths $a = BC$, $b = AC$, and $c = AB$ and angles $\alpha = m(\angle BAC)$, $\beta = m(\angle ABC)$, and $\gamma = m(\angle ACB)$, then

$$c^2 = a^2 + b^2 - 2a\, b\, \cos(\gamma)$$

It is clear that if two pairs of sides in two triangles are proportional and if the included angles are equal, then by the Law of Cosines the third pair of sides must have the same constant of proportionality. Then, using the Law of Cosines again with all three sides proportional, it is clear that the other angles must be equal.

It is left as an exercise to show that the Law of Cosines can be proven from within the Birkhoff model.

We have thus proved that Birkhoff's Postulates are satisfied within the model of analytic geometry. Since Birkhoff showed that classical Euclidean geometry can be derived from his set of postulates, we see that Euclidean geometry can be derived from within an analytic model, just by using the properties of ordered pairs of numbers and analytic equations of lines and angles. Birkhoff's system does not suffer the foundational problems of Euclid's original axiomatic system. For example, there is no need to axiomatize the idea of "betweeness" since it can be analytically defined. However, the system derives its power and elegance at a price—the assumption of the existence of the real numbers. To be complete, we would have to develop the real numbers *before* using Birkhoff's system, by axiomatically creating the reals from within another system.

Exercise 3.6.1. Finish the verification that the Ruler Postulate is satisfied by the analytic geometry model. That is, show that $|x_A - x_B| = d(A, B)$ for all points A, B on l, where l is represented as $(x, y) = (x_0, y_0) + t(dx, dy)$.

Exercise 3.6.2. We have to take care when using a result, such as the Law of Cosines, to prove that Birkhoff's SAS Similarity Postulate holds for the analytic geometry model. We have to ensure that the Law of Cosines can be proved *solely* within the model itself. Derive the Law of Cosines by using the vector equation for the cosine of an angle and by using general properties of vectors. [Hint: $c^2 = \|\vec{B} - \vec{A}\|^2 = \|(\vec{B} - \vec{C}) - (\vec{A} - \vec{C})\|^2$.]

Exercise 3.6.3. Using the vector definition of sine and cosine, prove that $\sin^2(\theta) + \cos^2(\theta) = 1$.

Exercise 3.6.4. We have seen in this chapter two approaches to constructing analytic geometry. In the first approach, we construct analytic geometry from a basis of synthetic Euclidean geometry, building it from prior work on triangles, parallels, perpendiculars, angles, and so on, in the style of Euclid and Hilbert. In the second approach, we start with Birkhoff's axioms for geometry and then consider analytic geometry as a *model* for this axiom set. Which approach do you think is better pedagogically?

Exercise 3.6.5. The invention of analytic geometry has often been described as the "arithmetization" of geometry. What is meant by this?

Chapter 4

Constructions

Geometry is the science of correct reasoning on incorrect figures.
—George Pólya (1887–1985) (from [35, page 208])

4.1 Euclidean Constructions

The quote by Pólya is somewhat tongue-in-cheek, but contains an important nugget of wisdom that is at the heart of how the Greeks viewed geometric constructions. For Euclid a geometric figure drawn on paper was only an *approximate* representation of the abstract, *exact* geometric relationship described by the figure and established through the use of axioms, definitions, and theorems.

When we think of drawing a geometric figure, we typically imagine using some kind of straightedge (perhaps a ruler) to draw segments and a compass to draw circles. Euclid, in his first three axioms for planar Euclidean geometry, stipulates that there are *exact*, *ideal* versions of these two tools that can be used to construct *perfect* segments and circles. Euclid is making an abstraction of the concrete process of drawing geometric figures so that he can provide logically rigorous proofs of geometric results.

But Euclid was paradoxically quite *concrete* in his notions of what constituted a proof of a geometric result. It was not enough to show a figure or result *could* be constructed—the *actual* construction had to be demonstrated, using an ideal straightedge and compass or other constructions that had already been proved valid.

In this section we will follow Euclid and assume the ability to construct a segment connecting two given points and the ability to construct a circle with a given point as center and a constructed segment as radius. From these two basic constructions, we will develop some of the more useful constructions that appear in Euclidean geometry.

Many of these constructions are well known from high school geometry and proofs of validity will be developed in the exercises.

Construction 4.1. *(Copying an Angle) To copy the angle defined by two rays* $\overrightarrow{AB}$ *and* $\overrightarrow{AC}$ *to ray* $\overrightarrow{DE}$*, we*

- *Construct the circle at* A *of radius* AB*. This will intersect* $\overrightarrow{AC}$ *at point* F*.*
- *Copy segment* $\overline{AB}$ *to* $\overrightarrow{DE}$*, getting segment* $\overline{DG}$*.*
- *Copy a circle centered at* B *of radius* BF *to point* G*. Let* H *be an intersection of this circle with the arc used in part 2. Then* $\angle HDG$ *will have the same angle measure as* $\angle BAC$ *(Fig. 4.1).*

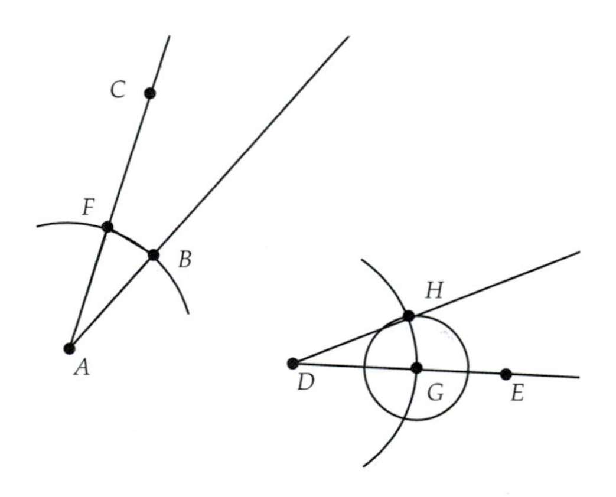

Fig. 4.1 Copying an Angle

Proof: The proof is left as an exercise. □

Note in this construction the implied ability to copy a segment from one place to another. This assumes the existence of a *non-collapsing* compass, that is, a compass that perfectly holds the relative position of its dividers as you move it from one place to another.

Euclid, in the second proposition of Book I of *The Elements*, proves that one can copy the length of a given segment to another point, using only the two axiomatic constructions and Proposition 1 (the construction of an equilateral triangle). It can be inferred that Euclid does not assume that a compass preserves its position as it is moved, for if he did, then there would be no need for the second proposition.

Why would Euclid make the assumption of a *collapsing* compass? Just as he strove to make his set of axioms as simple and economical as possible, so too he sought to make the *ideal* tools of construction as simple and free of ambiguity as possible. Thus, if one could not guarantee that the dividers of a compass would stay perfectly fixed when lifted from a drawing, then Euclid did not want to assume this property for his ideal compass.

It is interesting to note, however, that Proposition 2 implies that one can assume either a collapsing compass or a non-collapsing compass with equivalent results. For Proposition 2 implies that a collapsing compass and straightedge allow one to copy segment lengths from one position to another. Because of this equivalence, we will assume compasses are non-collapsing for the rest of this chapter.

Two quite elegant and simple constructions are those of bisecting a segment and bisecting an angle.

Construction 4.2. *(Perpendicular Bisector of a Segment) To bisect segment $\overline{AB}$ we set the compass center at A and construct a circle of radius AB. Likewise we set the compass center at B and draw a circle of radius AB. These two circles will intersect at two points C, D. The line through C, D will be a perpendicular bisector of $\overline{AB}$ and will intersect $\overline{AB}$ at the midpoint M of $\overline{AB}$ (Fig. 4.2).*

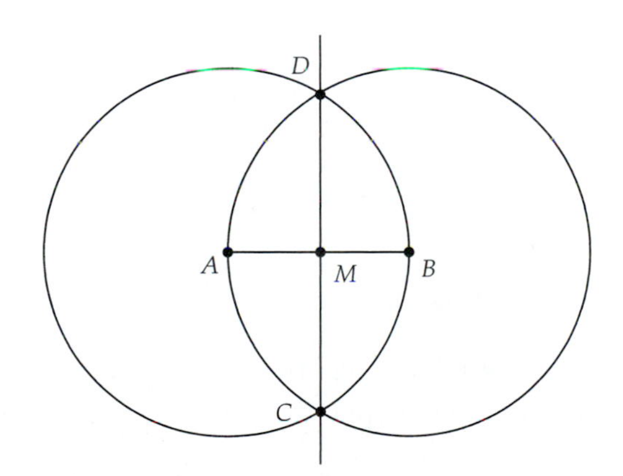

Fig. 4.2 Perpendicular Bisector of a Segment

Proof: The proof is left as an exercise. □

Construction 4.3. *(Bisecting an Angle) To bisect the angle defined by rays $\overrightarrow{AB}$ and $\overrightarrow{AC}$:*

- *At A construct a circle of radius AB, intersecting $\overrightarrow{AC}$ at D.*

- *At B construct a circle of radius BD and at D construct a circle of radius BD. These will intersect at E.*

- *Construct ray $\overrightarrow{AE}$. This will be the bisector of the angle (Fig. 4.3).*

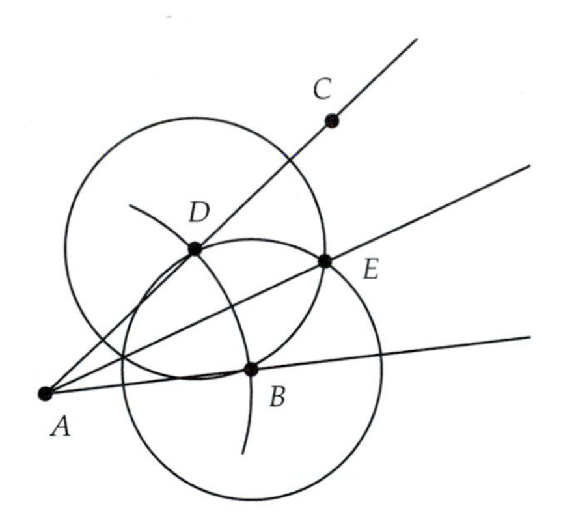

Fig. 4.3 Bisector of an Angle

Proof: The proof is left as an exercise. □

To construct a line perpendicular to a given line l at a point A, there are two possible constructions, depending on whether A is on l or off l.

Construction 4.4. *(Perpendicular to a Line through a Point on the Line) To construct a perpendicular to line l at a point A on l, we first set our compass center at A and draw a circle of some positive radius. This creates a segment $\overline{BC}$ where the circle intersects l. Then do the construction for the perpendicular bisector (Fig. 4.4).*

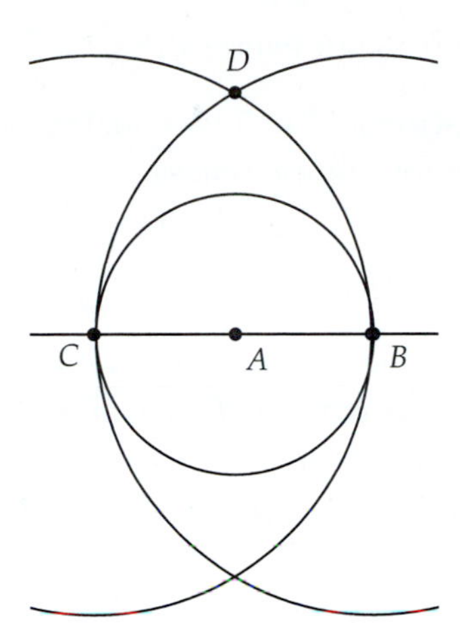

Fig. 4.4 Perpendicular to a Line through a Point on the Line

Proof: The proof is similar to the proof of the perpendicular bisector construction. □

Construction 4.5. *(Perpendicular to a Line through a Point Not on the Line) To construct a perpendicular to line l at a point A that is not on l, we first select a point B on l and set our compass center at A and draw a circle of some radius AB. Either l will be tangent to the circle at B (in which case $\overline{AB}$ will be perpendicular to l) or the circle will intersect l at two points B, C. If the circle intersects at two points, construct the angle bisector to $\angle BAC$. This will intersect $\overline{BC}$ at right angles (Fig. 4.5).*

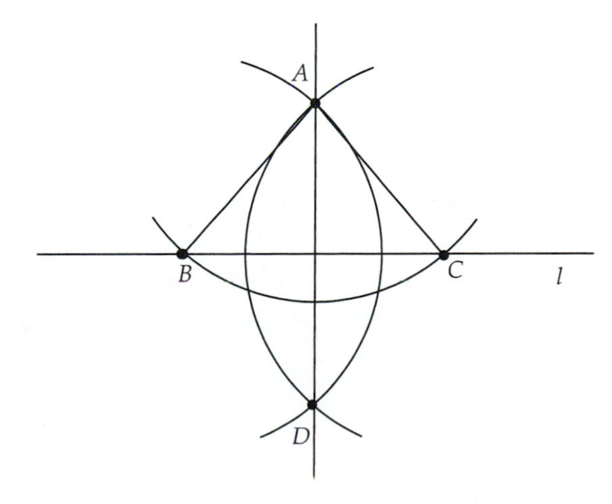

Fig. 4.5 Perpendicular to a Line through a Point Not on the Line

Proof: The proof is left as an exercise. □

We have already mentioned Euclid's construction of an equilateral triangle. We include it here for completeness.

Construction 4.6. *(Equilateral Triangle) To construct an equilateral triangle on a segment $\overline{AB}$, we construct two circles, one with center A and one with center B, with both having radius AB. Let C be an intersection of these circles. Then ΔABC is equilateral (Fig. 4.6).*

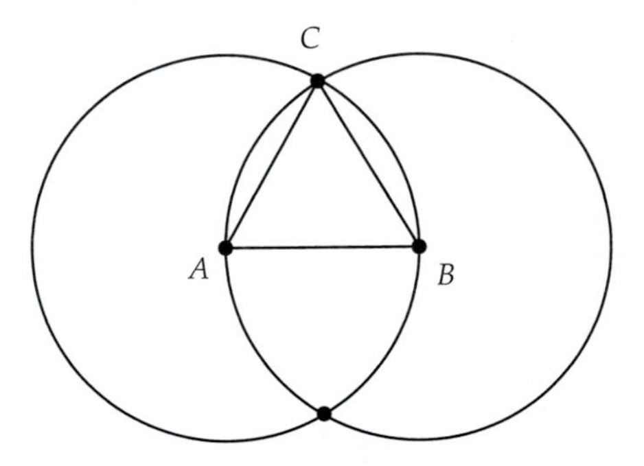

Fig. 4.6 Equilateral Triangle

Proof: The proof is left as an exercise. □

For the next set of constructions, we will not include all of the helping marks from earlier constructions but will instead just outline the major construction steps. The proofs of the correctness of these constructions can be found in Project 2.3 from Chapter 2.

Construction 4.7. *(Circumcenter) To construct the circumcenter P of ΔABC, we find the intersection of two perpendicular bisectors of the triangle's sides (Fig. 4.7).*

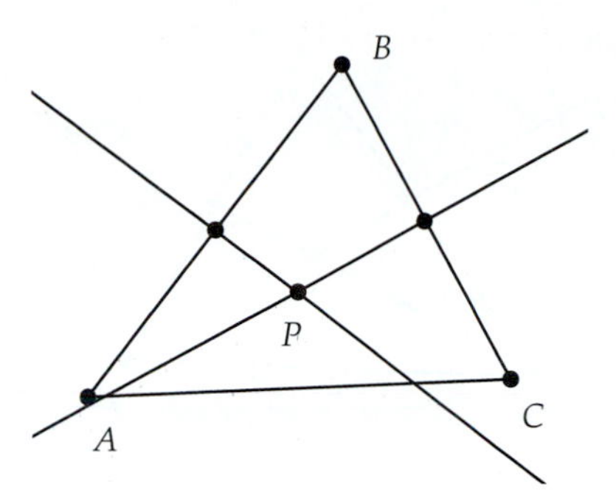

Fig. 4.7 Triangle Circumcenter

Construction 4.8. *(Orthocenter) To construct the orthocenter* O *of* ΔABC*, we find the intersection of two altitudes of the triangle. An altitude is a perpendicular to a side of the triangle that passes through the opposite vertex (Fig. 4.8).*

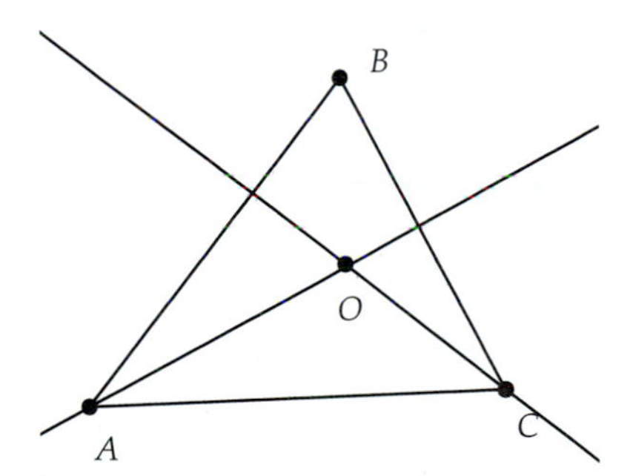

Fig. 4.8 Triangle Orthocenter

Construction 4.9. *(Incenter) To construct the incenter* I *of* ΔABC*, we find the intersection of two angle bisectors of the angles of the triangle (Fig. 4.9).*

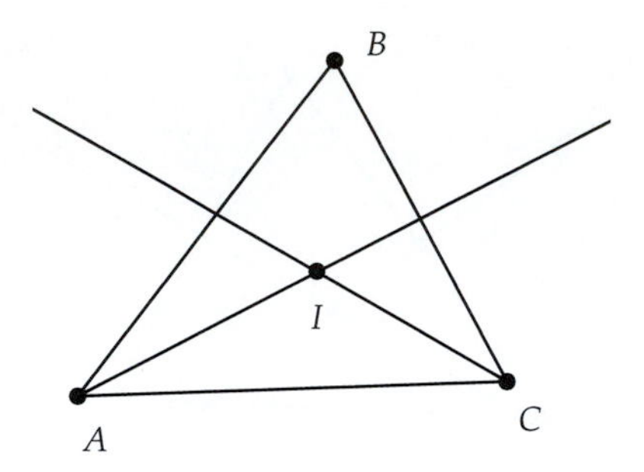

Fig. 4.9 Triangle Incenter

Construction 4.10. *(Centroid) To construct the centroid Q of ΔABC, we find the intersection of two medians of the triangle. A median is a segment from the midpoint of a side to the opposite vertex (Fig. 4.10).*

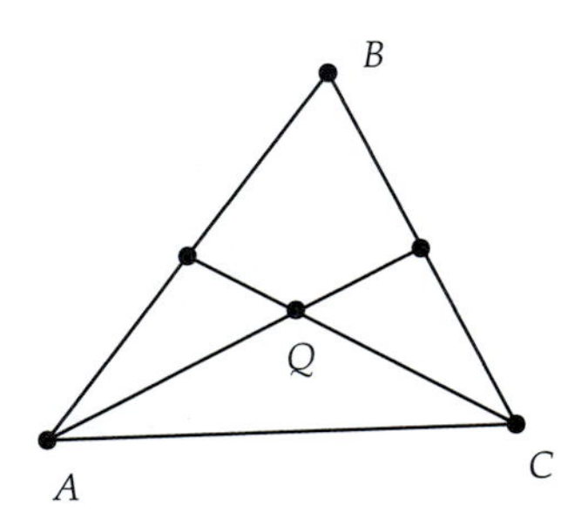

Fig. 4.10 Triangle Centroid

We will now look at some circle constructions. Again we will not put in all of the helping construction marks that are based on constructions covered earlier. The proofs of the correctness of these constructions can be found in the section on circle geometry in Chapter 2.

Construction 4.11. *(Tangent to a Circle at a Point on Circle) To construct the tangent to a circle c at a point P on the circle, we construct the line through the center O of the circle and P and then find the perpendicular to this line at P (Fig. 4.11).*

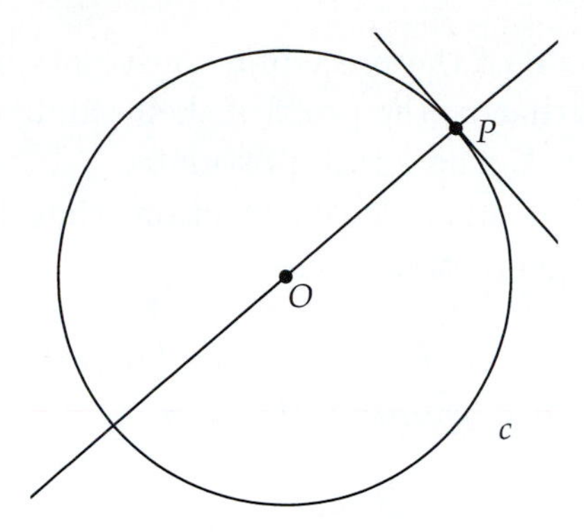

Fig. 4.11 Tangent to a Circle at a Point on Circle

Construction 4.12. *(Tangent to a Circle at a Point Not on Circle) To construct the two tangents to a circle c, with center O, at a point P not on the circle, we*

- *Construct segment $\overline{OP}$ and find the midpoint M of $\overline{OP}$.*
- *Construct the circle at M of radius OM.*
- *Let T_1 and T_2 be the two intersections of this new circle with c.*
- *Construct two lines: one through P and T_1 and the other through P and T_2. These will be tangents to c through P (Fig. 4.12).*

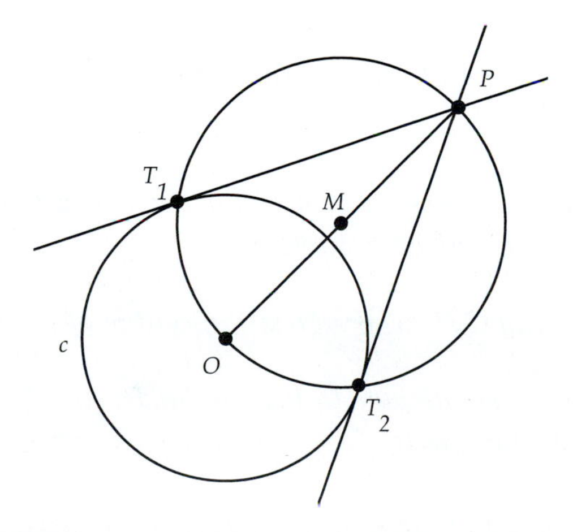

Fig. 4.12 Tangent to a Circle at a Point Not on Circle

We note here that all of the preceding constructions are independent of the parallel postulate; that is, the proof of their validity does not depend on any result derived from Euclid's fifth postulate.

Here are three Euclidean circle constructions that do rely on the parallel property of Euclidean geometry.

Construction 4.13. *(Circle through Three Points) To construct the circle through three non-collinear points A, B, C, we*

- *Construct the segments $\overline{AB}$ and $\overline{BC}$.*
- *Construct the perpendicular bisectors of these segments. Let O be the intersection of these bisectors.*
- *Construct the circle with center O and radius OA. This is the desired circle (Fig. 4.13).*

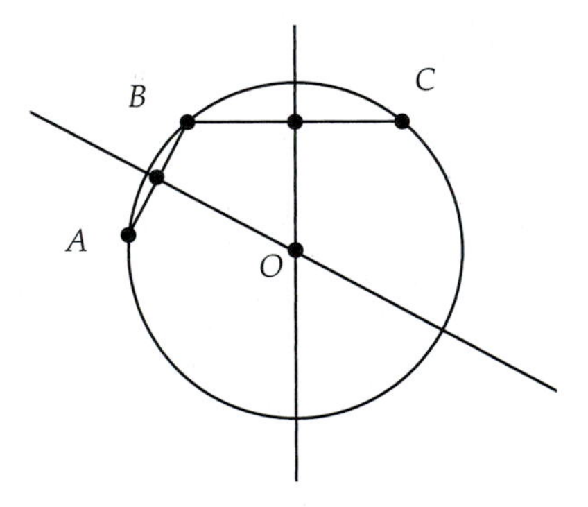

Fig. 4.13 Circle through Three Points

Construction 4.14. *(Inversion of a Point through a Circle) To construct the inverse of a point P inside a circle c, we*

- *Construct the ray $\overrightarrow{OP}$, where O is the center of c.*
- *Construct the perpendicular to this ray at P. Let T be an intersection of this ray with the circle.*
- *At T, construct the tangent to the circle. The intersection P' of this tangent with $\overrightarrow{OP}$ will be the inverse point to P through c (Fig. 4.14).*

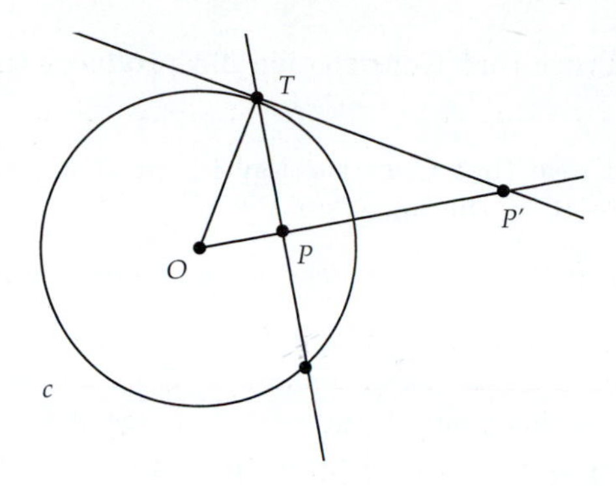

Fig. 4.14 Inversion of a Point through a Circle

Construction 4.15. *(Orthogonal Circles) To construct a circle orthogonal to a circle c through two points P, Q inside c, we*

- *Construct the inverse point P' to P through c.*
- *Construct the circle through the three points P, P', and Q (Fig. 4.15).*

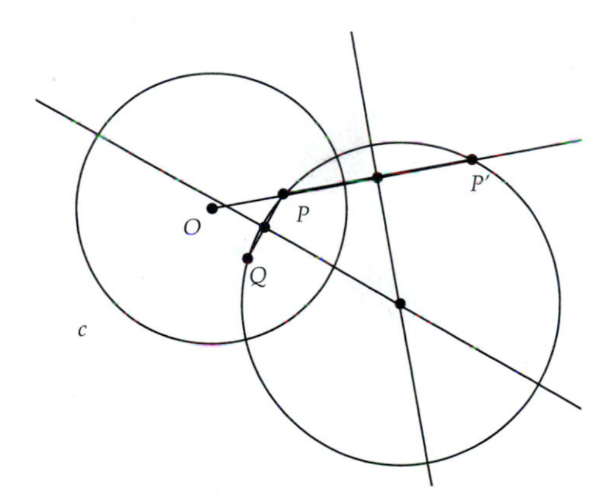

Fig. 4.15 Orthogonal Circles

This concludes our review of some of the most basic Euclidean constructions. Other basic constructions involving parallels and triangles are covered in the exercises of this section.

Exercise 4.1.1. Prove that Construction 4.1 creates a new angle congruent to the original angle.

Exercise 4.1.2. Prove that Construction 4.2 produces the perpendicular bisector of a segment.

Exercise 4.1.3. Prove that Construction 4.3 produces the angle bisector of an angle.

Exercise 4.1.4. Prove that Construction 4.5 produces the perpendicular to a line through a point not on the line.

Exercise 4.1.5. Prove that Construction 4.6 produces an equilateral triangle.

Exercise 4.1.6. In this exercise we will investigate a construction that will allow us to copy a circle with a *collapsing* compass. Given a circle c with center O and radius point A and another point B, we wish to construct a circle centered at B of radius OA. It suffices to prove the result for the case where B is outside c. (Why?) First, construct a circle centered at O of radius OB. Then construct a circle at B of radius OB. Let C and D be the intersection points of these circles. Let E be an intersection of the circle centered at B with the original circle c.

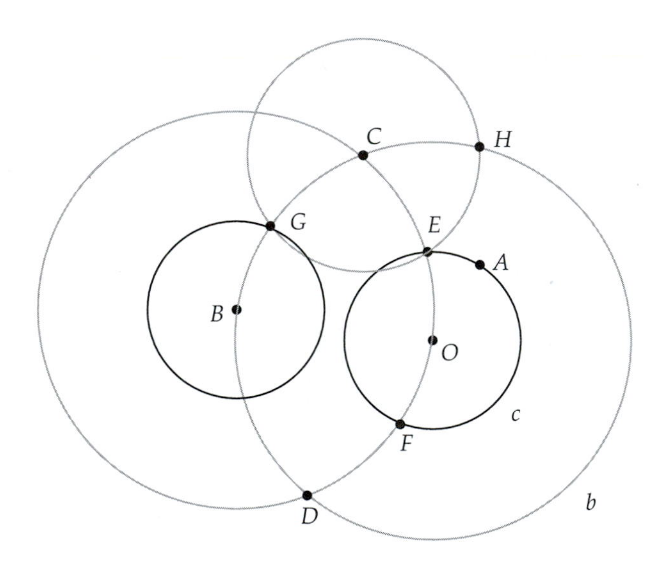

Fig. 4.16 Copying a Circle

At intersection point C, construct a circle of radius CE. This circle will intersect the circle at B of radius OB at a point G. Show that the circle with center B and radius point G is the desired circle. [Hint: Show that ΔCGO and ΔCEB are congruent, and then consider ΔGBO and ΔEOB.] Why is this construction valid for a collapsing compass?

Exercise 4.1.7. Using the perpendicular construction, show how one can construct a line parallel to a given line through a point not on the line.

Exercise 4.1.8. Draw a segment $\overline{AB}$ and devise a construction of the isosceles right triangle with $\overline{AB}$ as a base.

Exercise 4.1.9. Show how to construct a triangle given two segments a and b and an angle $\angle ABC$ that will be the included angle of the triangle. To what triangle congruence result is this construction related?

Exercise 4.1.10. Given a segment $\overline{AB}$ and a positive integer n, devise a construction for dividing $\overline{AB}$ into n congruent sub-segments.

4.2 Project 6 - Euclidean Eggs

So far we have looked at constructions of fairly traditional geometric figures—lines, circles, parallels, perpendiculars, tangents, and so on. In this project we will look at a way of joining circular arcs together in a smooth fashion to make interesting shapes.

The idea of joining curves together so that they look "smooth" across the point of attachment can be traced back to at least the time of the ancient Romans and their construction of arches and oval tracks. The fifteenth-century artist Albrecht Dürer makes great use of this technique in his design of alphabets for the printing press.

In his book *Mathographics* [12], Robert Dixon describes how to make a variety of curves and oval shapes using a simple method of smoothly joining circular arcs.

To see how this method works, we'll try a little experiment.

Start *Geometry Explorer* and create a segment $\overline{AB}$. Attach a point C to the segment and create a circle c_1 with center A and radius point C, and a circle c_2 with center B and radius point C.

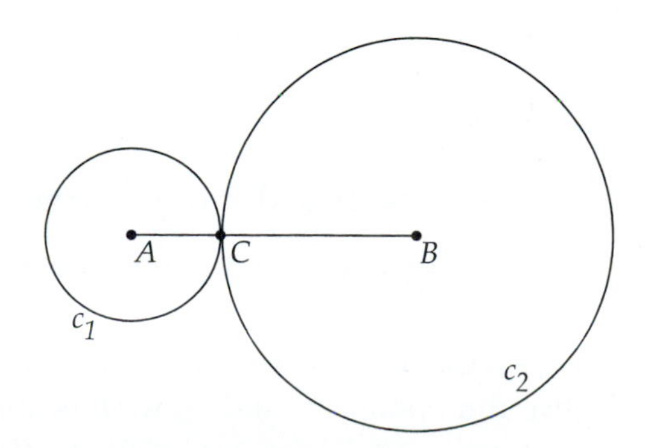

Now attach two points D and E on circle c_1 and two points F and G on c_2 as shown.

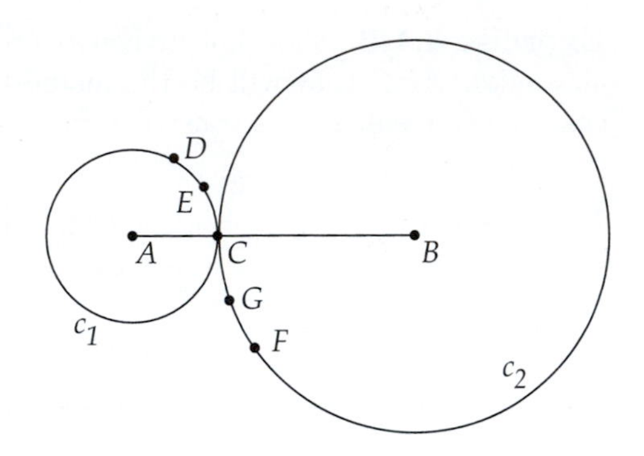

Hide the two circles and multi-select points C, D, and E. Click on the Arc construction tool to create an arc passing through these three points. Similarly, construct the arc on C, F, and G.

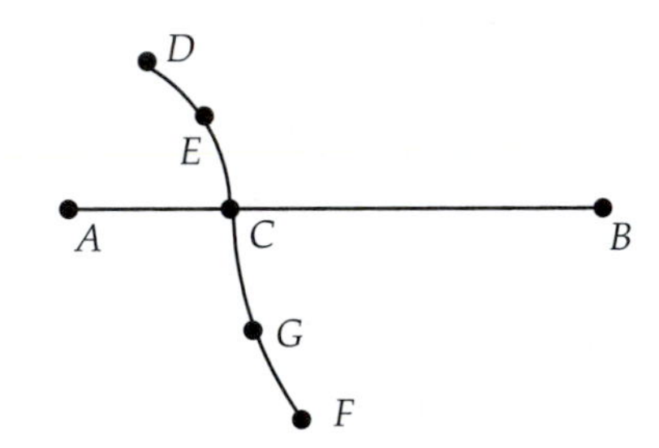

Note how smoothly these two different arcs join together at C, even though they are constructed from circles of different radii.

Exercise 4.2.1. Show that two circular arcs joined together in this fashion will always be smooth across the join point (that is, their tangents will coincide) if the following condition holds: a straight line drawn through the centers of the two arcs passes through the point where they are joined. (In the preceding figure, this would refer to $\overleftrightarrow{AB}$ passing through point C.)

We will use this idea to construct an oval, or more precisely an *egg*.

Clear the screen and create segment $\overline{AB}$. Construct the midpoint M of $\overline{AB}$ and construct a perpendicular to $\overline{AB}$ at M (refer to Fig. 4.17). Next create three circles: c_1 with center M and radius point A, c_2 with center A and radius point B, and c_3 with center B and radius point A.

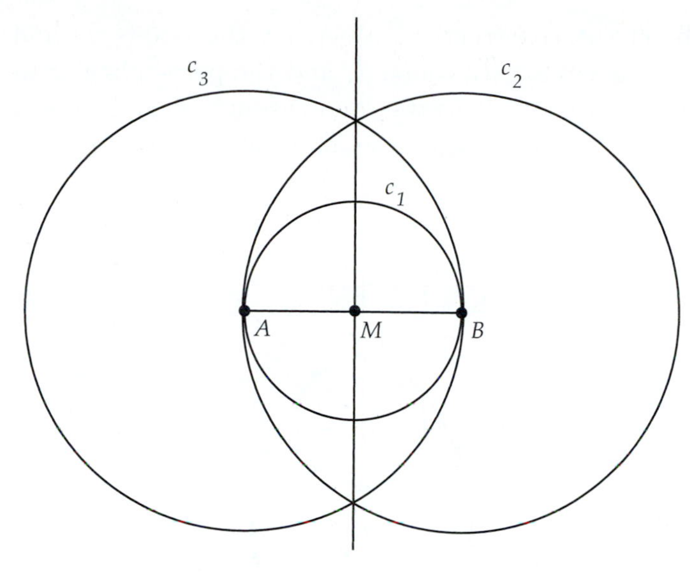

Fig. 4.17

Next construct the intersection points I_1 and I_2 of the perpendicular with c_1, and then create two rays, one from B through I_1, and one from A through I_1 (refer to Fig. 4.18). Construct the intersection points C and D of these rays with circles c_2 and c_3. Then create a circle with center I_1 and radius point C. Can you see an egg emerging from the figure?

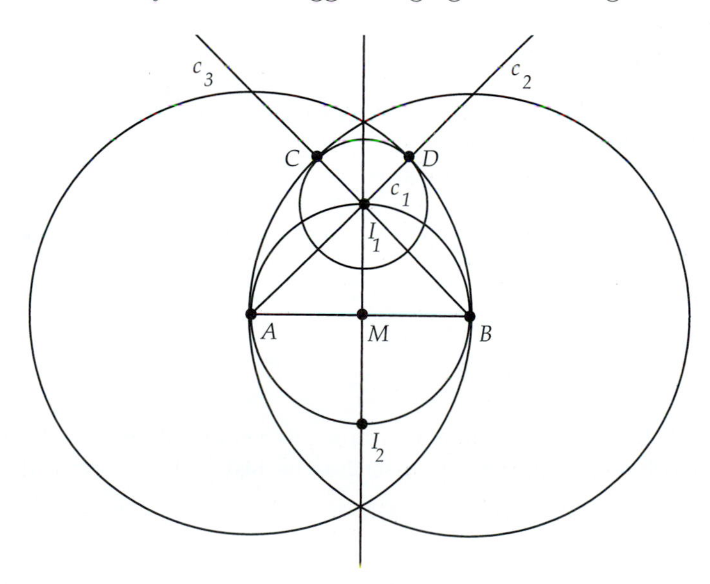

Fig. 4.18

To finish off the construction, construct the points I_3 and I_4 of the intersection of the circle with center I_1 and the perpendicular to $\overline{AB}$ (refer to Fig. 4.19). Construct two rays: one through A and I_4 and the other though B and I_4. Then construct the intersection points E and F of these rays with c_2 and c_3.

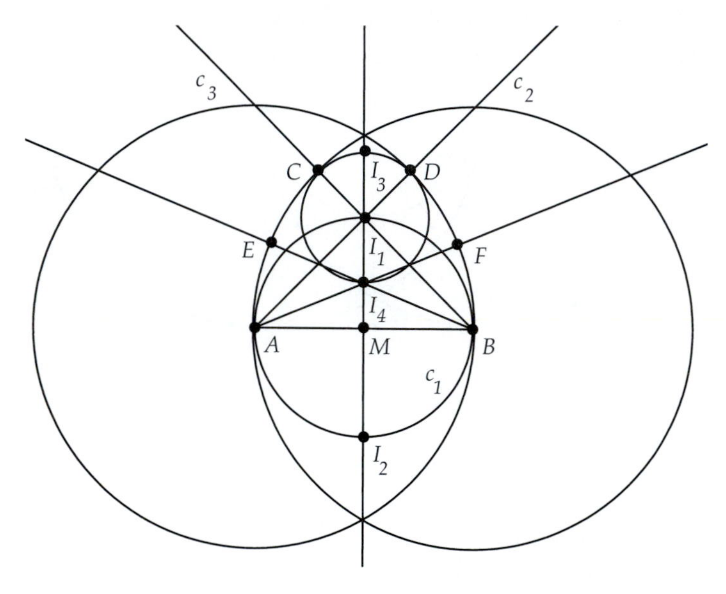

Fig. 4.19

The four arcs $\{A, E, C\}$, $\{C, I_3, D\}$, $\{D, F, B\}$, and $\{B, I_2, A\}$, will define our egg. Construct these arcs and hide all helper circles and lines.

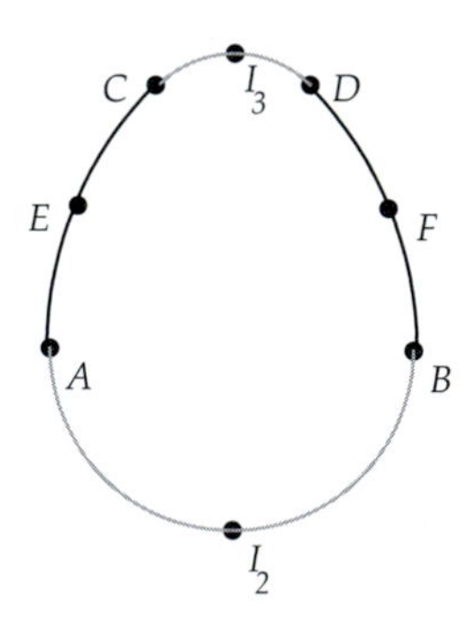

Exercise 4.2.2. Verify that these four arcs meet smoothly across the join points, using the smoothness definition discussed in the last exercise (coincident tangent lines).

Exercise 4.2.3. Here is another interesting oval shape from Dixon's book. It is made of four circular arcs. Study Fig. 4.20 and describe the construction steps necessary to construct this oval.

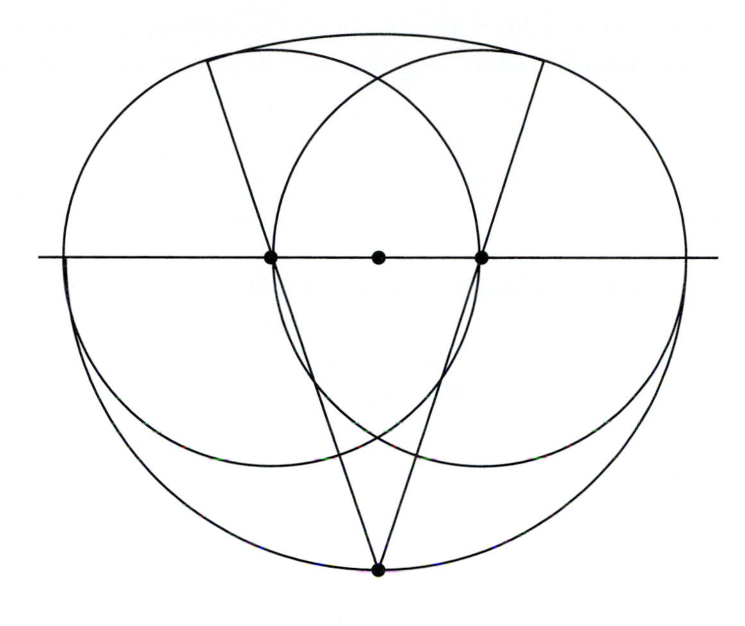

Fig. 4.20

The ability to construct complex figures from simple curves, such as circular arcs, is one of the most significant practical uses for geometry. Everything from the fonts used on a computer screen to the shapes of airframes depends on the ability to smoothly join together simple curves. In Project 3.3 we saw how quadratic and cubic Bézier curves could be used to model curved shapes. Interestingly enough, it is not possible to perfectly model a circle using Bézier curves, although we can construct approximations that will fool the eye. Thus, for shapes that require the properties of a circle (at least in small sections), we can use the joining method above to create such shapes.

In your project report you should include a discussion of the construction steps you used to create the preceding egg shapes and also include complete answers to each of the exercises.

4.3 Constructibility

In the last two sections we looked at some of the basic construction techniques of Euclidean geometry. Constructions were fundamental to Euclid's axiomatic approach to geometry, as is evident by the fact that the first two

Propositions of Book I of *Elements* deal with the construction of equilateral triangles and the transferring of segment lengths (the question of collapsing versus non-collapsing compasses).

The notion of constructibility was also fundamental to the way Euclid viewed numbers. To Euclid a number existed only in relation to a particular geometric figure—the length of a *constructed* segment. A number was the end result of a series of straightedge and compass constructions, starting with a given segment that was assumed to be of unit length.

Euclid did allow the independent existence of some numbers—the positive integers. The number theory found in the *Elements* can be traced back to the mystical beliefs of the Pythagoreans. Integers could be "perfect" or "prime." Euclid also allowed the consideration of the *ratio* of integers independent of any geometric context.

However, any other number not expressible as a ratio of positive integers (that is, *irrational*) could only be discussed as a purely geometric quantity—the end product of a sequence of geometric constructions. For example, the irrational number $\sqrt{2}$ is constructible, as it is possible to construct a right triangle with two base sides of unit length, given an existing segment of unit length. The hypotenuse of this triangle would be a segment of length $\sqrt{2}$.

Definition 4.1. A number α is *constructible* if a segment of length α can be constructed by a finite sequence of straightedge and compass constructions, starting with a given segment of unit length.

Given Euclid's insistence on the construction of numbers, it is not surprising that constructibility puzzles—riddles asking whether certain numbers could be constructed—came to be a celebrated, almost mythic, aspect of Euclidean geometry.

Three of these puzzles have occupied the attention of mathematicians from the time of Euclid (300 BC) until the work of Niels Henrik Abel (1802–1829) and Evariste Galois (1811–1832) in the early 1800s:

Doubling the Cube Given an already constructed cube, construct another cube with twice the volume of the given cube.

Angle Trisection Given an arbitrary constructed angle, construct an angle that divides the given angle into three congruent parts.

Squaring the Circle Given an already constructed circle, construct a square of area equal to the area of the circle.

The solution of these puzzles involves a deeper understanding of how constructibility relates to solutions of *algebraic* equations. The connection with algebra comes from embedding the construction of geometric figures within the context of analytic geometry.

As was mentioned previously, all constructions start with a segment assumed to be of unit length. Let $\overline{OI}$ be this segment. We know by the constructions of the first section of this chapter that we can construct a perpendicular to the line $\overleftrightarrow{OI}$ at O, and thus we can construct a Cartesian coordinate system with origin O and with $x = 1$ at the point I (Fig. 4.21).

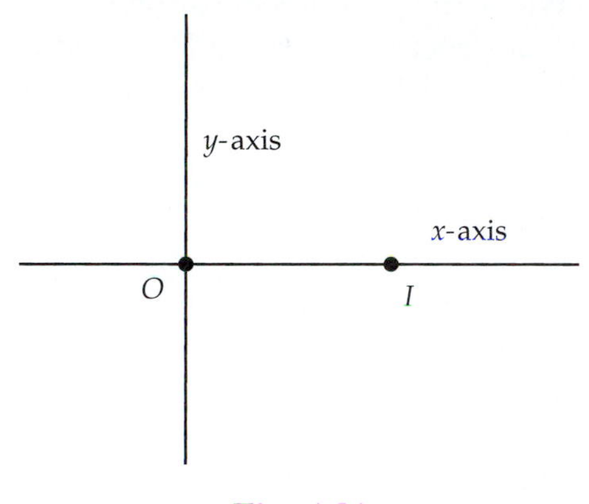

Fig. 4.21

One of the great insights of Descartes was that straightedge and compass constructions were equivalent to the solution of linear and quadratic equations.

Theorem 4.1. *Let* $P_1 = (x_1, y_1)$, $P_2 = (x_2, y_2)$, . . . , $P_n = (x_n, y_n)$ *be a set of points that have been constructed from the initial segment* $\overline{OI}$. *Then a point* $Q = (\alpha, \beta)$ *can be constructed from these points if and only if the numbers* α *and* β *can be obtained from* x_1, x_2, . . . , x_n *and* y_1, y_2, . . . , y_n *by simple arithmetic operations of* $+$, $-$, $\cdot$, *and* $\div$ *and the solution of a finite set of linear and quadratic equations.*

Proof: Since straightedge and compass constructions involve the construction of segments (and lines by extension) and circles, then a point in the plane is constructible if and only if it is either the intersection of two lines, a line and a circle, or two circles. (Or, it could be one of the two initial points O or I.)

From our work in Chapter 3, we know that a line through two constructed points, say P_1 and P_2, has the equation

$$(y - b_1)(a_2 - a_1) = (x - a_1)(b_2 - b_1)$$

This can be transformed into the form

$$Ay + Bx + C = 0$$

where A, B, and C are simple arithmetic combinations of the numbers a_1, a_2, b_1, and b_2 of the type described in the statement of the theorem.

A circle through a constructed point, say P_1, of radius equal to the length of a constructible segment, say $\overline{P_1P_2}$, will have the equation

$$(x - a_1)^2 + (y - b_1)^2 = ((a_2 - a_1)^2 + (b_2 - b_1)^2)$$

where $((a_2 - a_1)^2 + (b_2 - b_1)^2) = r^2$, r being the radius of the circle.

Again, this can be written as

$$x^2 + Bx + y^2 + Cy + D = 0$$

where the coefficients are again simple arithmetic combinations of the x and y coordinates of the constructed points.

If we are given two lines, say $A_1y + B_1x + C_1 = 0$ and $A_2y + B_2x + C_2 = 0$, then the intersection of these two lines will be given by a simple formula involving the addition, subtraction, multiplication, and division of the coefficients (proved as an exercise).

If we are given a line $Ay + Bx + C = 0$ and a circle $x^2 + Dx + y^2 + Ey + F = 0$, then if the lines intersect, we know from elementary algebra that the solution will involve solving the quadratic formula for an expression involving simple arithmetic combinations of the coefficients. However, a new operation, the square root, will be introduced in the solution.

To find the point of intersection of two circles, say $x^2 + B_1x + y^2 + C_1y + D_1 = 0$ and $x^2 + B_2x + y^2 + C_2y + D_2 = 0$, we first subtract the two equations yielding $(B_2 - B_1)x + (C_2 - C_1)y + (D_2 - D_1) = 0$. We solve this for x or y (depending on which terms are non-zero) and substitute into one of the original equations to get a quadratic equation in a single variable. The solution then follows from the quadratic formula.

Thus, we have proved that if a point can be constructed by a sequence of straightedge and compass constructions on a given set of already constructed points, then the coordinates of the new point can be obtained from simple arithmetic operations of $+$, $-$, $\cdot$, $\div$, and $\sqrt{}$ used in combination on the coordinates of the existing points.

The converse is also true. That is, any combination of using $+$, $-$, $\cdot$, $\div$, and $\sqrt{}$ on the coordinates of a set of already constructed points, that results in a pair of numbers α and β, can be realized as the coordinates of a point that arises from a sequence of straightedge and compass constructions on the given points.

To show this, it is enough to show that if $a \neq 0$ and $b \neq 0$ are given constructible numbers, then the numbers

$$a + b,\ a - b,\ ab, \frac{a}{b},\ \sqrt{a}$$

are also constructible.

It is fairly trivial to show that the numbers $a + b$ and $a - b$ are constructible. For $a + b$ we lay out a segment $\overline{OA}$ of length a on the x-axis and extend $\overline{OA}$ to a point B such that $AB = b$ (this is possible since we can transfer segment lengths). Then $\overline{OB}$ has length $a + b$. For $a - b$ we just lay off a segment for b in the negative direction from A on the x-axis.

For the product of two numbers, ab, consider the following construction:

On the x-axis, lay off the length a so that $OA = a$. Likewise, on the y-axis, lay off the unit length and b to get points I_y and B, with $OB = b$. Construct the segment from I_y to A and construct a parallel to this line through B, cutting the x-axis at C. Let $c = OC$ (Fig. 4.22).

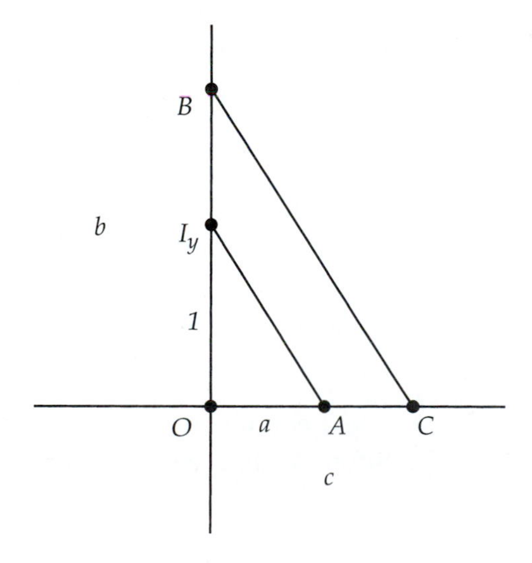

Fig. 4.22 Division

By similar triangles we have that

$$\frac{a}{1} = \frac{c}{b}$$

Clearly, $c = ab$ and we have constructed the product of a and b.

The construction of the ratio of two numbers is left as an exercise, as is the construction of the square root of a positive number. □

Starting with a unit length segment, what kinds of numbers are constructible? Clearly, all non-zero integers can be constructed. Using the construction of the ratio of two numbers, we see that all rational numbers are constructible, a rational number being a fraction of two integers. In fact, if we restrict our constructions to those that involve only the intersection of lines, it is clear that the set of rational numbers contains all such constructed numbers (if we include 0).

The set K of all constructible numbers (together with 0) forms a special type of algebraic structure called a *field*. A field is a set of elements having two operations (like $+$ and $\cdot$) that satisfies a set of properties such as associativity and commutativity for the operations. A field also satisfies the distributive property for the two operations and is *closed* under the operations and their inverses.

The prime example of a field is perhaps the set of real numbers with ordinary addition and multiplication. Given two (non-zero) real numbers, a and b, the sum $a + b$ is again a real number, as is the additive inverse -a, the product ab, and the multiplicative inverse $\frac{1}{a}$. This shows that the real numbers are closed under each of these operations.

By Theorem 4.1 we know that the set of constructible numbers K is a field, as this set is closed under the algebraic operations $+$, $-$, $\cdot$, and $\div$.

What about the taking of square roots? How does this operation affect the algebraic structure of a given set of constructible numbers? Let F be the field of rational numbers. Suppose we construct the number $\sqrt{2}$ and then consider all possible algebraic combinations of rational numbers with this new number, for example, $\sqrt{2} + 1$, $\frac{1+\sqrt{2}}{10-\sqrt{2}}$, and so on.

It turns out that all such combinations can be written in a simpler form, as we will see next.

Definition 4.2. Let F be a field contained in the real numbers and let k be a positive number in F. Suppose that $\sqrt{k}$ does not belong to F. Then the set

$$F(k) = \{x + y\sqrt{k}\ |x, y \in F\}$$

is called a *quadratic extension of F*.

Theorem 4.2. *Let F be a field contained in the real numbers, let k be an element of F, and suppose $\sqrt{k}$ does not belong to F. Then the set F' of all possible simple algebraic combinations (ones using $+$, $-$, $\cdot$, and $\div$) of elements of F and the number $\sqrt{k}$ is a field and is equal to $F(k)$.*

Proof: We first note that F' contains $F(k)$ by definition. Let z be an element of F'. Then z is obtained by algebraic operations using elements of F and the number $\sqrt{k}$. If we can show that each of these operations can be represented in $F(k)$, then we will have shown that $F(k)$ contains F' (which implies $F(k) = F'$), and we will also have shown that $F(k)$ is a field, as it will be closed under the basic operations and their inverses.

Clearly, any sum or difference of elements of F or $\sqrt{k}$ can be represented in $F(k)$. What about products? Any product of elements in F is again in F, as F is a field. Also, if a, b, c are in F, then $a(b + c\sqrt{k}) = ab + ac\sqrt{k}$ is an element in $F(k)$, as is $a(c\sqrt{k})$.

In fact, the only algebraic operation that is not obviously represented in $F(k)$ is division. Suppose a and b are in F and suppose $a + b\sqrt{k} \neq 0$. Then

$$\begin{aligned} \frac{1}{a + b\sqrt{k}} &= \frac{1}{a + b\sqrt{k}} \left(\frac{a - b\sqrt{k}}{a - b\sqrt{k}} \right) \\ &= \frac{a - b\sqrt{k}}{a^2 - b^2 k} \end{aligned}$$

which is again an element in $F(k)$. □

As an application of this notion of field extensions, we will consider cubic polynomials.

Theorem 4.3. *Given a cubic polynomial*

$$p(z) = z^3 + az^2 + bz + c = 0$$

with coefficients in a field F (contained in the reals), if $p(w) = 0$, where w is an element of a quadratic extension $F(k)$ but not an element of F, then the polynomial has another root in F.

Proof: We know that every polynomial with real coefficients can be factored into $p(z) = (z - z_1)(z - z_2)(z - z_3)$ with possibly some of the z_i being complex numbers. Equivalently, we have

$$p(z) = z^3 - (z_1 + z_2 + z_3)z^2 + (z_1 z_2 + z_1 z_3 + z_2 z_3)z - z_1 z_2 z_3 = 0$$

The coefficients for a polynomial can only be represented one way, and so

$$-(z_1 + z_2 + z_3) = a$$

Now suppose the root w of $p(z)$ is represented by $z_1 = x + y\sqrt{k}$, with x and y in F. Then $x - y\sqrt{k}$ is also a root. This can be proved by showing that $x - y\sqrt{k}$ acts like the complex conjugate when compared to $x + y\sqrt{k}$. Then we use the fact that roots come in conjugate pairs. We can assume that $z_2 = x - y\sqrt{k}$. Then

$$-((x + y\sqrt{k}) + (x - y\sqrt{k}) + z_3) = a$$

which implies that $z_3 = -(a + 2x)$ is in F. □

Corollary 4.4. *Let F be a field contained in the reals. If a cubic polynomial has a root w that is in a field F_n that is the result of a series of quadratic extensions*

$$F = F_0,\ F_1,\ F_2, \ldots, F_n$$

where each F_i is a quadratic extension of the previous F_{i-1}, then the polynomial must have a root in F.

Proof: The proof of this result is just a repeated application of the preceding theorem. □

We are now in a position to tackle two of the classic constructibility puzzles.

Duplication of the Cube

Given a cube constructed from a segment $\overline{AB}$, is it possible to construct another segment $\overline{CD}$ such that the cube on $\overline{CD}$ has volume double that of the cube on $\overline{AB}$?

If it is possible, then we have the algebraic relationship

$$(CD)^3 = 2(AB)^3$$

Or

$$\left(\frac{CD}{AB}\right)^3 = 2$$

This implies that if it is possible to double the original volume of the cube, then it must be possible to construct $\overline{CD}$ and thus it must be possible to construct the fraction $\frac{CD}{AB}$. This fraction is then a root of the cubic polynomial $z^3 - 2 = 0$. Thus, if we can carry out a sequence of straightedge and compass constructions that yield $\frac{CD}{AB}$, then this construction sequence would be mirrored in an algebraic sequence of larger and larger quadratic extensions of the rational numbers. By Corollary 4.4, letting F be the rationals, we would have that $z^3 - 2 = 0$ has a root in the rationals. But from algebra we know that the only possible rational roots of this polynomial are 1, -1, 2, and -2. Since none of these are actually roots, then no element in a quadratic extension can be a root, and the construction is impossible.

Trisection of an Angle

Given an arbitrary angle, is it always possible to construct an angle that is $\frac{1}{3}$ the measure of the given angle?

Note that a solution to this problem would imply that *all* angles can be trisected and, in particular, a 60 degree angle. We will show that it is impossible to construct a 20 degree angle and thus that the trisection puzzle has no solution.

If it is possible to construct a 20 degree angle, then it must be possible to construct the number $\cos(20^o)$, as this will be the base of a right triangle with angle of 20 degrees and hypotenuse equal to 1.

We make use of several of the trigonometric formulas from Chapter 3 to consider the formula for the cosine of 3θ for a given angle θ:

$$\begin{aligned} \cos(3\theta) &= \cos(2\theta + \theta) \\ &= \cos(2\theta)\cos(\theta) - \sin(2\theta)\sin(\theta) \\ &= (\cos^2(\theta) - \sin^2(\theta))\cos(\theta) - 2\sin(\theta)\cos(\theta)\sin(\theta) \\ &= (2\cos^2(\theta) - 1)\cos(\theta) - 2(1 - \cos^2(\theta))\cos(\theta) \\ &= 4\cos^3(\theta) - 3\cos(\theta) \end{aligned}$$

Letting $\theta = 20^o$ and using the fact that $\cos(60^o) = \frac{1}{2}$, we have

$$\frac{1}{2} = 4\cos^3(20^o) - 3\cos(20^o)$$

which is equivalent to $8\cos^3(20^o) - 6\cos(20^o) - 1 = 0$.

We conclude that if we can construct $\cos(20^o)$, then it must be a root of the polynomial $8z^3 - 6z - 1 = 0$. We will simplify the analysis of this polynomial by using the observation that a construction of $\cos(20^o)$ would imply the construction of $2\cos(20^o)$. Making the substitution $x = 2z$ in the polynomial $8z^3 - 6z - 1$, we get that $2\cos(20^o)$ is a root of $x^3 - 3x - 1 = 0$. The only possible rational roots of this polynomial are 1 and -1, and neither are actually roots. So the number $2\cos(20^\circ)$ is not constructible, and general angle trisection is impossible.

Squaring the Circle

Given a circle of radius r, can we construct a square whose area is equal to the area of the circle? Since the area of the circle is πr^2, then we are looking for a segment whose length is $\sqrt{\pi}r$. Since r is assumed constructible, then if $\sqrt{\pi}r$ is constructible, we would be able to divide by r and have a construction for $\sqrt{\pi}$. Multiplying this number by itself, we would have a construction for π.

The proof of the impossibility of the construction of π is beyond the level of this text. While the previous two puzzles could be resolved by considering roots of polynomials with rational coefficients, no such analysis will prove the impossibility of squaring the circle. This is because the number π is *transcendental*; that is, it is not the root of a polynomial with rational coefficients. This was first proved by Carl Louis Ferdinand von Lindemann (1852–1939) in 1882. An interesting historical note about Lindemann is that David Hilbert was one of his doctoral students in Germany.

In this section we have just scratched the surface as to the connection between constructibility and the algebraic theory of fields. For a more detailed review of this connection, see Chapter 19 of Moise's text [32] or the excellent book by Robin Hartshorne [19].

Exercise 4.3.1. Prove that if two lines $A_1y + B_1x + C_1 = 0$ and $A_2y + B_2x + C_2 = 0$ intersect, then the coordinates of the intersection point will be given by a simple formula involving the addition, subtraction, multiplication, and division of the coefficients of the lines.

Exercise 4.3.2. Show that the number $\sin(22\frac{1}{2}^\circ)$ is constructible. [Hint: Use a trigonometric formula for 45°.]

Exercise 4.3.3. Devise a construction for the *ratio* of two numbers.

Exercise 4.3.4. In this exercise we see how to construct square roots of positive numbers. Let $\overline{AB}$ be a segment representing a length a (Fig. 4.23). Extend $\overrightarrow{AB}$ beyond B by the unit length to a point C. Let M be the midpoint of $\overline{AC}$ and construct a circle centered at M of radius AM. Construct the perpendicular to $\overline{AC}$ at B and let D be an intersection point of this perpendicular with the circle. Show that $BD = \sqrt{a}$. [Hint: Use right triangles.]

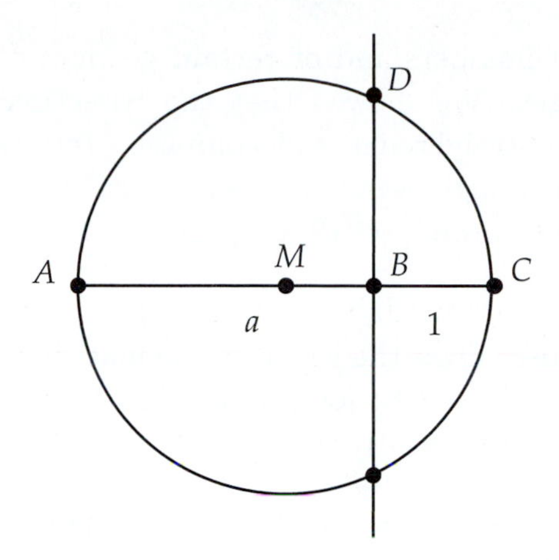

Fig. 4.23 Square Root

Exercise 4.3.5. Devise constructions for the numbers $\sqrt{3}$ and $\sqrt{5}$ that are different than the one given in the previous exercise.

Exercise 4.3.6. Show that there are an infinite number of non-constructible numbers.

Exercise 4.3.7. Show that there is at least one non-constructible number in every interval $[0, a]$ for $a > 0$.

Exercise 4.3.8. Show that every circle centered at the origin contains at least two points which are not constructible. [Hint: Use the preceding exercise.]

Exercise 4.3.9. Show that every circle contains at least two points which are not constructible. [Hint: Use the preceding exercise.]

Exercise 4.3.10. Galois was one of the more colorful figures in the history of mathematics. His work on the solvability of equations revolutionized algebra. Research this area and prepare a short report on the significance of Galois's work.

4.4 Mini-Project - Origami Construction

So far we have looked at constructions where the constructing tools are primarily a straightedge and (collapsing) compass. We have seen that a non-collapsing compass can be substituted for a collapsing one, with equivalent capabilities for Euclidean construction. We have also discussed the issue of *constructibility* of numbers and of certain geometric figures such as the trisection of an angle. We showed that the trisection of a general angle was impossible with straightedge and compass. Interestingly enough, the trisection of a general angle is possible with a ruler (*marked* straightedge) and compass. (For the proof see [19, page 260].)

The question of constructibility is thus dependent on the tool set that one is permitted to use. Over the years, mathematicians have experimented with using other types of tools; for example, rusty compasses (ones where the divider length is permanently fixed).

One method of construction that has become popular in recent years is that of paper folding, or *origami.* The seemingly simple practice of folding paper can produce quite complex geometric configurations. In this project we will investigate the geometric constructions possible in origami by setting up a set of axioms for "perfect" paper folding. This is similar to the first few axioms of Euclid's geometry, where he postulates the ability to do perfect straightedge and compass constructions.

The axioms we will use for paper folding were first formulated by Humiaki Huzita in 1992 [26]. Huzita postulates six axioms for paper folding:

(Axiom O1) Given two constructed points P and Q, we can construct (fold) a line through them.

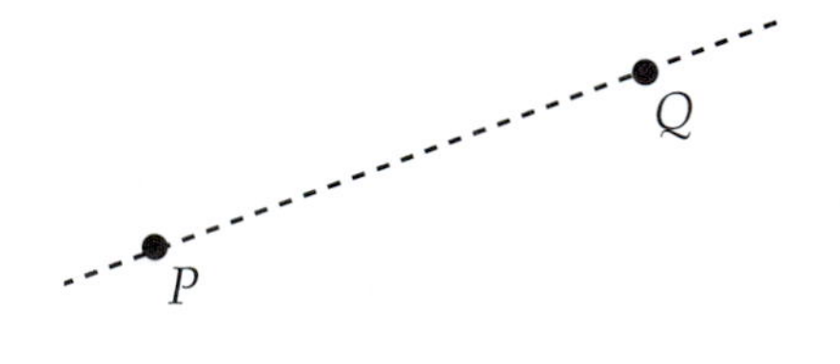

(Axiom O2) Given two constructed points P and Q, we can fold P onto Q.

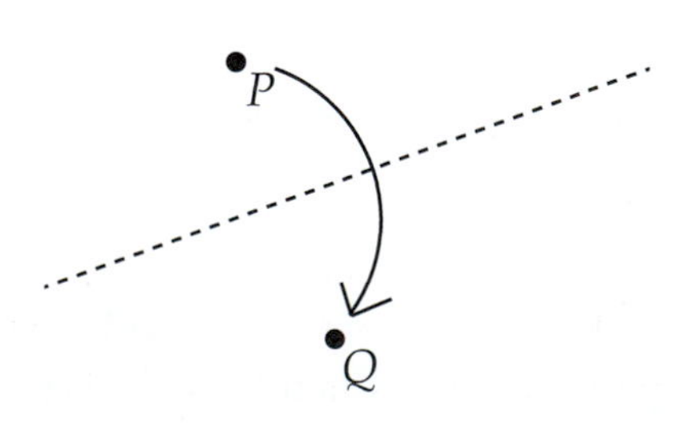

(Axiom O3) Given two constructed lines l_1 and l_2, we can fold line l_1 onto l_2.

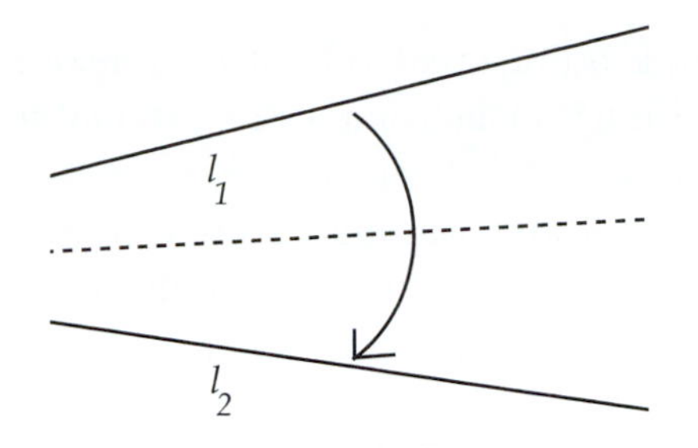

(Axiom O4) Given a constructed point P and a constructed line l, we can construct a perpendicular to l passing through P.

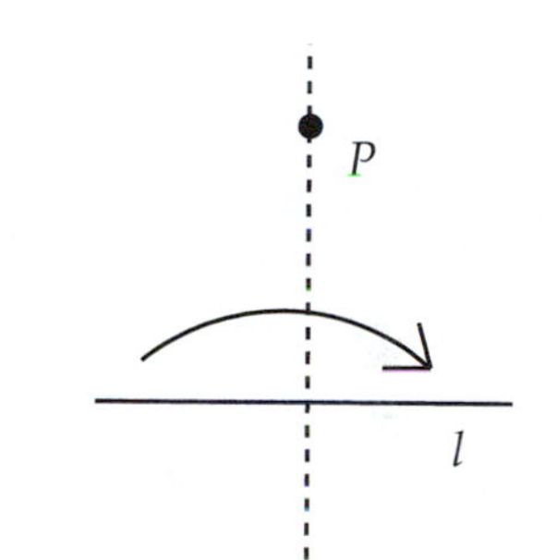

(Axiom O5) Given two constructed points P and Q and a constructed line l, then whenever possible, the line through Q, which reflects P onto l, can be constructed.

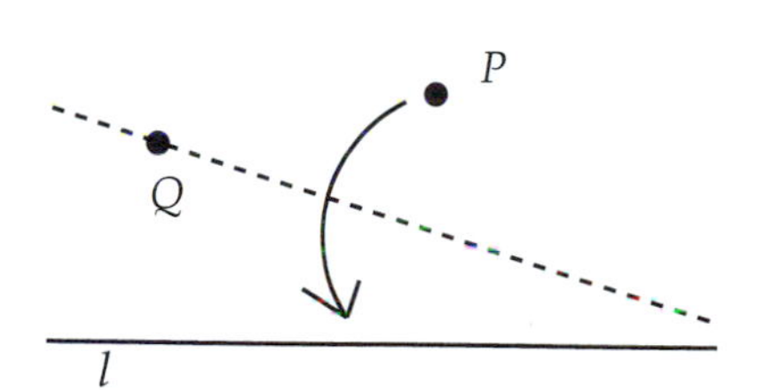

(Axiom O6) Given two constructed points P and Q and two constructed lines l_1 and l_2, then whenever possible, a line that reflects P onto l_1 and also reflects Q onto l_2 can be constructed.

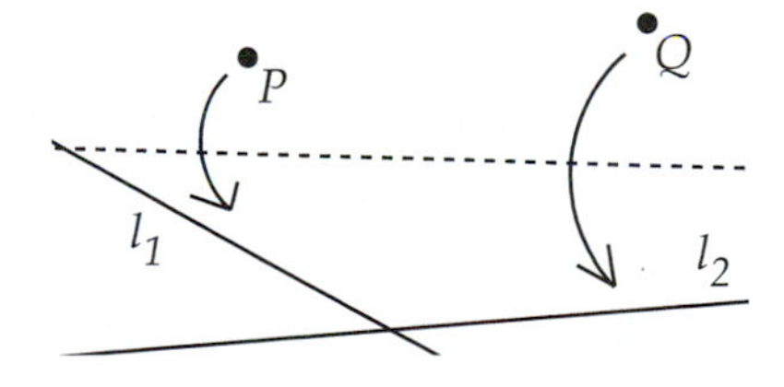

In the illustrations for the axioms, a fold line created by the axiom is indicated by a dotted line, and the direction of the fold is indicated by a curved arrow.

We are assuming in the statement of these axioms that a constructed line is either one of the original four edge lines of a square piece of origami paper or is a line created by using one or more of the folding axioms. A constructed point is either one of the original four vertices of the square paper or a point created by using one or more of the folding axioms. We will also assume that folds take lines to lines and preserve segment lengths and angles. We will prove this fact carefully in Chapter 5, where we will see that a fold is essentially a Euclidean *reflection* across the fold crease line. For now we will take this fact as a rule of reasoning, assumed without proof—thus the references to reflections in the axioms will be assumed without explanation or proof.

Another rather strange property of Axioms O5 and O6 is the phrase "whenever possible." All of the axiomatic systems we have studied up to this point have been quite definitive. For example, Euclid's third postulate states that circles are *always* constructible.

Before continuing on with this project, practice each of the six axiomatic foldings using a square sheet of paper, preferably origami paper or waxed paper.

Exercise 4.4.1. Using the six axiomatic foldings, devise a construction for the perpendicular bisector of a segment.

Exercise 4.4.2. Devise a folding construction for the parallel to a given line through a point not on the line.

Many of the axioms for paper folding are quite similar to the constructions of Euclidean geometry. Axiom O1 is essentially equivalent to Euclid's first axiom on the construction of a line joining two points. Axiom O2 mimics the construction of the perpendicular bisector of a segment. For the third axiom, if the given pair of lines intersect, then the axiom gives the angle bisector construction for the angle formed by the lines. If the lines are parallel, the construction is slightly more complicated but still possible (convince yourself of this fact). Axiom O4 is equivalent to the perpendicular to a line through a point.

What straightedge and compass construction has the equivalent effect of Axiom O5? After carrying out this construction, it must be the case that $PQ = P'Q$, where P' is the folding (reflection) of P onto l.. Thus, we are looking for an intersection point of a circle, centered at Q of radius PQ, with the line l, as depicted in Fig. 4.24.

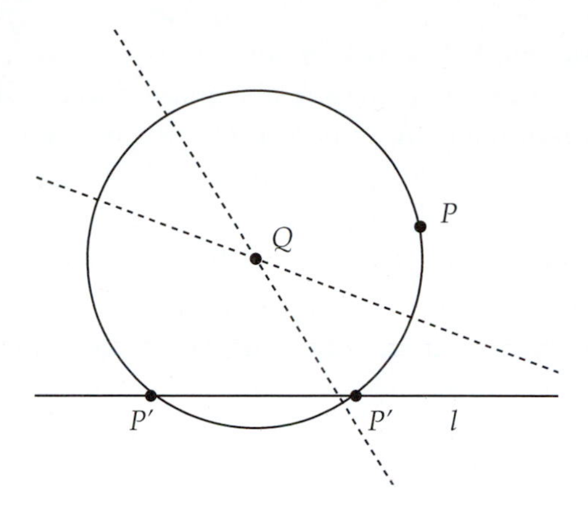

Fig. 4.24

It is clear that there are three possibilities: either the circle has no intersections with the line (in which case the construction is impossible), or there is one intersection (at a point of tangency), or there are two different intersections. (The third case is illustrated in Fig. 4.24, with the two different lines of reflection shown as dotted lines.)

In fact, Axiom O5 allows us to construct a parabola with focus P and directrix l. Recall that a parabola is the set of points that are equidistant from a given point (the focus) and a given line (the directrix).

Exercise 4.4.3. Show that Axiom O5 can be used to construct a parabola with focus P and directrix l by referring to Fig. 4.25. In this figure $\overleftrightarrow{P'R}$ is the perpendicular to l at the constructed point P' and R is the intersection of this perpendicular with the line of reflection t taking P to P'. [Hint: Use the distance-preserving properties of a folding (reflection).]

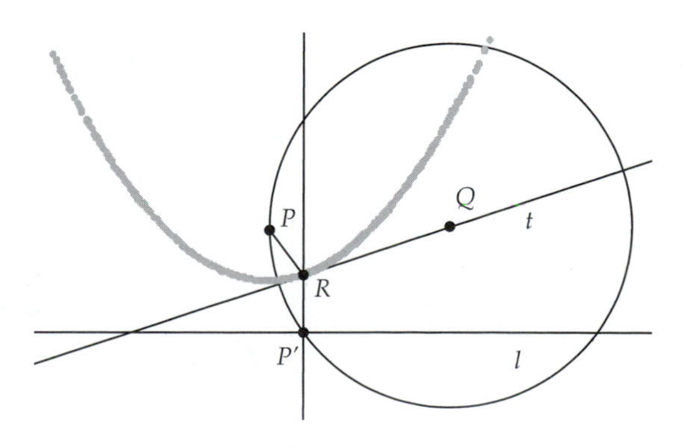

Fig. 4.25

The creation of point R involves the solution of a quadratic equation and thus allows for the construction of square roots. In fact, the set of constructible numbers using Axioms O1–O5 is the same as the set of constructible numbers using straightedge and compass (allowing for arbitrary large initial squares of paper). A proof of this can be found in [2].

We see, then, that almost all of the folding axioms can be carried out by simple straightedge and compass constructions. How about the last axiom? It turns out that the last axiom is not constructible using straightedge and compass. In fact, using the last axiom we can actually *trisect* a general angle.

Let's see how this is done.

Let the given angle ($\angle A$) be defined in the lower left corner of the paper square by line m as shown. Construct two lines l_1 and l_2 that are parallel to the bottom edge l_b with the property that l_1 is equidistant from l_2 and l_b. (What is an easy way to construct l_2 given just the initial four lines and four points?)

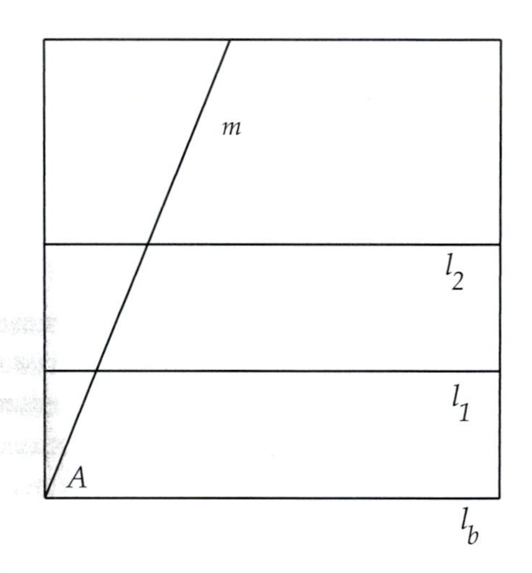

Let P be the lower left corner vertex of the square, and let Q be the intersection of l_2 with the left edge of the square. Then carry out the fold in Axiom O6 to place P on l_1 at P' and Q on m at Q'.

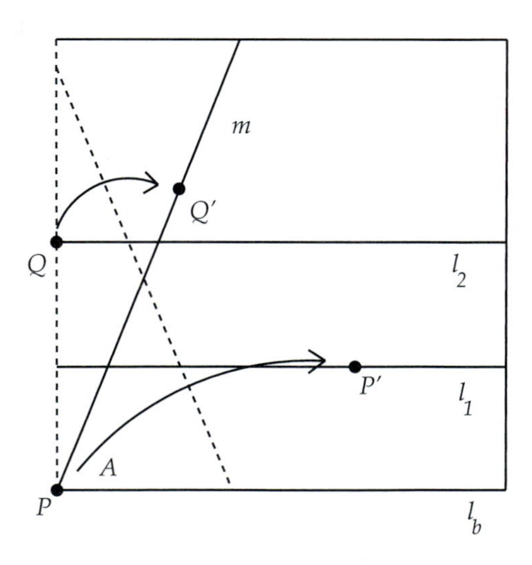

Leaving the paper folded, fold the paper once again along the folded-over portion of l_1. This will create line l_3. Unfold the paper. The claim is that line l_3 will make an angle with l_b of $\frac{2}{3}$ the angle A.

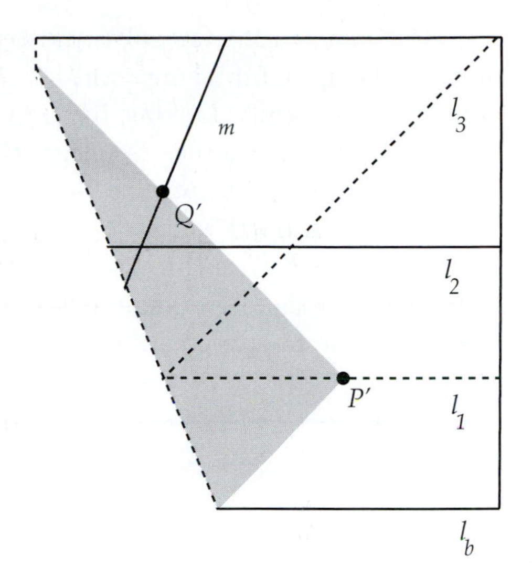

Exercise 4.4.4. Prove that the preceding construction actually does give an angle that is $\frac{2}{3}$ the angle A. [Hint: Prove that the three triangles $\Delta PQ'R$, $\Delta PP'R$, and $\Delta PP'S$ are congruent in Fig. 4.26.]

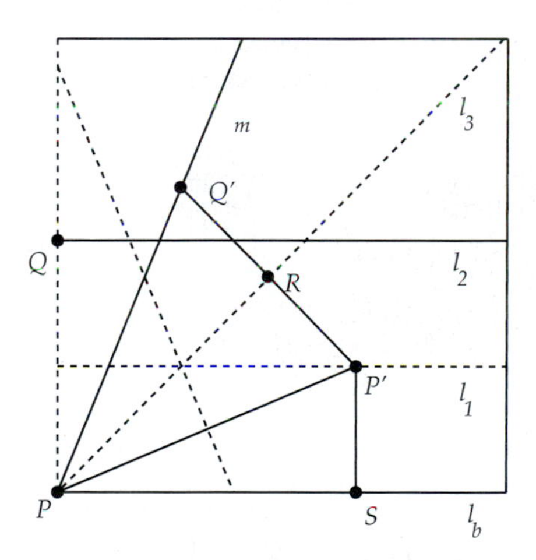

Fig. 4.26 Origami Trisection

Since we have constructed an angle that is $\frac{2}{3}$ of angle A, we can easily bisect this new angle to get the trisection of angle A. It is clear that the new folding axiom construction (Axiom O6) *cannot* be equivalent to a series of straightedge and compass constructions since we know it is impossible to trisect a general angle with straightedge and compass alone.

How can Axiom O6 solve the trisection problem? Since Axiom O6 is essentially a simultaneous solution to two Axiom O5 constructions, then what we are really looking for is a reflection line that is simultaneously tangent to two parabolas (see the "Hint" at the end of Exercise 4.4.3). The solution of this simultaneous tangent problem leads to a cubic equation of the form developed in the last section when we looked at the trisection puzzle in detail. Thus, Axiom O6 guarantees that such cubic equations are solvable, and therefore angle trisection is possible. For complete details on the connection between Axiom O6 and cubic polynomials, see [2].

Chapter 5

Transformational Geometry

> Geometry is the study of those properties of a set which are preserved under a group of transformations on that set.
>
> —Felix Klein (1849–1925)

Classical Euclidean geometry, such as the material covered in Chapter 2, is primarily concerned with *static* properties of objects. To expand this static geometry to a more *dynamic* geometry, we need to explore what it means to *transform* objects.

A transformation will be some *function* on points in the plane. That is, it will be some process whereby points are transformed to other points. This process could be the simple movement of points or could be a more complex alteration of the points.

Transformations are basic to both a practical and theoretical understanding of geometry. Object permanence, the idea that we can move an object to a different position, but the object itself remains the same, is one of the first ideas that we learn as infants.

Felix Klein, one of the great geometers of the late nineteenth century, gave an address at Erlanger, Germany, in 1872, in which he proposed that geometry should be *defined* as the study of transformations and of the objects that transformations leave unchanged, or *invariant*. This view has come to be known as the *Erlanger Program*.

If we apply the Erlanger Program to Euclidean geometry, what kinds of transformations characterize this geometry? That is, what are the transformations that leave basic Euclidean figures, such as lines, segments, triangles, and circles, invariant? Since segments are the basic building blocks of many geometric figures, Euclidean transformations must, at least, preserve the "size" of segments; that is, they must preserve *length*.

5.1 Euclidean Isometries

Definition 5.1. A function f on the plane is called a *Euclidean isometry* (or a *Euclidean motion*) if f has the property that for all points A and B the segment $\overline{AB}$ and the transformed segment $\overline{f(A)f(B)}$ have the same length.

This simple definition has important implications.

Theorem 5.1. *Let f be a Euclidean isometry. Then*

(i) f is one-to-one. That is, if $f(A) = f(B)$, then $A = B$.

(ii) If $f(A) = A'$ and $f(B) = B'$, then f maps all points between A and B to points between A' and B'. That is, $f(\overline{AB}) = \overline{A'B'}$.

(iii) f maps lines to lines.

(iv) f preserves angles.

(v) f is onto the plane. That is, for all points P', there is a point P such that $f(P) = P'$.

(vi) f preserves parallel lines.

Proof: (i) Suppose $f(A) = f(B)$. Then, $f(A)f(B) = 0$ and by the definition of an isometry, $AB = 0$. Then, $A = B$ and f is one-to-one.

(ii) Let C be a point between A and B and let $C' = f(C)$. We need to show that C' is on the line through A', B' and that C' is between A' and B'. Since f is one-to-one, C' cannot be A' or B'. Now, $AB = AC + CB$. Since f is an isometry we have that

$$A'B' = A'C' + C'B'$$

This implies that C' is on the line through A', B'. For if it were not on this line, then the triangle inequality would imply that $A'B' < A'C' + C'B'$.

Now either A' is between B' and C', or B' is between A' and C', or C' is between A' and B'. In the first case, we would get

$$B'C' = B'A' + A'C'$$

If we subtract this from the equation above, we would get

$$A'B' - B'C' = C'B' - B'A'$$

and

$$2A'B' - 2B'C' = 0$$

So $A'B' = B'C'$, which would contradict the fact that $A'B' < B'C'$ if A' is between B', C'. Likewise, we cannot have B' between A', C', and so C' must be between A', B'.

(iii) Let A, B be points on a line l. By part (ii) of this theorem, we know that segment $\overline{AB}$ gets mapped to segment $\overline{f(A)f(B)}$. Let D be a point on the ray $\overrightarrow{AB}$ not on segment $\overline{AB}$. Then B is between A, D on $\overrightarrow{AB}$ and since f preserves betweenness, $f(B)$ will be between $f(A)$ and $f(D)$ and so will be on the ray $\overrightarrow{f(A)f(B)}$. Thus, we have that ray $\overrightarrow{AB}$ gets mapped to ray $\overrightarrow{f(A)f(B)}$ and similarly ray $\overrightarrow{BA}$ gets mapped to ray $\overrightarrow{f(B)f(A)}$. This implies that the line through A, B gets mapped to the line through $f(A), f(B)$.

(iv) Let $\angle ABC$ be an angle with vertex B. Since f preserves length, by SSS triangle congruence, ΔABC and $\Delta f(A)f(B)f(C)$ will be congruent and their angles will be congruent.

(v) We know that f is one-to-one. Thus, given P' we can find two points A, B such that $f(A) \neq f(B) \neq P'$. Let $f(A) = A'$ and $f(B) = B'$. There are two cases for A', B', P': either they lie on the same line or not.

If A', B', P' are collinear, then P' is either on the ray $\overrightarrow{A'B'}$ or on the opposite ray. Suppose P' is on $\overrightarrow{A'B'}$. Let P be a point on $\overrightarrow{AB}$ such that $AP = A'P'$. Since $AP = f(A)f(P) = A'f(P)$ and since P' and $f(P)$ are on the same ray $\overrightarrow{A'B'}$, then $P' = f(P)$. If P' is on the opposite ray to $\overrightarrow{A'B'}$, we would get a similar result.

If A', B', P' are not collinear, then consider $\angle P'A'B'$. On either side of the ray through A, B, we can find two points P, Q such that $\angle P'A'B' \cong \angle PAB \cong \angle QAB$ (Fig. 5.1).

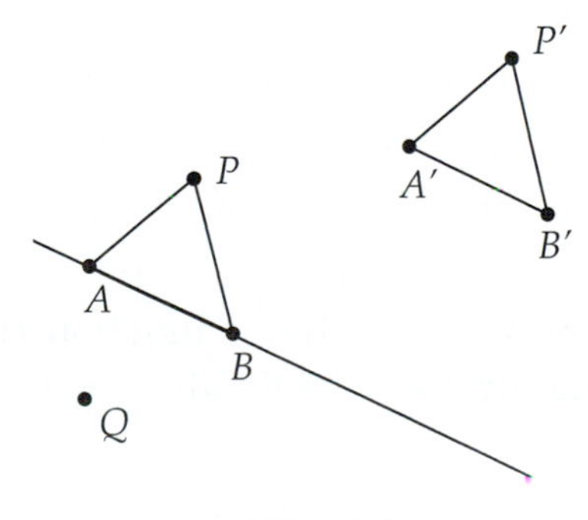

Fig. 5.1

We can also choose these points such that $AP = AQ = A'P'$. Since f preserves betweenness (proved in statement (ii)) we know that one of $\overline{A'f(P)}$ or $\overline{A'f(Q)}$ will be on the same side as $\overline{A'P'}$. We can assume that $\overline{A'f(P)}$ is on this same side. Then, since f preserves angles, we have that $\angle P'A'B' \cong \angle f(P)A'B'$ and thus $\overrightarrow{A'P'} \cong \overrightarrow{A'f(P)}$. Since f preserves lengths, we have that $A'P' = AP = f(A)f(P) = A'f(P)$ and thus $P' = f(P)$.

(vi) Let l, n be parallel lines. Suppose that $f(l), f(m)$ were not parallel. Then, for some P on l and Q on m, we would have $f(P) = f(Q)$. But, we know that $PQ \neq 0$ as l and m are parallel. As f is an isometry, we then have that $f(P)f(Q) \neq 0$. Thus, it cannot be true that $f(P) = f(Q)$, and $f(l), f(m)$ must be parallel. □

We have shown that isometries are

- length-preserving
- one-to-one
- onto

Isometries are a special type of transformation, but we have not yet explicitly defined what we mean by a "transformation." We have said that a transformation is a function on points in the plane, but this definition is too general. In the spirit of Klein's Erlanger Program, we want to consider "reasonable" functions that leave Euclidean figures invariant. Functions that map lines to points, or areas to segments, are not reasonable geometric equivalences. However, functions that are one-to-one and onto do not have such pathological behavior. Thus, we will define transformations as follows:

Definition 5.2. A function f on the plane is a *transformation* of the plane if f is a one-to-one function that is also onto the plane.

An isometry is then a length-preserving transformation. One important property of any transformation is that it is *invertible.*

Definition 5.3. Let f, g be functions on a set S. We say that g is the *inverse* of f if $f(g(s)) = s$ and $g(f(s)) = s$ for all s in S. That is, the composition of g and f (f and g) is the identity function on S. We denote the inverse by f^{-1}.

It is left as an exercise to show that a function that is one-to-one and onto must have a unique inverse. Thus, all transformations have unique inverses.

A nice way to classify transformations (isometries) is by the nature of their fixed points.

Definition 5.4. Let f be a transformation. P is a *fixed point* of f if $f(P) = P$.

How many fixed points can an isometry have?

Theorem 5.2. *If points A, B are fixed by an isometry f, then the line through A, B is also fixed by f.*

Proof: We know that f will map the line $\overleftrightarrow{AB}$ to the line $\overleftrightarrow{f(A)f(B)}$. Since A, B are fixed points, then $\overleftrightarrow{AB}$ gets mapped back to itself.

Suppose that P is between A and B. Then, since f preserves betweenness, we know that $f(P)$ will be between A and B. Also

$$AP = f(A)f(P) = Af(P)$$

This implies that $P = f(P)$.

A similar argument can be used in the case where P lies elsewhere on $\overleftrightarrow{AB}$. □

Definition 5.5. The isometry that fixes all points in the plane will be called the *identity* and will be denoted as *id*.

Theorem 5.3. *An isometry f having three non-collinear fixed points must be the identity.*

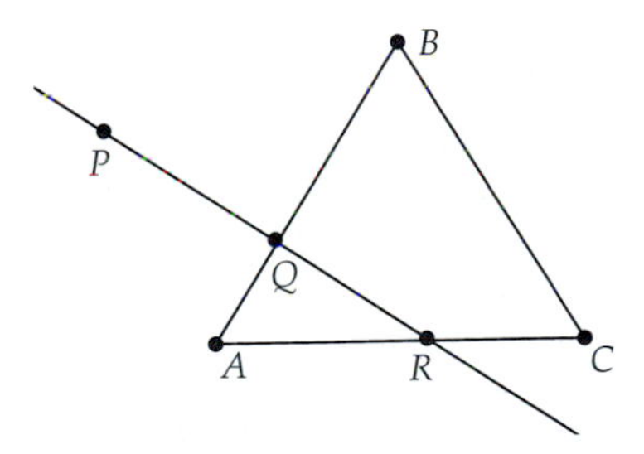

Fig. 5.2

Proof: Let A, B, C be the three non-collinear fixed points. From the previous theorem we know that f will fix lines $\overleftrightarrow{AB}$, $\overleftrightarrow{AC}$, and $\overleftrightarrow{BC}$.

Let P be a point not on one of these lines. Let Q be a point between A, B (Fig. 5.2). Consider the line through P, Q. By Pasch's axiom, this line will intersect one of $\overline{AC}$ or $\overline{BC}$ at some point R. By the previous theorem, f fixes the line $\overleftrightarrow{QR}$ and thus fixes P. Since P was chosen arbitrarily, then f fixes all points in the plane and is the identity. □

Corollary 5.4. *If two isometries f, g agree on any three non-collinear points, then the two isometries must agree everywhere, that is, $f = g$.*

The proof of this result is left as an exercise.

It is clear from this theorem that we can classify isometries into three non-trivial (non-identity) types: those with two fixed points, those with one fixed point, and those with no fixed points. In the following sections we will study the properties of isometries with two, one, or zero fixed points. We will make extensive use of techniques from both synthetic and analytic geometry in our proofs and development.

It is interesting to note that the preceding results on isometries do not depend on Euclid's fifth postulate, the parallel postulate. They are part of *neutral geometry* (or *absolute geometry*). We will make use of this fact in Chapter 7, where we explore *non-Euclidean* geometry.

Exercise 5.1.1. Prove that every function f on a set S that is one-to-one and onto has a unique inverse. [Hint: First, define f^{-1} using f and show that it is a valid function. Then, show that $f \circ f^{-1} = id$ and $f^{-1} \circ f = id$, where id is the identity on S. Finally, show that the inverse is unique.]

Exercise 5.1.2. Prove that the inverse of an isometry is again an isometry. (This implies that the set of isometries is closed under the inverse operation.)

Exercise 5.1.3. Let f, g be two invertible functions from a set S to itself. Let $h = f \circ g$; that is, h is the composition of f and g. Show that $h^{-1} = g^{-1} \circ f^{-1}$.

Exercise 5.1.4. Let f, g be two isometries. Show that the composition $f \circ g$ is again an isometry. (This says the set of isometries is closed under composition.)

Exercise 5.1.5. Show that isometries map circles of radius r to circles of radius r. That is, isometries preserve circles.

Exercise 5.1.6. Prove that the image of a triangle under an isometry is a new triangle congruent to the original.

Exercise 5.1.7. Given an equilateral triangle ABC, show that there are exactly six isometries that map the triangle back to itself. [Hint: Consider how the isometry acts on the vertices of the triangle.]

Exercise 5.1.8. Prove Corollary 5.4.

Exercise 5.1.9. Consider points in the plane as ordered pairs (x, y) and consider the function f on the plane defined by $f(x, y) = (kx + a, ky + b)$, where k, a, b are real constants, and $k \neq 0$. Is f a transformation? Is f an isometry?

Exercise 5.1.10. Define a *similarity* to be a transformation on the plane that preserves the betweenness property of points and preserves angle measure. Prove that under a similarity, a triangle is mapped to a *similar* triangle.

Exercise 5.1.11. Use the previous exercise to show that if f is a similarity, then there is a positive constant k such that

$$f(A)f(B) = k\,AB$$

for all segments $\overline{AB}$.

Exercise 5.1.12. Consider points in the plane as ordered pairs (x, y) and consider the function f on the plane defined by $f(x, y) = (kx, ky)$, where k is a non-zero constant. Show that f is a similarity.

5.2 Reflections

Definition 5.6. An isometry with two different fixed points, and that is not the identity, is called a *reflection*.

What can we say about a reflection? By Theorem 5.2 if A, B are the fixed points of a reflection, then the reflection also fixes the line through A, B. This line will turn out to be the equivalent of a "mirror" through which the isometry reflects points.

Theorem 5.5. *Let r be a reflection fixing A and B. If P is not collinear with A, B, then the line through A and B will be a perpendicular bisector of the segment connecting P and $r(P)$.*

Proof: Drop a perpendicular from P to $\overleftrightarrow{AB}$, intersecting at Q. At least one of A or B will not be coincident with Q; suppose B is not. Consider ΔPQB and $\Delta r(P)QB$. Since we know that Q and B are fixed points of r, then $PQ = r(P)Q$, $BP = Br(P)$, and the two triangles are congruent by SSS. Since the two congruent angles at Q make up a straight line, $\angle r(P)QB$ will be a right angle and $\overleftrightarrow{AB}$ will be a perpendicular bisector of the segment $\overline{Pr(P)}$. □

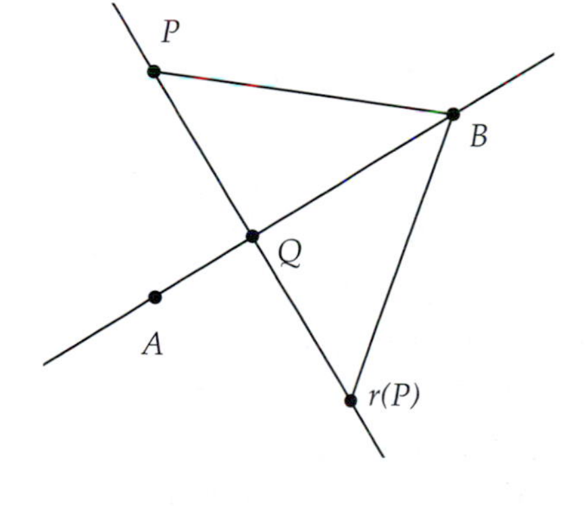

We call the line through A, B the *line of reflection* for r.

Theorem 5.6. *Let P, P' be two points. Then there is a unique reflection taking P to P'. The line of reflection will be the perpendicular bisector of $\overline{PP'}$.*

Proof: Let $\overleftrightarrow{AB}$ be the perpendicular bisector of PP' (Fig. 5.3). Define a function r on the plane as follows: If a point C is on $\overleftrightarrow{AB}$, let $r(C) = C$. If C is not on this line, drop a perpendicular from C to $\overleftrightarrow{AB}$ intersecting at Q, and let $r(C)$ be the unique point on this perpendicular such that $r(C) \neq C$ and $\overline{r(C)Q} \cong \overline{CQ}$.

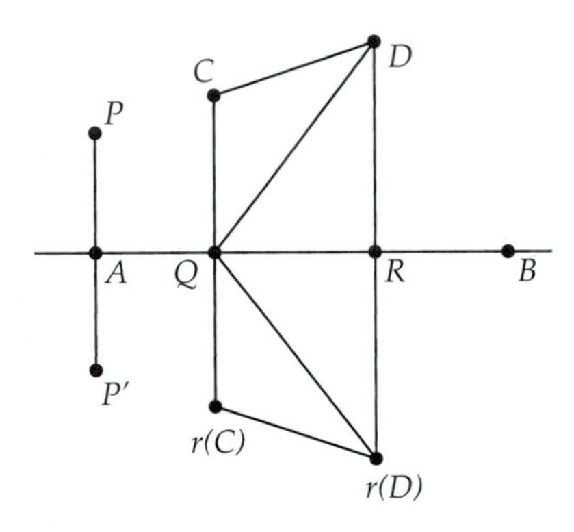

Fig. 5.3

Will r be an isometry? We need to show that for all $C \neq D$, $r(C)r(D) = CD$. Let C be a point not on $\overleftrightarrow{AB}$ and D a point on the same side of $\overleftrightarrow{AB}$ as C. Consider Fig. 5.3. By SAS, $\Delta QRD \cong \Delta QR\, r(D)$. Again using SAS congruence, we have $\Delta CQD \cong \Delta r(C)\, Q\, r(D)$. Thus, $CD = r(C)\, r(D)$. Similar arguments using congruent triangles can be used if D is on $\overleftrightarrow{AB}$ or on the other side of $\overleftrightarrow{AB}$ as C. (The proof is left as an exercise.)

If C is a point on $\overleftrightarrow{AB}$ and if D is also on $\overleftrightarrow{AB}$, then clearly $CD = r(C)\, r(D)$. If D is not on $\overleftrightarrow{AB}$, then a simple SAS argument will show that $CD = r(C)\, r(D)$.

Thus, r is an isometry. Is the reflection r unique? Suppose there was another reflection r' taking P to P'. By the previous theorem we know that the fixed points of r' are on the perpendicular bisector of $\overline{PP'}$. Since the perpendicular bisector is unique, we have that the fixed points of r' are on $\overleftrightarrow{AB}$. Thus, r and r' have the same values on three non-collinear points P, P', and B and so $r = r'$. $\square$

5.2.1 Mini-Project - Isometries through Reflection

In the first part of this chapter, we discussed Felix Klein's idea of looking at geometry as the study of figures that are invariant under sets of transformations. In the case of transformations that are isometries, invariance means that lengths and angles are preserved, and lines get mapped to lines. Thus, isometries must not only map triangles to triangles, but must map triangles to *congruent* triangles.

We can reverse this idea and ask whether, given two congruent triangles, there is an isometry that maps one to the other.

We will start with an easy case. Clearly, if two triangles are the same, then the identity isometry will map the triangle to itself.

Now, suppose you have two congruent triangles that share two points in common. Let ΔABC and ΔPQR be congruent with $A = P$ and $B = Q$.

Exercise 5.2.1. Show that either the triangles are the same or that there is a reflection that takes ΔABC to ΔPQR.

Now suppose you have two congruent triangles that share only one point in common. Let $\Delta ABC \cong \Delta PQR$ with $A = P$. Let l_1 be the angle bisector of $\angle BAQ$ and r_1 the reflection across l_1.

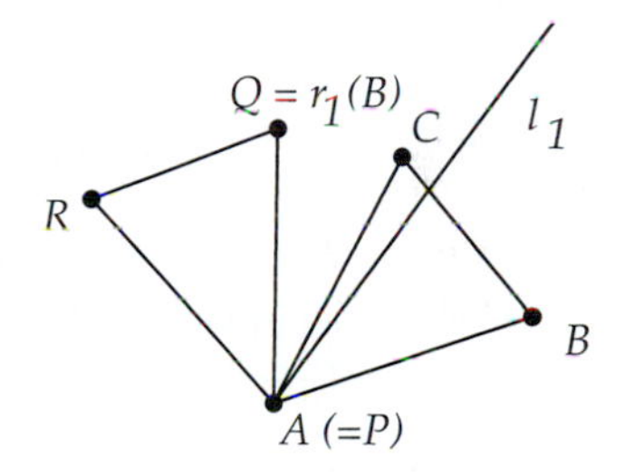

Exercise 5.2.2. Show that $r_1(B) = Q$. Then, use the preceding exercise to argue that there is a sequence of at most two reflections that will take ΔABC to ΔPQR, if the two triangles share one point in common.

Finally, suppose two congruent triangles share no point in common.

Exercise 5.2.3. Show that there is a sequence of at most three reflections that will take ΔABC to ΔPQR in this case.

We have now proved the following theorem:

Theorem 5.7. *Let $\Delta ABC \cong \Delta PQR$. Then there is an isometry composed of at most three reflections that takes ΔABC to ΔPQR.*

This theorem has the following amazing corollary:

Corollary 5.8. *Every isometry can be written as the product of at most three reflections.*

Proof: Let f be an isometry and consider any triangle ΔABC. Then $\Delta f(A)\, f(B)\, f(C)$ is a triangle congruent to ΔABC and, by the preceding theorem, there is an isometry g composed of at most three reflections taking ΔABC to $\Delta f(A)\, f(B)\, f(C)$. Since two isometries that agree on three non-collinear points must agree everywhere, then f must be equal to g. □

Exercise 5.2.4. Let two triangles be defined by coordinates as follows: ΔABC with $A = (-3, 2)$, $B = (-3, 6)$, $C = (-6, 2)$ and ΔDEF with $D = (1, -4)$, $E = (4, -4)$, $F = (1, -8)$. Verify that these two triangles are congruent and then find a sequence of three (or fewer) reflection lines such that ΔABC can be transformed to ΔDEF.

We note here that the results in this project, and in the preceding section on reflections, are *neutral*—they do not depend on Euclid's fifth postulate, the parallel postulate. We will make use of this fact in Chapter 7.

5.2.2 Reflection and Symmetry

The word *symmetry* is usually used to refer to objects that are in *balance.* Symmetric objects have the property that parts of the object look similar to other parts. The symmetric parts can be interchanged, thus creating a visual balance to the entire figure. How can we use transformations to mathematically describe symmetry?

Perhaps the simplest definition of mathematical symmetry is the one that most dictionaries give: an arrangement of parts equally on either side of a dividing line. While this type of symmetry is not the only one possible, it is perhaps the most basic in that such symmetry pervades the natural world. We will call this kind of symmetry *bilateral* symmetry.

Definition 5.7. A figure F in the plane is said to have a *line of symmetry* or *bilateral symmetry* if there is a reflection r that maps the figure back to itself having the line as the line of reflection. For example, the line l in Fig. 5.4 is a line of symmetry for ΔABC since if we reflect the triangle across this line, we get the exact same triangle back again.

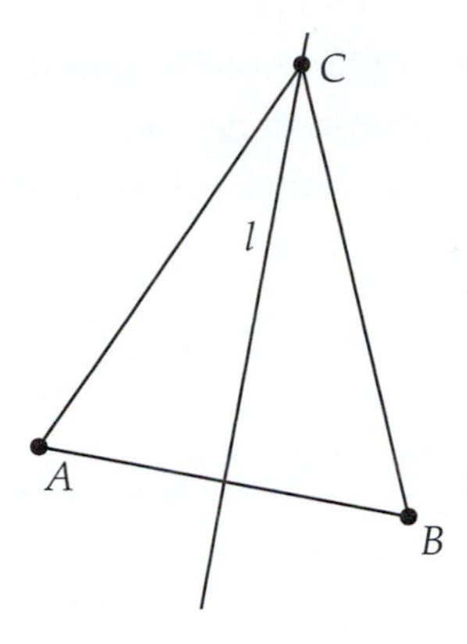

Fig. 5.4

Where is bilateral symmetry found in nature? Consider the insect body types in Fig. 5.5. All exhibit bilateral symmetry. In fact, most animals, insects, and plants have bilateral symmetry. Why is this the case? Living creatures need bilateral symmetry for *stability*. Consider an animal that needs to be mobile, that needs to move forward and backward. To move with the least expenditure of energy, it is necessary that a body shape be balanced from side to side so that the creature does not waste energy keeping itself upright. Likewise, an immobile living creature, such as a tall pine tree, needs to be bilaterally symmetric in order to keep itself in a vertical equilibrium position.

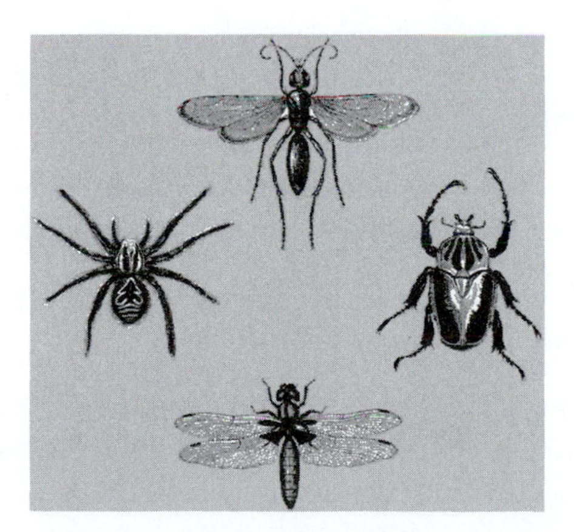

Fig. 5.5 Insect Symmetry, Shelter Online, www.shelterpub.com

We know that a reflection will map its line of symmetry back to the same line of symmetry. Lines perpendicular to the line of symmetry are also mapped back to themselves. (The proof is one of the exercises that follow.) These lines are perhaps the simplest figures that are bilaterally symmetric.

In general, we can ask which lines are preserved under the action of an arbitrary transformation.

Definition 5.8. Lines that are mapped back to themselves by a transformation f are called *invariant lines* of f.

Note that we do not require that *points* on the line get mapped back to themselves, only that the line as a set of points gets mapped back to itself. Thus, a line may be invariant under f, but the points on the line need not be fixed by f.

In the next chapter we will use invariant lines extensively to classify different sets of symmetries in the plane.

Exercise 5.2.5. Find examples of five objects in nature that have two or more lines of bilateral symmetry. Draw sketches of these along with their lines of symmetry.

Definition 5.9. A polygon is a *regular polygon* if it has all sides congruent and all interior angles congruent.

Exercise 5.2.6. Show that the angle bisectors of a regular pentagon are lines of symmetry. Would your proof be extendable to show that the angle bisectors of *any* regular polygon are lines of symmetry?

Exercise 5.2.7. Show that the perpendicular bisector of a side of a regular pentagon is a line of symmetry. Would your proof be extendable to show that the perpendicular bisectors of the sides of *any* regular polygon are lines of symmetry?

Exercise 5.2.8. Show that if a parallelogram has a diagonal as a line of symmetry, then the parallelogram must be a rhombus (i.e., have all sides congruent).

Exercise 5.2.9. Show that if a parallelogram has a line of symmetry parallel to a side, then the parallelogram must be a rectangle.

Exercise 5.2.10. Finish the proof of Theorem 5.6. That is, prove that the function r defined in the proof is length-preserving for the case where C is not on $\overleftrightarrow{AB}$ and D is either on $\overleftrightarrow{AB}$ or on the other side of $\overleftrightarrow{AB}$.

Exercise 5.2.11. Prove that the composition of a reflection with itself is always the identity. Thus, a reflection is its own inverse.

Exercise 5.2.12. Show that the lines invariant under a reflection r_m, where m is the line of reflection for r_m, consist of the line m and all lines perpendicular to m.

Exercise 5.2.13. Let r_l and r_m be two reflections with lines of reflection l and m, respectively. Show that the composition $r_m \circ r_l \circ r_m = r_{l'}$, where l' is the reflection of l across m. [Hint: Let A, B be distinct points on l. Show that $r_m(A)$ and $r_m(B)$ are fixed points of $r_m \circ r_l \circ r_m$.]

Exercise 5.2.14. Show that the product of four reflections can be written as the product of two reflections.

Exercise 5.2.15. An object at point O is visible in a mirror from a viewer at point V. What path will the light take from O to the mirror to V? Light always travels through a homogeneous medium to minimize total travel distance. Consider a possible light ray path from the object that hits the mirror at P and then travels to V. Show that the total length of this path is the same as the path from O' to P to V, where O' is the reflection of O across the mirror. Use this to find the *shortest* path for the light ray from O to the mirror to V. Describe this path in terms of the two angles made at P by the light ray.

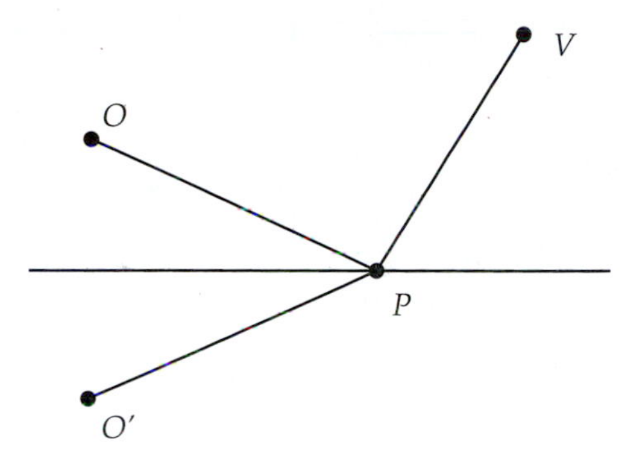

5.3 Translations

Definition 5.10. An isometry that is made up of two reflections, where the lines of reflection are parallel, or identical, is called a *translation*.

What can we say about a translation?

Theorem 5.9. *Let T be a translation that is not the identity. Then, for all points $A \neq B$, if A, B, $T(A)$, and $T(B)$ form a quadrilateral, then that quadrilateral is a parallelogram.*

Proof: Let r_1, r_2 be the two reflections comprising T and let l_1, l_2 be the two lines of reflection. Since T is not the identity, we know that l_1 is parallel to l_2. If we set up a coordinate system where l_1 is the x-axis, then $r_1(x, y) = (x, -y)$.

Since l_1 is parallel to l_2, we can assume l_2 is the line at $y = -K$, $K \neq 0$ (Fig. 5.6).

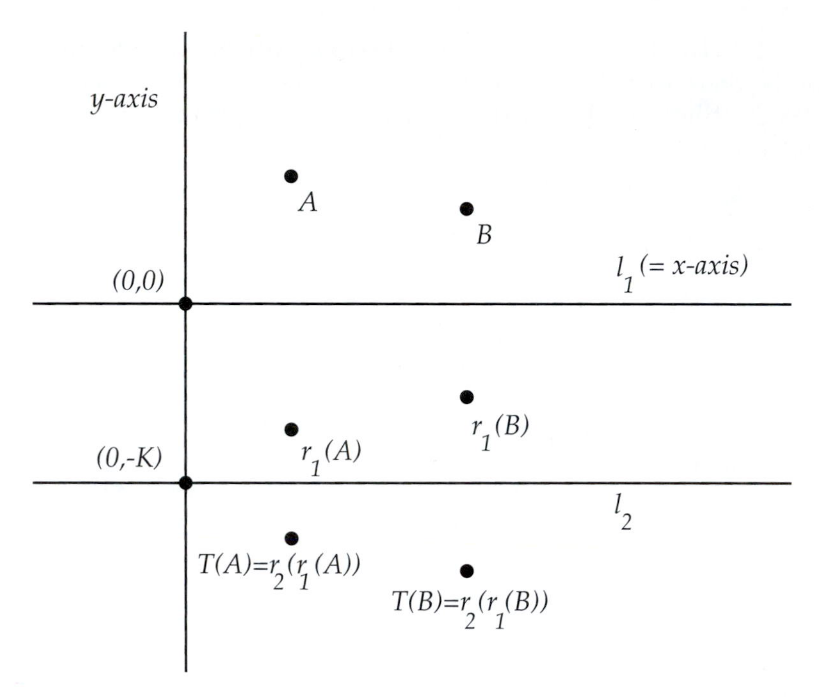

Fig. 5.6

A point (x, y) reflected across l_2 must be transformed to a point at a distance $y + K$ below l_2. Thus, the y-coordinate must be $-K - (y + K)$ and $r_2(x, y) = (x, -2K - y)$. Note that K is the distance between l_1, l_2.

Now, if $A = (x, y)$, then $A\,T(A) = A\,r_2(r_1(A))$ will be just the difference in the y values of A and $r_2(r_1(A))$; that is, $y - (-2K - y) = 2K$.

So, $A\,T(A) = 2K$ and $B\,T(B) = 2K$. Since T is an isometry, we also have that $T(A)\,T(B) = AB \neq 0$, as $A \neq B$. Thus, the sides of quadrilateral $A\,B\,T(B)\,T(A)$ are pair-wise congruent. Also, since $\overleftrightarrow{A\,T(A)}$ is perpendicular to l_1 and $\overleftrightarrow{B\,T(B)}$ is perpendicular to l_1, then these two sides of the quadrilateral are parallel. By constructing the diagonal of the quadrilateral, and using a triangle congruence argument, we see that the other pair of sides in the quadrilateral are also parallel, and thus the quadrilateral must be a parallelogram. □

From this theorem we see that for every point A in the plane, a translation T (not equal to the identity) will map A to $T(A)$ in such a way that the length of the segment from A to $T(A)$ will be constant and the direction of this segment will also be constant. Thus, a translation is determined by a *vector*, the vector from A to $T(A)$ for any A. This will be called the *displacement vector* of the translation.

Since the displacement vector is the vector from A to $T(A)$, this vector is given by $T(A) - A$ (considering the points as vectors from the origin). Thus, if $A = (x, y)$, then a translation T with displacement vector $v = T(A) - A = (v_1, v_2)$ will have the coordinate equation

$$T(x, y) = (x, y) + (v_1, v_2)$$

On the other hand, suppose that we are given a function defined by the coordinate equation $T(x, y) = (x, y) + (v_1, v_2)$. Let l_1 be a line through the origin that is perpendicular to the vector v and l_2 a line parallel to l_1, but passing through the midpoint of the segment along vector v (Fig. 5.7).

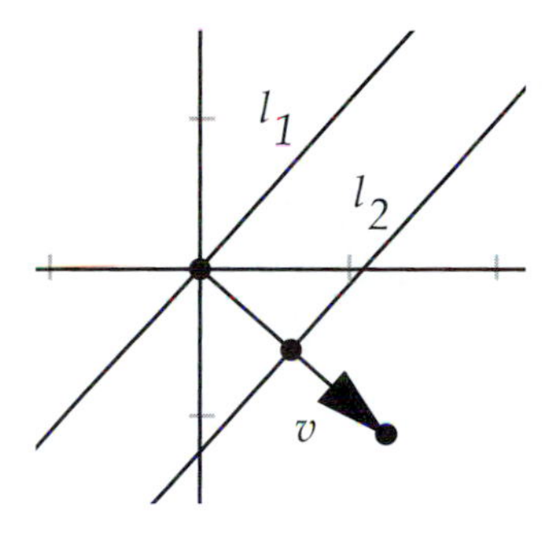

Fig. 5.7

Then the translation T' defined by successive reflection across l_1 and l_2 will map the origin to the head of v and thus will have a translation vector equal to v. The coordinate equation for T' will then be $T'(x, y) = (x, y) + (v_1, v_2)$. Thus, the translation T' and the function T must be the same function.

Corollary 5.10. *Every translation T can be expressed in rectangular coordinates as a function $T(x, y) = (x, y) + (v_1, v_2)$ and, conversely, every coordinate function of this form represents a translation.*

We also have the following result that expresses the translation vector in terms of the original lines of reflection.

Corollary 5.11. *Let a translation be defined by reflection across two parallel lines l_1, l_2 (in that order). Let m be a line perpendicular to both lines at A on l_1 and B on l_2. Then the displacement vector is given by $2\overrightarrow{AB}$.*

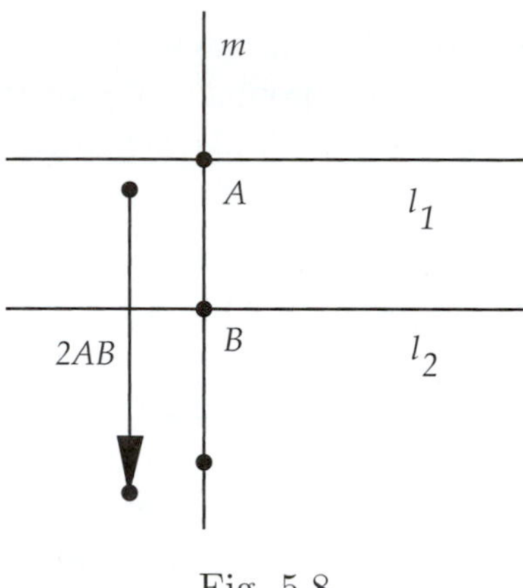

Fig. 5.8

Proof: Set up a coordinate system with the x-axis on l_1. Then, as in the proof of the last theorem, the translation will move a point a distance of twice the distance between the lines l_1, l_2. The direction will be perpendicular to the lines. Thus, the vector will be $2\overrightarrow{AB}$, as this vector has the right length and direction. □

How many fixed points will a translation have? If the two lines of reflection defining the translation are not coincident, then the translation always moves points a finite distance and thus there are *no* fixed points. On the other hand, if the two lines are the same, then the translation is the identity. Thus, a non-trivial translation (one that is not the identity) has no fixed points.

5.3.1 Translational Symmetry

When we discussed bilateral symmetry in the last section, we saw that many objects in nature exhibited bilateral symmetry. Such objects, when reflected across a line of symmetry, remain unchanged or invariant. Bilateral symmetry is a property that a single, *finite* object can exhibit.

What can we say about an object that is invariant under translation, that is, an object that has translational symmetry? To remain unchanged under translation, an object must repeat its form at regular intervals, defined by the translation vector. When we take a portion of the object and translate it, we must overlap the exact same shape at the new position. Thus, an object that is invariant under translation is necessarily *infinite* in extent. As Hermann Weyl stated in his foundational work on symmetry:

> A figure which is invariant under a translation t shows what in

> the art of ornament is called "infinite rapport," i.e. repetition in a regular spatial rhythm [40, page 47].

Translational symmetry, being infinite in extent, cannot be exhibited by finite living creatures. However, we can find evidence of a limited form of translation symmetry in some animals and plants. For example, the millipede has leg sections that are essentially invariant under translation (Fig. 5.9).

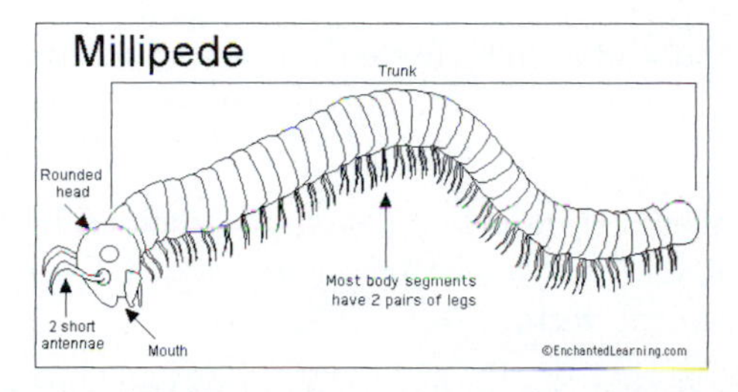

Fig. 5.9 Millipede, Copyright EnchantedLearning.com

Also, many plants have trunks (stems) and branching systems that are translation invariant.

Whereas translational symmetry is hard to find in nature, it is extremely common in human ornamentation. For example, wallpaper must have translational symmetry in the horizontal and vertical directions so that when you hang two sections of wallpaper next to each other the seam is not noticeable. Trim patterns called *friezes*, which often run horizontally along tops of walls, also have translational invariance (Fig. 5.10).

Fig. 5.10 Frieze Patterns

Exercise 5.3.1. Find and sketch two examples of translational symmetry in nature.

Exercise 5.3.2. Find and sketch three examples of wallpaper or frieze patterns that have translational invariance. Indicate on your sketch the translation vector(s) for each pattern.

Exercise 5.3.3. Given a translation $T = r_1 \circ r_2$ defined by two reflections r_1, r_2 and with displacement vector v, show that the inverse of T is the translation $T^{-1} = r_2 \circ r_1$ having displacement vector $-v$.

Exercise 5.3.4. Show that the composition of two translations T_1 and T_2 is again a translation. Find the translation vector for $T_1 \circ T_2$.

Exercise 5.3.5. Show that the composition of translations is a commutative operation. That is, if T_1 and T_2 are translations, then $T_1 \circ T_2 = T_2 \circ T_1$.

Exercise 5.3.6. Given a reflection r across a line l and a translation T in the same direction as l, show that $r \circ T = T \circ r$. [Hint: Choose a "nice" setting in which to analyze r and T.] Will this result hold if T is not in the direction of l?

Exercise 5.3.7. In the section on reflections, we saw that a simple reflection across the x-axis, which we will denote by r_x, could be expressed as $r_x(x, y) = (x, -y)$. Let r be the reflection across the line $y = K$. Let T be the translation with displacement vector of $v = (0, -K)$. Show that the function $T^{-1} \circ r_x \circ T$ is equal to r and find the coordinate equation for r. [Hint: Show that $T^{-1} \circ r_x \circ T$ has the right set of fixed points.]

Exercise 5.3.8. Let T be a (non-identity) translation. Show that the set of invariant lines for T are all pair-wise parallel and that each is parallel to the displacement vector of T. [Hint: Use Theorem 5.9.]

Exercise 5.3.9. Let T be a translation with (non-zero) displacement vector parallel to a line l. Let m be any line perpendicular to l. Show that there is a line n perpendicular to l such that $T = r_n \circ r_m$. [Hint: Suppose m intersects l at P. Choose n to be the perpendicular bisector of $P\,T(P)$. Show that $r_n \circ T$ fixes m.]

Exercise 5.3.10. Let r_l be a reflection across l and T be a translation with displacement vector parallel to l. Show that $r_l \circ T \circ r_l = T$.

5.4 Rotations

Definition 5.11. An isometry that is made up of two reflections where the lines of reflection are *not* parallel will be called a *rotation*.

To analyze rotations we will make use of the following lemma.

Lemma 5.12. *If two non-coincident lines l_1, l_2 intersect at O, and if m is the bisector of $\angle Q_1OQ_2$, with Q_1 on l_1 and Q_2 on l_2, then $r_m(l_1) = l_2$ and $r_m \circ r_{l_1} = r_{r_m(l_1)} \circ r_m = r_{l_2} \circ r_m$.*

Proof: We can assume that Q_1, Q_2 were chosen such that $\overline{OQ_1} = OQ_2$ (Fig 5.11). Let P be the intersection of the bisector m with $\overline{Q_1Q_2}$.

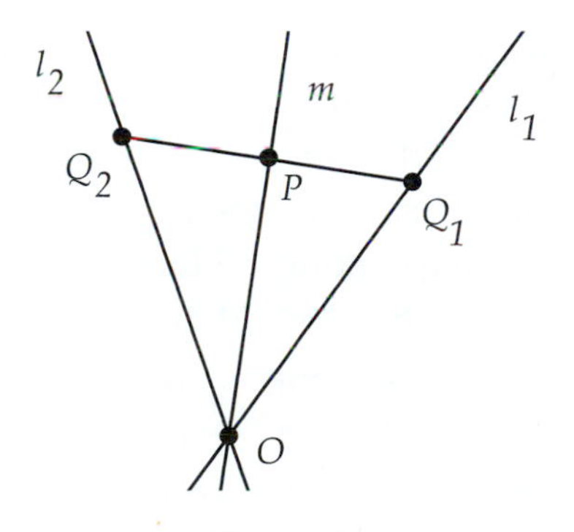

Fig. 5.11

Then, by SAS we know that P is the midpoint of $\overline{Q_1Q_2}$ and m is the perpendicular bisector of $\overline{Q_1Q_2}$. Thus, $r_m(Q_1) = Q_2$. Since $r_m(O) = O$, the line l_1 (defined by O and Q_1) must get mapped to l_2 (defined by O and Q_2), or $r_m(l_1) = l_2$.

For the second part of the lemma, we know by Exercise 5.2.13 that $r_m \circ r_{l_1} \circ r_m = r_{r_m(l_1)} = r_{l_2}$. Thus, $r_m \circ r_{l_1} = r_{l_2} \circ r_m$. □

Rotations are characterized by the fact that they have a single fixed point.

Theorem 5.13. *An isometry $R \neq id$ is a rotation iff R has exactly one fixed point.*

Proof: Suppose R has a single fixed point, call it O (Fig. 5.12). Let A be another point with $A \neq O$.

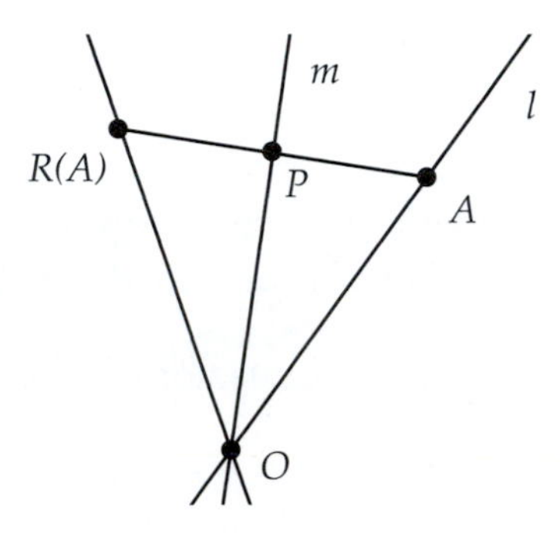

Fig. 5.12

Let l be the line through O and A and let m be the bisector of $\angle AOR(A)$ (or the perpendicular bisector of $\overline{A\,R(A)}$ if the three points are collinear). If $A, O, R(A)$ are non-collinear, we know from the previous lemma that $r_m \circ R$ will fix point A. If they are collinear, $r_m \circ R$ will also fix A, as m is the perpendicular bisector of $\overline{A\,R(A)}$. Since $r_m \circ R$ also fixes O, $r_m \circ R$ is either the identity or a reflection. If it is the identity, then $R = r_m$. But, then R would have more than one fixed point. Thus, $r_m \circ R$ must be a reflection fixing O and A, and thus $r_m \circ R = r_l$ and $R = r_m \circ r_l$.

Conversely, let $R = r_m \circ r_l$ for lines l, m intersecting at O. If R had a second fixed point, say $B \neq O$, then $r_m(B) = r_l(B)$. Clearly, B cannot be on m or l. But, then the segment joining B to $r_m(B)$ is perpendicular to m, and this same segment would also be perpendicular to l, as $r_m(B) = r_l(B)$. This is impossible. □

The next lemma describes triples of reflections about coincident lines.

Lemma 5.14. *Let l, m, n be three lines intersecting at a point O. Then, $r_l \circ r_m \circ r_n$ is a reflection about a line p passing through O. Also, $r_l \circ r_m \circ r_n = r_n \circ r_m \circ r_l$.*

Proof: Let $f = r_l \circ r_m \circ r_n$, and let A be a point on n not equal to O (Fig. 5.13). For $A' = f(A)$, either $A' \neq A$ or $A' = A$.

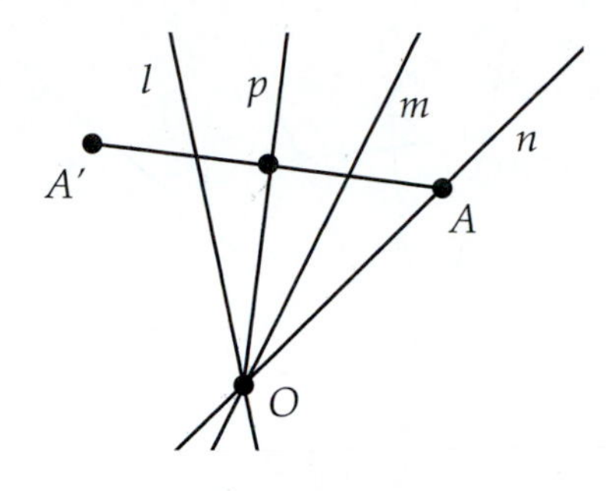

Fig. 5.13

If $A' \neq A$, let line p be the angle bisector of $\angle AOA'$. A simple triangle argument shows that p is also the perpendicular bisector of $\overline{AA'}$ and thus r_p maps A to A'. If $A' = A$ choose $p = n$.

In either case, $r_p \circ f$ will fix both A and O and thus fixes n. Thus, $r_p \circ f$ is either equal to r_n or is the identity. If $r_p \circ f = r_p \circ r_l \circ r_m \circ r_n = r_n$, then $r_p \circ r_l \circ r_m = id$ and $r_p = r_m \circ r_l$. Since m and l intersect, then r_p would be either a rotation or the identity. Clearly, r_p cannot be a rotation or the identity, and thus $r_p \circ f = id$; that is, $r_p = f$.

We conclude that f is a reflection about a line p passing through O, and $r_p = f = r_l \circ r_m \circ r_n$.

For the second part of the theorem, we note that $(r_p)^{-1} = (r_n)^{-1} \circ (r_m)^{-1} \circ (r_l)^{-1}$. Since reflections are their own inverses, we have $r_p = r_n \circ r_m \circ r_l$.

□

The next theorem tells us how rotations transform points through a fixed angle.

Theorem 5.15. *Let R be a rotation about a fixed point O. For any point $A \neq O$, there is a unique line m passing through O such that $R = r_m \circ r_l$, where l is the line through O and A. Also, if $m\angle AOR(A)$ is θ degrees, then for any point $B \neq O$, we have $m\angle BOR(B)$ is also θ degrees.*

Proof: Let m be the angle bisector of $\angle AOR(A)$. By the first part of the proof of Theorem 5.13, we know that $R = r_m \circ r_l$.

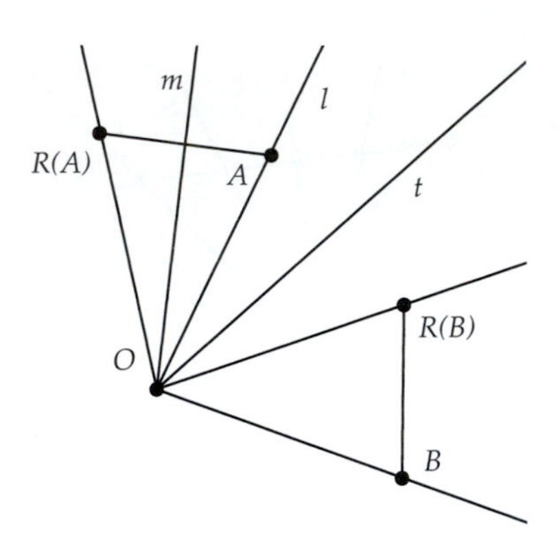

Fig. 5.14

For the second part of the theorem, we can assume B has the property that $\overline{OA} \cong \overline{OB} \cong \overline{O\,R(B)}$ (Fig. 5.14).

Let t be the angle bisector of $\angle R(B)\,OA$. Then, $r_t(A) = R(B) = r_m(r_l(B))$. Thus, $A = r_t(r_m(r_l(B)))$ and $B = r_l(r_m(r_t(A)))$. By the previous lemma we also have $B = r_t(r_m(r_l(A)))$. Since $r_m(r_l(A)) = R(A)$, then $B = r_t(R(A))$.

Since $R(B) = r_t(A)$ and $B = r_t(R(A))$, then $\overline{A\,R(A)} \cong \overline{B\,R(B)}$, and by SSS congruence $\Delta AOR(A) \cong \Delta BOR(B)$, and so $\angle AOR(A) \cong \angle BOR(B)$. Thus, the two angles have the same measure. □

From this theorem we can see that the construction of a rotation about O of a specific angle θ requires the choice of two lines that meet at O and make an angle of $\frac{\theta}{2}$.

We note here that the preceding theorems on rotations do not depend on Euclid's fifth postulate and are thus part of *neutral geometry*. We will use this fact to consider non-Euclidean rotations in Chapter 7.

Definition 5.12. The point O of intersection of the reflection lines of a rotation R is called the *center of rotation*. The angle ϕ defined by $\angle AO\,R(A)$ for $A \neq O$ is called the *angle of rotation*.

Now let's consider the coordinate form of a rotation R.

Given a point (x, y) in a coordinate system, we know that we can represent the point as $(r\ \cos(\theta), r\ \sin(\theta))$. A rotation of (x, y) through an angle of ϕ about the origin $O = (0, 0)$ is given by

$$Rot_\phi(x, y) = (r\ \cos(\theta + \phi), r\ \sin(\theta + \phi))$$

From the trigonometric formulas covered in Chapter 3, we know that the right side of this equation can be written as

$$(r\ \cos(\theta)\cos(\phi) - r\ \sin(\theta)\sin(\phi), r\ \sin(\theta)\cos(\phi) + r\ \cos(\theta)\sin(\phi))$$

Therefore,

$$Rot_{\phi}(x, y) = (x\ \cos(\phi) - y\ \sin(\phi), x\ \sin(\phi) + y\ \cos(\phi))$$

This is the coordinate form for a rotation about the origin by an angle of ϕ.

We note for future reference that a rotation about a point O of 180 degrees is called a *half-turn* about O.

5.4.1 Rotational Symmetry

Definition 5.13. A figure is said to have *rotational symmetry* (or *cyclic symmetry*) of angle ϕ if the figure is preserved under a rotation about some center of rotation with angle ϕ.

Rotational symmetry is perhaps the most widespread symmetry in nature.

Rotational symmetry can be found in the very small, such as this radiolarian illustrated by Ernst Haeckel in his book *Art Forms in Nature* [18], to the very large, as exhibited by the rotationally symmetric shapes of stars and planets.

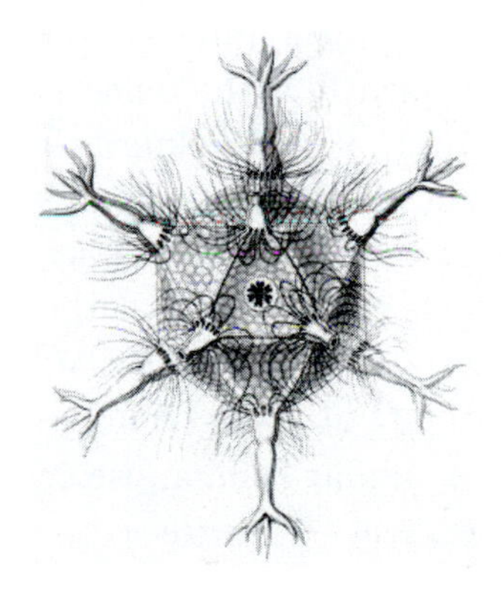

Many flowers exhibit five-fold symmetry, the rotational symmetry of the regular pentagon. Let's prove that the regular pentagon has five-fold symmetry.

Theorem 5.16. *The regular pentagon has rotational symmetry of 72 degrees.*

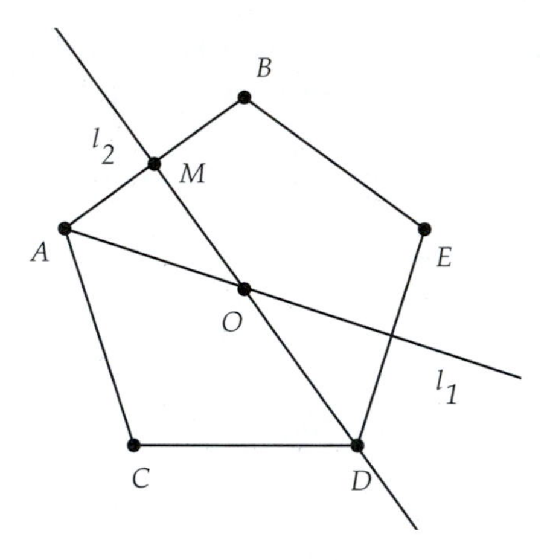

Fig. 5.15

Proof: In earlier exercises it was shown that the angle bisectors of the pentagon are lines of symmetry and the perpendicular bisectors of the sides are also lines of symmetry. Let l_1 be the angle bisector of $\angle CAB$ and l_2 be the perpendicular bisector of side $\overline{AB}$ at M in the pentagon shown in Fig. 5.15. Then l_1 and l_2 must intersect. For, suppose that they were parallel. Then $\angle CAB$ must be a straight angle, that is, C, A, and B are collinear, which is clearly impossible.

Let O be the intersection of l_1, l_2 and ϕ be the measure of $\angle AOB$. The composition R of the reflections through l_1, l_2 is a rotation through an angle of $2\angle AOM = \phi$. We know that the pentagon is invariant under R, since it is invariant under the component reflections. If C, D, E are the other vertices, we have that iterated application of the rotation R on A will cycle through these other vertices, and thus $OC = OA = OD = OE$, making all of the interior triangles ΔAOB, ΔCOA, and so on, congruent. Since there are five angles at O of these triangles, then $\phi = \frac{360}{5} = 72$. □

In a similar fashion, we could show that the regular n-gon has rotational symmetry of $\frac{360}{n}$ degrees.

Exercise 5.4.1. Suppose we are given a coordinate system centered at a point O. Let Rot_ϕ be a rotation about O of angle ϕ. Let $C = (x, y)$ be a point not equal to O and let T be the translation with vector $v = (-x, -y)$. Show that $T^{-1} \circ Rot_\phi \circ T$ is a rotation about C of angle ϕ.

Exercise 5.4.2. Given a coordinate system centered at a point O and a line l that does not pass through O, find an expression for reflection across l, by using translations, rotations, and a reflection across the x-axis.

Exercise 5.4.3. Find three examples in nature of each of the following rotational symmetries: 90 degrees (square), 72 degrees (pentagon), and 60 degrees (hexagon). Sketch your examples and label the symmetries of each.

Exercise 5.4.4. Show that if R is a rotation of $\theta \neq 0$ degrees about O, and l is a line not passing through O, then $R(l) \neq l$. That is, if a (non-identity) rotation has an invariant line, it must pass through the center of rotation. [Hint: Drop a perpendicular from O to l at A. If $\theta < 180$ consider the triangle $AO\,R(A)$. If $\theta = 180$ use a different argument.]

Exercise 5.4.5. Show that if a rotation $R \neq id$ has an invariant line, then it must be a rotation of 180 degrees. Also, the invariant lines for such a rotation are all lines passing through the center of rotation O. [Hint: Use the preceding exercise.]

Exercise 5.4.6. Let R be a rotation about a point O and m be a line. Show that if $R(m) \parallel m$, then the rotation angle for R is 180 degrees. [Hint: Suppose the angle is less than 180 and consider $\Delta AO\,R(A)$ where A is a point on m.]

Exercise 5.4.7. Show, using reflections, that the inverse to a rotation about a point of ϕ degrees is a rotation about the same point of $-\phi$ degrees.

Exercise 5.4.8. Show that the composition of two rotations centered at the same point is again a rotation centered at that point.

Exercise 5.4.9. Suppose that two rotations R, R' centered at O have the same effect on a point $A \neq O$. Show that $R = R'$.

Exercise 5.4.10. Suppose that the composition of a rotation $R \neq id$ with a reflection r_1 is again a reflection r_2. That is, suppose that $r_1 \circ R = r_2$. Show that r_1 and r_2 must pass through the center of rotation for R.

Exercise 5.4.11. Let l be perpendicular to m at a point A on l (or m). Let $H = r_l \circ r_m$. Show that H is a rotation about A of 180 degrees; that is, a half-turn about A. [Hint: Show that for all $B \neq A$ that A is the midpoint of $\overline{B\,H(B)}$.]

Exercise 5.4.12. Let A, B be distinct points. Let H_A,H_B be half-turns about A, B, respectively. Show that $H_B \circ H_A$ is a translation in the direction of the vector from A to B.

Exercise 5.4.13. Let f be any isometry and H_A a half-turn about a point A. Show that $f \circ H_A \circ f^{-1}$ is a half-turn. [Hint: Show it is a rotation and that it maps lines through a point back to themselves.]

5.5 Project 7 - Quilts and Transformations

Before we look at the last type of isometries, those composed of three reflections, we will take a break to have a little fun with our current toolbox of isometries (reflections, translations, and rotations).

One of the uniquely American craft forms is that of quilting. In making a quilt, we start with a basic block, usually a square, made up of various pieces of cloth. This basic block is then copied to form a set of identical blocks that are sewn together to make a quilt.

For example, here is a quilt block we'll call "square-in-square" that is made up of a piece of white cloth on top of a black background.

If we translate this block up and down (equivalently, sew copies of this block together), we will get the quilt shown here.

Note that the square-in-square block can be built up from a simpler shape, that of the triangular-divided square found in the bottom left corner of the square-in-square block. The square-in-square block is built of four copies of this basic shape, using various reflections.

In the first part of this project, we will see how to use *Geometry Explorer's* built-in transformation capability to construct the triangular-divided square, and then use this to construct the square-in-square and a quilt based on this block.

We note here that when we refer to a "block" of a quilt, we are referring to a square region of the quilt that can cover the entire quilt when repeatedly transformed via reflections, rotations, and translations. Thus, in the quilt shown in the middle figure in the preceding series of figures, either the square-in-square shape, or the triangular-divided square, or the entire quilt itself, could be considered basic quilting blocks.

Start *Geometry Explorer*. For this project we will use the Transform panel of buttons. In this panel are several transformation buttons, as well as three buttons labeled **Mark**, **Custom**, and **Base**. These three are *pop-up* menus. When you click on them, a menu will pop up, allowing you to select an option. In the rest of this project, you will be instructed to use these pop-up menus to define transformations.

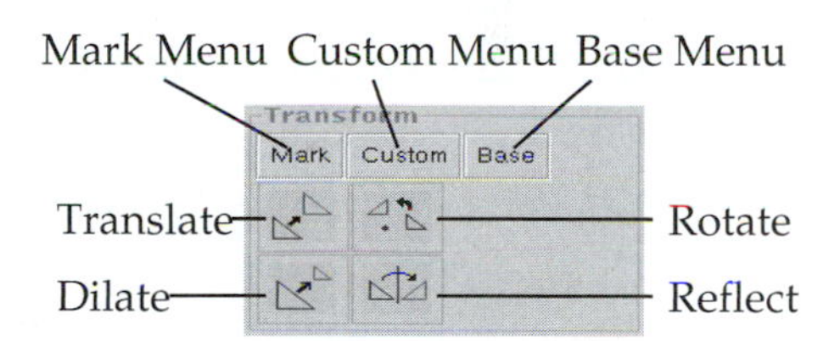

Create a segment $\overline{AB}$, select A, and choose **Center** under the **Mark** pop-up menu. This sets point A to be a center of rotation. We are going to rotate point B 90 degrees about point A. To do this we need to define the rotation. Choose **Rotation** from the **Custom** pop-up menu. A dialog box will pop up. Type in "90" and hit return. Then, rotate B by selecting B and then clicking on the Rotate button in the Transform panel (second button in first row).

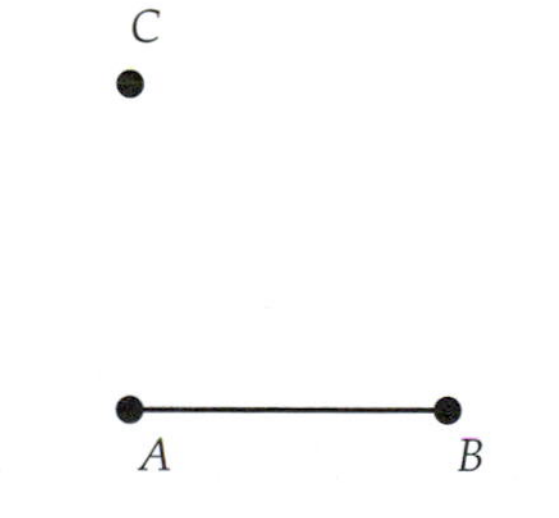

Now, set C as the center of rotation and rotate point A by 90 degrees to point D. (You will need to use the **Mark** pop-up menu as you did before.) Then, connect the vertices with segments to form the square shown.

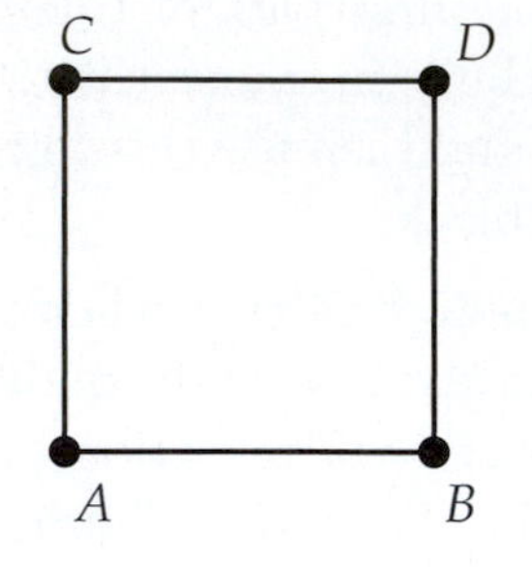

To construct the triangular-divided square, we will construct the figure shown. Multi-select points B, A, and C, and then click on the Filled-polygon button (third button in third row of the Construct panel).

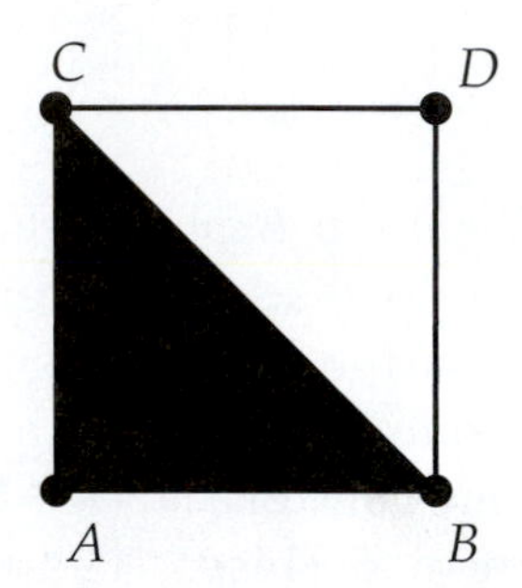

We can finish the creation of the square-in-square block either by reflection or rotation about D. Carry out whichever transformations you wish to get the block shown. (We will ignore the extra lines and points for now.)

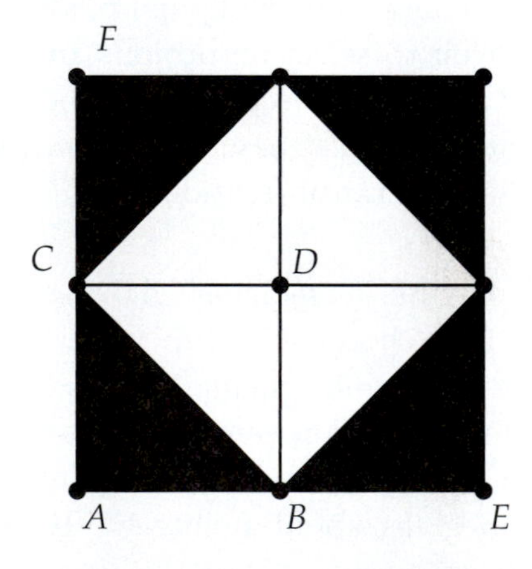

To create a larger quilt from this block, we will translate the block vertically and horizontally. For example, to translate horizontally, we will want to shift the block to the right by the vector $\overrightarrow{AE}$.

Define this horizontal translation by multi-selecting A and E and choosing **Vector** from the **Mark** pop-up menu. A dialog box will pop up, asking whether this vector is defined as a "polar" or "rectangular" vector. Choose "rectangular." To translate the block, click and drag the selection arrow to create a selection box around the block, and then click on the Translate button (first button in first row of the Transform panel) to get the image shown.

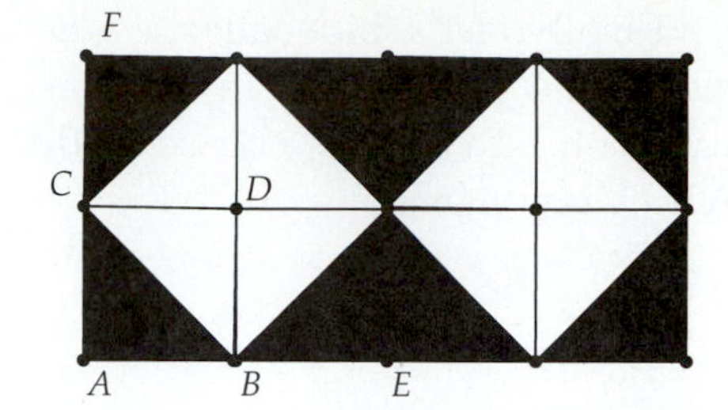

We can likewise set $\overrightarrow{AF}$ as a vector and translate the previous figure in a vertical direction multiple times (just keep hitting the Translate button).

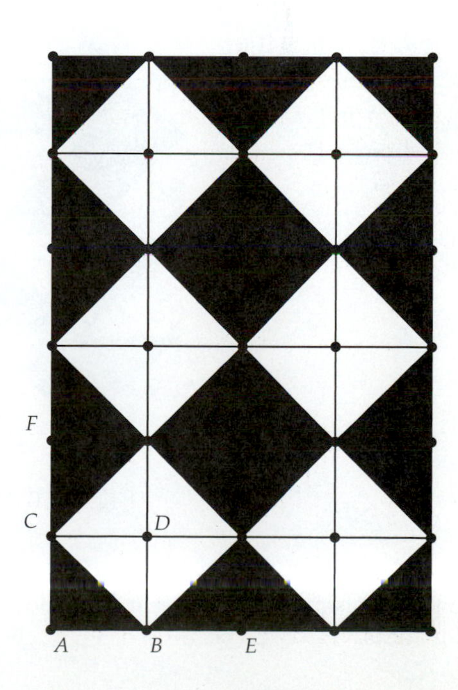

Finally, let's hide all the segments and points used in the construction. To do this choose **Hide All** (**View** menu).

Exercise 5.5.1. Using the same triangular-divided square and appropriate transformations, construct the quilt Yankee Puzzle shown in Fig. 5.16. Feel free to use different colors for the various shapes in the quilt. Describe the steps (i.e., sequence of transformations) that you took to build the quilt.

rot sym

Fig. 5.16 Yankee Puzzle

Exercise 5.5.2. Design your own pattern, based on triangular-divided squares or simple squares, and use it to build a quilt. Below are a few quilts you can use for ideas.

Exercise 5.5.3. Which of the four quilt patterns in the previous exercise have

- bilateral symmetry (Specify the lines of reflection.)
- rotational symmetry (Specify the rotation angle and center of rotation.)
- both rotational and bilateral symmetry

Exercise 5.5.4. Why must a quilt having two perpendicular lines of reflection have a rotational symmetry? What is the rotation angle?

The quilt patterns in this project, and many other intriguing quilt patterns, can be found in [5, pages 305–311] and also in [36].

5.6 Glide Reflections

Now we are ready to look at the last class of isometries—those made up of three reflections. Such isometries will turn out to be equivalent to either a reflection or a *glide reflection.*

Definition 5.14. An isometry that is made up of a reflection and a translation parallel to the line of the reflection is called a *glide reflection.*

A glide reflection is essentially a flip across a line and then a glide (or translate) along that line. If $\overrightarrow{AB}$ is a vector with T_{AB} translation by this vector and if l is a line parallel to $\overrightarrow{AB}$ with r_l reflection across l, then the glide reflection defined by these isometries is

$$G_{l,AB} = T_{AB} \circ r_l$$

Our first theorem about glide reflections says that it doesn't matter if you glide and then reflect or reflect and then glide. You always end up at the same place.

Theorem 5.17. *Let l be a line and $\overrightarrow{AB}$ a vector parallel to l.*

- $G_{l,AB} = T_{AB} \circ r_l = r_l \circ T_{AB}$
- $G_{l,AB}^{-1} = T_{BA} \circ r_l$

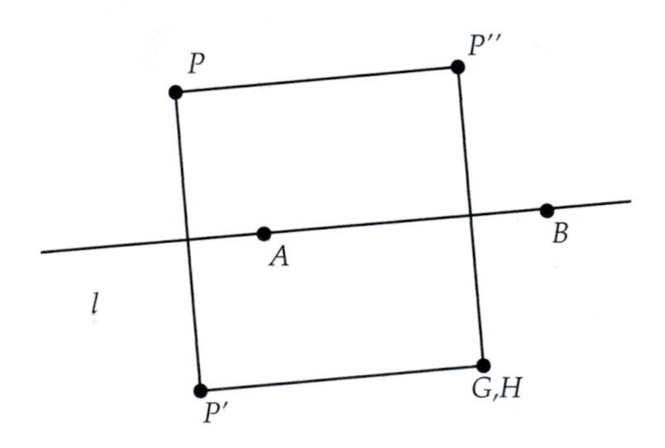

Fig. 5.17

Proof: For the first statement of the theorem, let P be a point not on l. Let $P' = r_l(P)$ and $P'' = T_{AB}(P)$ (Fig. 5.17). Let $G = T_{AB}(r_l(P))$ and $H = r_l(T_{AB}(P))$. We know that $\overline{PP'}$ is perpendicular to l. We also know that $APP''B$ and $AP'GB$ will be parallelograms, by Theorem 5.9. Thus, $\angle P'PP''$ and $\angle PP'G$ are right angles, as $\overline{PP''}$ and $\overline{P'G}$ are both parallel to l and $\overline{PP'}$ crosses l at right angles. The angles at P'' and G in quadrilateral $PP''GP'$ are also right angles, as translation preserves angles. Thus, $PP''GP'$ is a rectangle.

A similar argument will show that $PP''HP'$ is also a rectangle and thus $G = H$, or $T_{AB}(r_l(P)) = r_l(T_{AB}(P))$.

If P lies on l, then since translation of $T_{AB}(P)$ will still lie on l, we have that $r_l(T_{AB}(P)) = T_{AB}(P) = T_{AB}(r_l(P))$.

For the second statement we reference one of the earlier exercises of the chapter, which said that if a function h was the composition of f and g ($h = f \circ g$), then $h^{-1} = g^{-1} \circ f^{-1}$. So

$$G_{l,AB}^{-1} = r_l^{-1} \circ T_{AB}^{-1}$$

Since the inverse of a reflection is the reflection itself and the inverse of a translation from A to B is the reverse translation from B to A, we get that

$$G_{l,AB}^{-1} = r_l \circ T_{BA} = T_{BA} \circ r_l$$

□

Now we are ready to begin the classification of isometries that consist of three reflections.

Theorem 5.18. *Let l_1, l_2, l_3 be three lines such that exactly two of them meet at a single point. Then the composition of reflections across these three lines ($r_3 \circ r_2 \circ r_1$) is a glide reflection.*

Proof: Let r_1, r_2, r_3 be the reflections across l_1, l_2, l_3. There are two cases to consider. Either the first two lines l_1, l_2 intersect or they are parallel.

Suppose that l_1, l_2 intersect at P (Fig. 5.18). Drop a perpendicular from P to l_3 intersecting at Q.

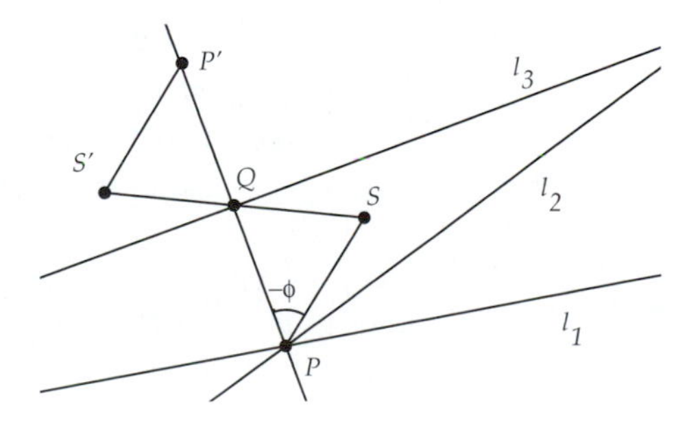

Fig. 5.18

We know that the composition of r_1 and r_2 will be a rotation about P of some non-zero angle, say ϕ degrees. Let S be the rotation of Q about P of $-\phi$ degrees. Since $r_2(r_1(S)) = Q$ and $r_3(Q) = Q$, we have

$$Q = r_3(r_2(r_1(S)))$$

Now, let P' be the reflection of P across l_3. Let $S' = r_3(r_2(r_1(Q)))$. We know that $S' \neq S$ as S' must be Q rotated about P by ϕ and then reflected across l_3 and thus must be on the other side of $\overleftrightarrow{QP}$ from S. Since $r_3 \circ r_2 \circ r_1$ is an isometry and since this composition maps ΔPSQ to $\Delta P'QS'$, we know that these two triangles are congruent. Thus, $\angle PQS \cong \angle P'S'Q$. But, both triangles must also be isosceles ($\overline{PS} \cong \overline{PQ}$). Thus, $\angle PQS \cong \angle P'QS'$. This means that S, Q, S' must lie on a line. Since isometries take lines to lines then $r_3 \circ r_2 \circ r_1$ must map $\overleftrightarrow{QS}$ back to itself, with a shift via the vector from S to Q.

Let $G = T_{SQ} \circ r_{SQ}$, where r_{SQ} is reflection across $\overleftrightarrow{QS}$ and T_{SQ} is translation from S to Q. It is left as an exercise to show that $G(P) = P'$. Then, G and $r_3 \circ r_2 \circ r_1$ match on three non-collinear points P, S, Q and so $G = r_3 \circ r_2 \circ r_1$, and thus the composition $r_3 \circ r_2 \circ r_1$ is a glide reflection.

Now, what about the second case, where l_1, l_2 are parallel? Then it must be the case that l_2 and l_3 intersect at a single point. Then, by the argument above, $r_1 \circ r_2 \circ r_3$ is a glide reflection. But, $r_1 \circ r_2 \circ r_3 = (r_3 \circ r_2 \circ r_1)^{-1}$. We know that the inverse of a glide reflection is again a glide reflection by the previous theorem. Thus, $((r_3 \circ r_2 \circ r_1)^{-1})^{-1} = r_3 \circ r_2 \circ r_1$ is a glide reflection.

□

We can now give a complete classification of isometries that consist of three reflections.

Theorem 5.19. *The composition of three different reflections is either a reflection or a glide reflection.*

Proof: There are three cases.

First, suppose that the three lines l_1, l_2, l_3 of reflection are parallel. We can suppose that there is a coordinate system set up so that each line is parallel to the x-axis. Then, as was discussed in the section on reflections, we know that the three reflections r_1, r_2, r_3 associated with l_1, l_2, l_3 can be given by

$$r_1(x, y) = (x, -y - 2K_1), r_2(x, y) = (x, -y - 2K_2), r_3(x, y) = (x, -y - 2K_3)$$

for some non-negative constants K_1, K_2, K_3. Then

$$r_3(r_2(r_1(x, y))) = (x, -y - 2K_1 + 2K_2 - 2K_3)$$

This clearly fixes the line at $y = -K_1 + K_2 - K_3$ and thus is a reflection.

Secondly, suppose that only two of the lines meet at exactly one point. By the previous theorem we have that $r_3 \circ r_2 \circ r_1$ is a glide reflection.

Finally, suppose that all three meet at a single point P. Then, by Lemma 5.14 we have that $r_3 \circ r_2 \circ r_1$ is a reflection about some line through the common intersection point.

□

5.6.1 Glide Reflection Symmetry

Definition 5.15. A figure is said to have *glide symmetry* if the figure is preserved under a glide reflection.

Where does glide symmetry appear in nature? You may be surprised to discover that your feet are creators of glide symmetric patterns! For example, if you walk in a straight line on a sandy beach, your footprints will create a pattern that is invariant under glide reflection (footsteps created by Preston Nichols).

Many plants also exhibit glide symmetry in the alternating structure of leaves or branches on a stem.

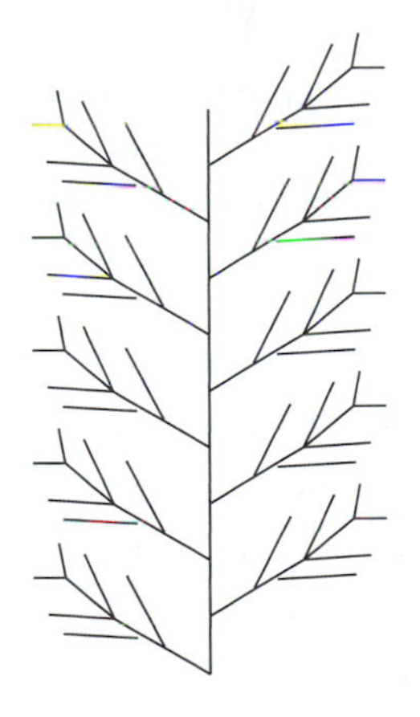

Exercise 5.6.1. Find examples of two objects in nature that exhibit symmetries of a glide reflection (other than the ones we have given). Draw sketches of these and illustrate the glide reflection for each object on your sketch.

Exercise 5.6.2. Finish the proof of Theorem 5.18; that is, show in the proof that $G(P) = P'$. [Hint: Let P'' be the reflection of P across $\overleftrightarrow{QS}$. Show that $\Delta P'P''Q \cong \Delta QSP$ and use this to show the result.]

Exercise 5.6.3. Show that the only invariant line under a glide reflection $T_{AB} \circ r_l$ (with $\overrightarrow{AB} \neq (0,0)$) is the line of reflection l. [Hint: If m is invariant, then it is also invariant under the glide reflection squared.]

Exercise 5.6.4. Show that if a glide reflection has a fixed point, then it is a pure reflection—it is composed of a reflection and a translation by the vector $v = (0,0)$. [Hint: Use a coordinate argument.]

Exercise 5.6.5. Show that the composition of a glide reflection with itself is a translation and find the translation vector in terms of the original glide reflection.

A *group* of symmetries is a set of Euclidean isometries that have the following properties:

1. Given any two elements of the set, the composition of the two elements is again a member of the set.

2. The composition of elements is an associative operation.

3. The identity is a member of the set.

4. Given any element of the set, its inverse exists and is an element of the set.

Note: Since function composition is associative, the fourth condition is true for all collections of isometries.

Exercise 5.6.6. Does the set of glide reflections form a group of symmetries? Why or why not?

Exercise 5.6.7. Show that the set of all reflections does not form a group of symmetries.

Exercise 5.6.8. Show that the set of *all* rotations does not form a group of symmetries. [Hint: Use half-turns.]

Exercise 5.6.9. Show that the set of rotations about a fixed center of rotation O does form a group of symmetries.

Exercise 5.6.10. Show that the set of translations forms a group of symmetries.

Definition 5.16. An isometry is called *direct* (or *orientation-preserving*) if it is a product of two reflections or is the identity. It is called *opposite* (or *orientation-reversing*) if it is a reflection or a glide.

Exercise 5.6.11. Given ΔABC, the ordering A, B, C is called a "clockwise" ordering of the vertices, whereas A,C,B would be a "counterclockwise" ordering. Describe, by example, how direct isometries preserve such orderings (i.e., map clockwise to clockwise and counterclockwise to counterclockwise) and how opposite isometries switch this ordering.

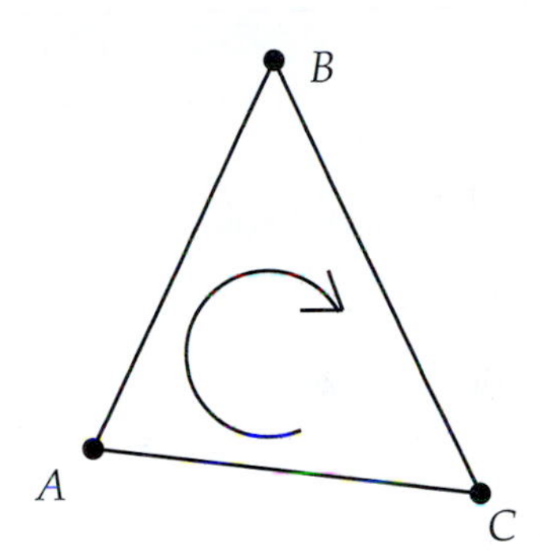

Exercise 5.6.12. Show that every isometry is either direct or opposite, but not both.

Definition 5.17. An isometry f will be called *even* if it can be written as the product of an even number of reflections. An isometry will be called *odd* if it can be written as the product of an odd number of isometries.

Exercise 5.6.13. Show that an isometry cannot be *both* even and odd. [Hint: Use Exercise 5.2.14.]

Exercise 5.6.14. Show that an isometry is even if and only if it is direct, and that it is odd if and only if it is indirect.

5.7 Structure and Representation of Isometries

We have now completely classified the possible isometries of the Euclidean plane. We can classify isometries in two ways—by the number of fixed points they have or by the number of reflections that make up isometries.

In the following table we have classified isometries by the number of fixed points they have.

Fixed Points	Isometry
0	translation, glide reflection
1	rotation
2	reflection
3 (not collinear)	identity

Table 5.1 Isometry Classification by Fixed Points

In the next table we classify an isometry by the number of reflections that comprise the isometry.

Number of Reflections	Isometry
1	reflection
2	identity, rotation, translation
3	glide reflection, reflection

Table 5.2 Isometry Classification by Reflections

We know how an individual isometry behaves, but how do isometries work when combined together? From Exercise 5.4.8 we know that the composition of two rotations about the same center is again a rotation about that center. But, what can we say about the composition of two rotations about *different* centers? We know that the composition of such rotations will again be an isometry, but what type? To answer such questions we will need a uniform way to represent all Euclidean isometries. This will be done through the use of matrices.

5.7.1 Matrix Form of Isometries

Recall that a rotation R_ϕ of a point (x, y) through an angle of ϕ degrees about the origin can be written as

$$R_\phi(x, y) = (x\ \cos(\phi) - y\ \sin(\phi), x\ \sin(\phi) + y\ \cos(\phi))$$

Note that this is equivalent to the matrix equation

$$R_\phi(x, y) = \begin{bmatrix} \cos(\phi) & -\sin(\phi) \\ \sin(\phi) & \cos(\phi) \end{bmatrix} \begin{pmatrix} x \\ y \end{pmatrix}$$

where $\binom{x}{y}$ is the column vector representing the point (x, y). Thus, the action of R_ϕ on points in the plane is the same as matrix multiplication (on the left) of the vector representing the point by the matrix $\begin{bmatrix} \cos(\phi) & -\sin(\phi) \\ \sin(\phi) & \cos(\phi) \end{bmatrix}$. We see, then, that any rotation about the origin can be identified with a $2x2$ matrix of the form above, and conversely any matrix of this form will give rise to a rotation about the origin.

Can we find a matrix form for a translation? Let T be a translation with vector $v = (v_1, v_2)$. Then

$$T(x, y) = (x, y) + (v_1, v_2) = (x + v_1, y + v_2)$$

Unfortunately, this cannot be written as the product of a $2x2$ matrix with $\binom{x}{y}$. However, consider the following matrix equation:

$$\begin{bmatrix} 1 & 0 & v_1 \\ 0 & 1 & v_2 \\ 0 & 0 & 1 \end{bmatrix} \begin{pmatrix} x \\ y \\ 1 \end{pmatrix} = \begin{pmatrix} x + v_1 \\ y + v_2 \\ 1 \end{pmatrix}$$

The vector $(x, y, 1)$ gets mapped to the vector $(x + v_1, v + v_2, 1)$. In the x and y coordinates, this is exactly what translation by v would produce. But, how can we use this fact about *three-dimensional* vectors in planar geometry?

Recall the discussion on *models* in the first chapter. If we can find a setting where the axioms of a geometry hold, then that setting will serve as a model for the geometry, and all theorems will hold equally well in that model. The normal model for planar geometry is the Euclidean (x, y) plane, which we can interpret as the plane at height 0 in three dimensions. But, the axioms of planar geometry work just as well for the set of points at height $z = 1$, that is, the set of points $(x, y, 1)$. Whatever operation we do on points in this new plane will be equal in effect to the corresponding operation in the abstract Euclidean plane, as long as we ensure that the third coordinate stays equal to 1.

For the rest of this section we will switch to this new model of planar geometry.

Definition 5.18. A *point* will be a vector $(x, y, 1)$. Distance and angle will be defined as the standard distance and angle of the x and y components.

In our new model the third coordinate does not carry any geometric significance. It is there just to make our representation of isometries easier.

We will define vector addition of two points $(x_1, y_1, 1)$ and $(x_2, y_3, 1)$ as $(x_1, y_1, 1) + (x_2, y_3, 1) = (x_1 + x_2, y_1 + y_2, 1)$. With this definition, vector addition carries the same geometric meaning as standard vector addition for points in the (x, y) plane. Also, if A is a matrix and v_1, v_2 are two vectors, then $A(v_1 + v_2) = Av_1 + Av_2$. This property of linearity will come in handy.

We have the following classification of "elementary" isometries.

- A rotation about the origin by an angle of ϕ is represented by

$$\begin{bmatrix} \cos(\phi) & -\sin(\phi) & 0 \\ \sin(\phi) & \cos(\phi) & 0 \\ 0 & 0 & 1 \end{bmatrix}$$

- A translation by a vector of $v = (v_1, v_2)$ is represented by

$$\begin{bmatrix} 1 & 0 & v_1 \\ 0 & 1 & v_2 \\ 0 & 0 & 1 \end{bmatrix}$$

- A reflection about the x-axis is represented by

$$\begin{bmatrix} 1 & 0 & 0 \\ 0 & -1 & 0 \\ 0 & 0 & 1 \end{bmatrix}$$

- A reflection about the y-axis is represented by

$$\begin{bmatrix} -1 & 0 & 0 \\ 0 & 1 & 0 \\ 0 & 0 & 1 \end{bmatrix}$$

Any other reflection, translation, rotation, or glide reflection can be built from these elementary isometries, as was shown in the exercises at the end of the last few sections. Thus, any isometry is equivalent to the product of $3x3$ matrices of the form above.

But, how exactly do compositions of isometries form new isometries? We know that the composition of two translations is again a translation, that two reflections form either a translation or rotation, and that two rotations about the same center form another rotation about that center. What about the composition of two rotations with different centers? Or the composition of a rotation and translation?

5.7.2 Compositions of Rotations and Translations

In the next two theorems, we will assume that any angle mentioned has been normalized to lie between 0 and 360.

Theorem 5.20. *Let $R_{a,\alpha}$ be rotation about point a by an angle $\alpha \neq 0$. Let $R_{b,\beta}$ be rotation about point $b \neq a$ by $\beta \neq 0$. Let T_v be translation by the vector $v = (v_1, v_2)$. Then*

(i) $R_{a,\alpha} \circ T_v = T_{R_{a,\alpha}(v)} \circ R_{a,\alpha}$.

(ii) $R_{a,\alpha} \circ R_{b,\beta}$ *is a translation iff* $\alpha + \beta = 0 (\text{mod } 360)$.

(iii) $R_{a,\alpha} \circ T_v$ *(or* $T_v \circ R_{a,\alpha}$*) is a rotation of angle* α.

(iv) $R_{a,\alpha} \circ R_{b,\beta}$ *is a rotation of angle* $\alpha + \beta$ *iff* $\alpha + \beta \neq 0 (\text{mod } 360)$.

Proof: For the first statement of the theorem, we can set our coordinate system so that a is the origin. Using the matrix forms for $R_{a,\alpha}$ and T_v, we get

$$\begin{aligned} R_{a,\alpha} \circ T_v &= \begin{bmatrix} \cos(\alpha) & -\sin(\alpha) & 0 \\ \sin(\alpha) & \cos(\alpha) & 0 \\ 0 & 0 & 1 \end{bmatrix} \begin{bmatrix} 1 & 0 & v_1 \\ 0 & 1 & v_2 \\ 0 & 0 & 1 \end{bmatrix} \\ &= \begin{bmatrix} \cos(\alpha) & -\sin(\alpha) & \cos(\alpha)v_1 - \sin(\alpha)v_2 \\ \sin(\alpha) & \cos(\alpha) & \sin(\alpha)v_1 + \cos(\alpha)v_2 \\ 0 & 0 & 1 \end{bmatrix} \end{aligned}$$

The x and y components of the third column of this product are precisely the x and y components of $R_{a,\alpha}(v)$. Let $R_{a,\alpha}(v) = (c, d)$. Then

$$\begin{aligned} T_{R_{a,\alpha}(v)} \circ R_{a,\alpha} &= \begin{bmatrix} 1 & 0 & c \\ 0 & 1 & d \\ 0 & 0 & 1 \end{bmatrix} \begin{bmatrix} \cos(\alpha) & -\sin(\alpha) & 0 \\ \sin(\alpha) & \cos(\alpha) & 0 \\ 0 & 0 & 1 \end{bmatrix} \\ &= \begin{bmatrix} \cos(\alpha) & -\sin(\alpha) & c \\ \sin(\alpha) & \cos(\alpha) & d \\ 0 & 0 & 1 \end{bmatrix} \end{aligned}$$

This finishes the proof of the first part of the theorem.

For the second part of the theorem, suppose that $R_{a,\alpha} \circ R_{b,\beta}$ is a translation T_v. We can assume that b is the origin. Let R_θ represent rotation about the origin by θ. We know that $R_{a,\alpha} = T_a \circ R_\alpha \circ T_{-a}$. Thus, using statement (i), we get that

$$\begin{aligned} T_v &= R_{a,\alpha} \circ R_{b,\beta} \\ &= T_a \circ R_\alpha \circ T_{-a} \circ R_\beta \\ &= T_a \circ T_{R_\alpha(-a)} \circ R_\alpha \circ R_\beta \\ &= T_{a+R_\alpha(-a)} \circ R_{\alpha+\beta} \end{aligned}$$

Thus, $T_{-a-R_\alpha(-a)} \circ T_v$ is a rotation ($R_{\alpha+\beta}$) and must have a fixed point. But, the only translation with a fixed point is the identity, and thus $R_{\alpha+\beta}$ must be the identity, and $\alpha + \beta$ must be a multiple of 360 degrees.

Conversely, if $\alpha + \beta$ is a multiple of 360, then $R_{a,\alpha} \circ R_{b,\beta} = T_{a+R_\alpha(-a)}$.

For the third statement of the theorem, we can assume that a is the origin and thus $R_{a,\alpha} = R_\alpha$. Consider the fixed points P of $R_\alpha \circ T_v$. Using the matrix form of these isometries, we get that if P is a fixed point, then

$$\begin{bmatrix} \cos(\alpha) & -\sin(\alpha) & 0 \\ \sin(\alpha) & \cos(\alpha) & 0 \\ 0 & 0 & 1 \end{bmatrix} \begin{bmatrix} 1 & 0 & v_1 \\ 0 & 1 & v_2 \\ 0 & 0 & 1 \end{bmatrix} P = P = \begin{bmatrix} 1 & 0 & 0 \\ 0 & 1 & 0 \\ 0 & 0 & 1 \end{bmatrix} P$$

Set $R_\alpha(v) = (e, f)$. After multiplying out the left side of the previous equation, we get

$$\begin{bmatrix} \cos(\alpha) & -\sin(\alpha) & e \\ \sin(\alpha) & \cos(\alpha) & f \\ 0 & 0 & 1 \end{bmatrix} P = \begin{bmatrix} 1 & 0 & 0 \\ 0 & 1 & 0 \\ 0 & 0 & 1 \end{bmatrix} P$$

If we subtract the term on the right from both sides, we get

$$\begin{bmatrix} \cos(\alpha) - 1 & -\sin(\alpha) & e \\ \sin(\alpha) & \cos(\alpha) - 1 & f \\ 0 & 0 & 0 \end{bmatrix} P = O$$

where O is the origin $(0, 0, 0)$. This equation has a unique solution iff the determinant of the $2x2$ matrix in the upper left corner is non-zero. (Remember that the third component is not significant.) This determinant is

$$(\cos(\alpha) - 1)^2 + \sin(\alpha)^2 = \cos(\alpha)^2 - 2\cos(\alpha) + 1 + \sin(\alpha)^2 = 2(1 - \cos(\alpha))$$

Since α is not a multiple of 360, then $\cos(\alpha) \neq 1$ and there is a unique fixed point. Thus, the composition is a rotation, say $R_{g,\gamma}$. Then $R_{-\alpha} \circ R_{g,\gamma}$

is a translation and by statement (ii) we have that $\gamma + (-\alpha) = 0 (\bmod 360)$ and thus $\gamma = \alpha (\bmod 360)$. A similar argument shows that $T_v \circ R_\alpha$ is a rotation of angle α.

For the fourth statement of the theorem, we know from the work above that $R_{a,\alpha} \circ R_\beta = T_{a+R_\alpha(-a)} \circ R_{\alpha+\beta}$. Since $\alpha + \beta \neq 0 (\bmod 360)$, by statement (iii) of the theorem, we know that the composition on the right is a rotation by an angle of $\alpha + \beta$. □

The only compositions left to consider are those involving reflections or glide reflections.

5.7.3 Compositions of Reflections and Glide Reflections

Theorem 5.21. *Let r_l be a reflection with line of symmetry l, $G_{m,v}$ a glide reflection along a line m with the vector v parallel to m, $R_{a,\alpha}$ a rotation with center a and angle $\alpha \neq 0$, and T_w translation along a non-zero vector w. Then*

(i) $r_l \circ R_{a,\alpha}$ (or $R_{a,\alpha} \circ r_l$) is a reflection iff l passes through a. If l does not pass through a, then the composition is a glide reflection.

(ii) $r_l \circ T_w$ (or $T_w \circ r_l$) is a reflection iff the vector w is perpendicular to l. If w is not perpendicular to l, then the composition is a glide reflection.

(iii) $G_{m,v} \circ R_{a,\alpha}$ (or $R_{a,\alpha} \circ G_{m,v}$) is a reflection iff m passes through the center of the rotation defined by $T_v \circ R_{a,\alpha}$ (or $R_{a,\alpha} \circ T_v$). If m does not pass through this center, then the composition is a glide reflection.

(iv) $G_{m,v} \circ T_w$ (or $T_w \circ G_{m,v}$) is a reflection iff the vector $v + w$ is perpendicular to m. If $v + w$ is not perpendicular to m, then the composition is a glide reflection.

(v) $r_l \circ G_{m,v}$ (or $G_{m,v} \circ r_l$) is a translation iff l is parallel to m. The composition is a rotation otherwise.

(vi) The composition of two different glide reflections is a translation iff the reflection lines for both are parallel. The composition is a rotation otherwise.

Proof: For the first two statements we note that the composition of a reflection and either a rotation or a translation will be equivalent to the composition of three reflections, and thus must be either a reflection or a glide reflection. Statements (i) and (ii) then follow immediately from Theorems 5.18 and 5.19 and their proofs.

For statement (iii) we note that $G_{m,v} = r_m \circ T_v$ and so $G_{m,v} \circ R_{a,\alpha} = r_m \circ T_v \circ R_{a,\alpha}$. The result follows from statement (iii) of the previous theorem and statement (i) of this theorem.

For statement (iv) we note that $G_{m,v} \circ T_w = r_m \circ T_v \circ T_w = r_m \circ T_{v+w}$. The result follows from statement (ii).

For statement (v) we note that $r_l \circ G_{m,v} = r_l \circ r_m \circ T_v$. If $r_l \circ G_{m,v}$ is a translation, say T_u, then $r_l \circ r_m = T_{u-v}$ and $l \parallel m$. Conversely, if $l \parallel m$, then $r_l \circ r_m$ is a translation and thus $r_l \circ G_{m,v}$ is a translation. If $l \not\parallel m$, then $r_l \circ r_m$ is a rotation and $r_l \circ G_{m,v}$ is also a rotation.

The last statement of the theorem follows immediately from looking at the structure of two glide reflections and is left as an exercise. □

We now have a complete characterization of how pairs of isometries act to form new isometries. In principle, any complex motion involving rotations, translations, and reflections can be broken down into a sequence of compositions of basic isometries. This fact is used to great effect by computer animators.

5.7.4 Isometries in Computer Graphics

Just as we can represent a point (x, y) as a point $(x, y, 1)$ on the plane at height 1 in three dimensions, we can represent a point in space (x, y, z) as a point $(x, y, z, 1)$ at "hyper" height 1 in four dimensions. Then, translations, rotations, and reflections in three dimensions can be represented as $4x4$ matrices acting on these points.

For example, to translate a point (x, y, z) by the vector $v = (a, b, c)$, we use the translation matrix

$$\begin{bmatrix} 1 & 0 & 0 & a \\ 0 & 1 & 0 & b \\ 0 & 0 & 1 & c \\ 0 & 0 & 0 & 1 \end{bmatrix}$$

and multiply this matrix (on the left) by the vector

$$\begin{bmatrix} x \\ y \\ z \\ 1 \end{bmatrix}$$

To animate an object on the computer screen, we need to do two things: represent the object in the computer and then carry out transformations on this representation. We can represent an object as a collection of polygonal patches that are defined by sets of vertices. To move an object we need only move the vertices defining the object and then redraw the polygons making up the object.

To realize movement of an object on the screen, we need to repeatedly carry out a sequence of rotations, reflections, and translations on the points defining the object. All of these isometries can be implemented using $4x4$ matrices as described above. Thus, a computer graphics system basically consists of point sets (objects) and sequences of $4x4$ matrices (motions) that can be applied to point sets.

One of the most popular graphics systems in use today is the OpenGL system [33]. OpenGL uses the notion of a *graphics pipeline* to organize how motions are carried out. For example, to rotate and then translate an object using OpenGL, you would define the rotation and then the translation and put these two $4x4$ matrices into a virtual pipeline. The graphics system then multiplies all the matrices in the pipeline together (in order) and applies the resulting matrix to any vertices that define the object. In a very real sense, computer graphics comes down to being able to quickly multiply $4x4$ matrices, that is, to quickly compose three-dimensional transformations.

5.7.5 Summary of Isometry Compositions

We summarize the theorems on compositions of isometries in the following table for future reference. (The table lists only non-trivial compositions; that is, ones where $l \neq m, a \neq b, v \neq w$.)

$\circ$	r_l	$R_{a,\alpha}$	T_v	$G_{l,v}$
r_m	T $(l \parallel m)$ R $(l \nparallel m)$	r $(a \in m)$ G $(a \notin m)$	r $(v \perp m)$ G $(v \not\perp m)$	T $(l \parallel m)$ R $(l \nparallel m)$
$R_{b,\beta}$	r $(b \in l)$ G $(b \notin l)$	$R_{c,\alpha+\beta}$ $(\alpha+\beta \neq 0 \pmod{360})$ T $(\alpha+\beta = 0 \pmod{360})$	$R_{c,\beta}$	r $(c_1 \in l)$ G $(c_1 \notin l)$
T_w	r $(w \perp l)$ G $(w \not\perp l)$	$R_{c,\alpha}$	T_{w+v}	r $((w+v) \perp l)$ G $((w+v) \not\perp l)$
$G_{m,w}$	T $(l \parallel m)$ R $(l \nparallel m)$	r $(c_2 \in m)$ G $(c_2 \notin m)$	r $((w+v) \perp m)$ G $((w+v) \not\perp m)$	T $(l \parallel m)$ R $(l \nparallel m)$

Table 5.3 Isometry Composition

In the table, c stands for an arbitrary center of rotation, c_1 is the center of rotation for $R_{b,\beta} \circ T_v$, and c_2 is the center of rotation for $T_w \circ R_{a,\alpha}$.

Exercise 5.7.1. Prove statement (vi) of Theorem 5.21. That is, show that the composition of two different glide reflections is a translation if the reflection lines for both are parallel and is a rotation otherwise.

An important algebraic operation on invertible functions is the idea of the *conjugate* of a function by another function. Given two invertible functions f, g, we construct the conjugate of g by f as $f \circ g \circ f^{-1}$. In the next five exercises, we look at how conjugation acts on types of isometries.

Exercise 5.7.2. Let f be an isometry and H_O a half-turn about point O (rotation by 180 degrees). Show that conjugation of H_O by f, that is $f \circ H_O \circ f^{-1}$, is equal to $H_{f(O)}$, a half-turn about $f(O)$.

Exercise 5.7.3. Let f be an isometry and r_m a reflection about line m. Show that $f \circ r_m \circ f^{-1} = r_{f(m)}$ (reflection about $f(m)$).

Exercise 5.7.4. Let f be an isometry and T_{AB} a translation with vector $\overrightarrow{AB}$. Show that $f \circ T_{AB} \circ f^{-1} = T_{f(A)f(B)}$ (translation with vector $\overrightarrow{f(A)f(B)}$).

Exercise 5.7.5. Let f be an isometry and $g = r_m \circ T_{AB}$ a glide reflection along m with translation vector $\overrightarrow{AB}$. Show that $f \circ g \circ f^{-1} = r_{f(m)} \circ T_{f(A)f(B)}$ (glide reflection along $f(m)$ with vector $\overrightarrow{f(A)f(B)}$).

Exercise 5.7.6. Let f be an isometry and $R_{A,\alpha}$ a rotation about point A by an angle of α. Show that $f \circ R_{A,\alpha} \circ f^{-1} = R_{f(A),\alpha}$ if f is a direct isometry (rotation or translation) and $f \circ R_{A,\alpha} \circ f^{-1} = R_{f(A),-\alpha}$ if f is an indirect isometry (reflection or glide reflection).

Exercise 5.7.7. Another model we can use for the Euclidean plane is the complex plane, where a point (x, y) is represented by a complex number $z = x + iy$ with $i = \sqrt{-1}$. Show that a rotation of a point (x, y) about the origin by an angle ϕ is equivalent to multiplication of $x + iy$ by $e^{i\phi}$. [Hint: Recall that $e^{i\phi} = \cos(\phi) + i\sin(\phi)$.] Show that translation of (x, y) by a vector $v = (v_1, v_2)$ is equivalent to adding $v = v_1 + iv_2$ to $x + iy$. Finally, show that reflection across the x-axis is the same as complex conjugation.

Exercise 5.7.8. Let $R_{a,\alpha}$ be rotation about a of angle α and R_β be rotation about the origin of β. Show that the center of $R_{a,\alpha} \circ R_\beta$ is the complex number $c = a\frac{1-e^{i\alpha}}{1-e^{i(\alpha+\beta)}}$.

Exercise 5.7.9. Let T_v be translation by v and R_β rotation about the origin by an angle of β. Show that the center of $T_v \circ R_\beta$ is the complex number $c = \frac{v}{1-e^{i\beta}}$.

Exercise 5.7.10. You are a computer game designer who is designing a two-dimensional space battle on the screen. You have set up your coordinate system so that the screen is a virtual world with visible coordinates running from -5 to 5 in the x direction and likewise in the y direction. Suppose that you want to have your ship start at the origin, move to the position $(2, 3)$, rotate 45 degrees there, and then move to the position $(-2, 3)$. Find the $3x3$ matrices that will realize this motion, and then describe the order in which you would put these matrices into a graphics pipeline to carry out the movement.

5.8 Project 8 - Constructing Compositions

At first glance the title of this project may seem a bit strange. We have been talking about compositions of isometries, which are essentially compositions of functions. How can you *construct* the composition of two functions? Function composition would seem to be primarily an *algebraic* concept. But composition of *geometric* transformations should have some geometric interpretation as well. In this project we will explore the construction of the composition of rotations and discover some beautiful geometry along the way.

Start *Geometry Explorer* and create a segment $\overline{AB}$. Then, create two rays $\overrightarrow{AC}$ and $\overrightarrow{BD}$ from A and B and construct the intersection point E of these two rays. Create a circle with center F and radius point G.

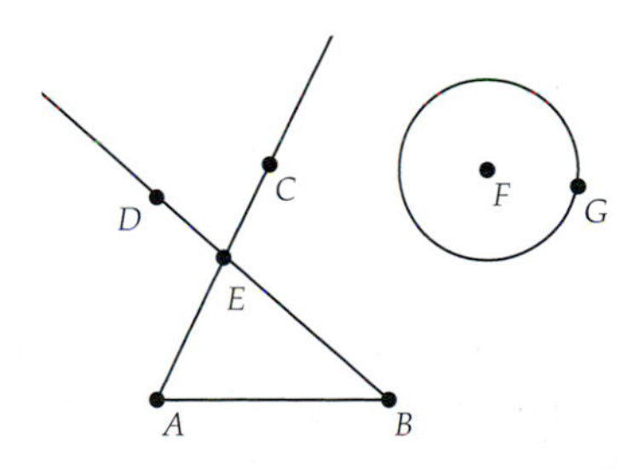

The two rotations we will compose are the rotation about A by $\angle EAB$, which we will denote by $R_{A,\angle EAB}$, and the rotation about B by $\angle ABE$, denoted by $R_{B,\angle ABE}$. Both angles will be *oriented* angles. That is, $\angle EAB$ will be directed clockwise, moving $\overrightarrow{AE}$ to $\overrightarrow{AB}$. $\angle ABE$ will also be directed clockwise. Note that this direction is opposite to the standard orientation used by the *Geometry Explorer* program.

Multi-select points E, A, B (in that order) and choose **Angle** from the **Mark** pop-up menu. Select A and choose **Center** from the **Mark** pop-up menu. These two actions define a rotation about A of $\angle EAB$. Select the circle and click on the Rotate button in the Transform panel to rotate the circle to a new circle, shown at right with center H.

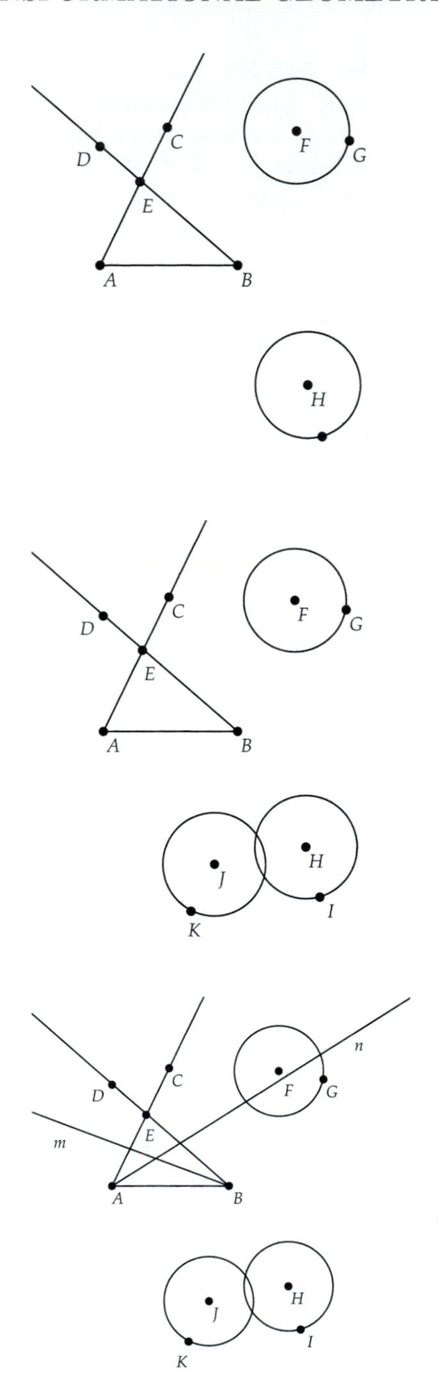

Now we will rotate the circle with center H by the rotation $R_{B,\angle ABE}$. As you did above, set $\angle ABE$ as a new angle of rotation (using the **Mark** pop-up menu) and set B as the new center of rotation. Then, select the circle centered at H and click on the Rotate button to get a new circle centered at J.

The circle at J is the result of applying $R_{B,\angle ABE} \circ R_{A,\angle EAB}$ to the original circle at F. To better analyze this composition, construct the angle bisector for $\angle BAE$ by multi-selecting B, A, and E (in that order) and clicking on the Angle Bisector tool in the Construct panel. Likewise, construct the angle bisector of $\angle EBA$. Label these bisectors "m" and "n."

Exercise 5.8.1. Show that the rotations $R_{A,\angle EAB}$ and $R_{B,\angle ABE}$ can be written as reflections using pairs of lines chosen from the three lines m, n, and $\overleftrightarrow{AB}$.

Find the intersection point O of m and n and note that O appears to be equidistant from the original circle at F and the twice transformed circle at J. Move point D around, thus changing the angles of rotation, and observe how O always seems to have this property. To convince ourselves that this holds true, let's measure the distance from O to F and from O to J. (To measure OF, multi-select O and F and choose **Distance** (**Measure** menu).)

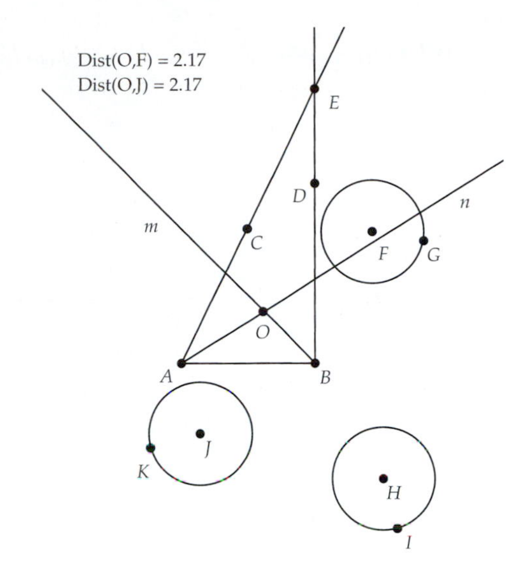

The distances do appear to stay the same.

Exercise 5.8.2. From what you have discovered so far, prove that the point O is fixed under the composite isometry $R_{B,\angle ABE} \circ R_{A,\angle EAB}$ and prove that the composite must be a rotation. [Hint: Use the results from the previous exercise.]

We conclude that $R_{B,\angle ABE} \circ R_{A,\angle EAB} = R_{O,\gamma}$, for some angle γ.

Exercise 5.8.3. Use the results of the last two exercises to show that $\gamma = (\angle EAB + \angle ABE)(\text{mod } 360)$. [Hint: Use ΔAOB and take care that angle orientation is calculated correctly.]

We have shown that the composition of two rotations about different centers is again a rotation, with new rotation angle the sum of the component angles of rotation (mod 360). This was also shown in the last section, but that proof was *algebraic* in nature. Here we have shown that the composition can be completely described by the *construction* of a certain triangle, namely ΔAOB. Transformational geometry has this interesting two-sided nature in that one can almost always explain a result either by algebraic manipulation or by geometric construction. This dual nature is what gives transformational geometry its great utility and power.

For what rotations will the preceding construction be valid? Clearly, we can construct the triangle only if $\frac{\angle EAB}{2} + \frac{\angle ABE}{2} < 180$ as this ensures that the triangle is well defined. But, this is equivalent to $(\angle EAB + \angle ABE)(\text{mod } 360) < 360$ and again we have confirmation of the condition from the last section for two rotations to create a new rotation.

What happens if $(\angle EAB + \angle ABE) (\text{mod } 360) = 360$ (or 0)?

Clear the screen and create $\overline{AB}$ and $\overrightarrow{AC}$ at some angle as shown. Create a circle centered at some point D.

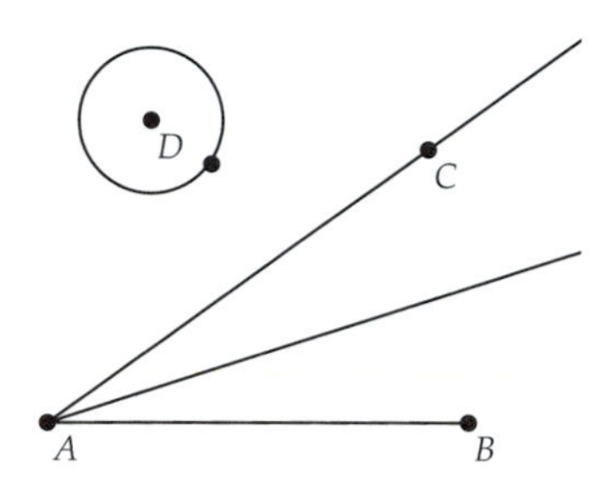

Define a rotation through $\angle CAB$, with center at A, and rotate the circle, yielding a new circle at F.

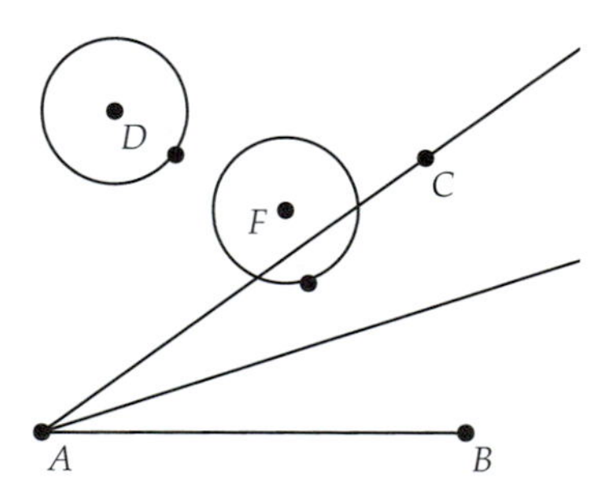

Now, we will rotate the circle at F through an angle β such that $\angle CAB + \beta$ is a multiple of 360. The simplest way to do this is merely to reverse the angle just defined. Multi-select points B, A, and C and define a rotation through $\angle BAC$ with center at A. (Note that the two angles of rotation together must be 360.) Rotate the circle at F to get a new circle at H.

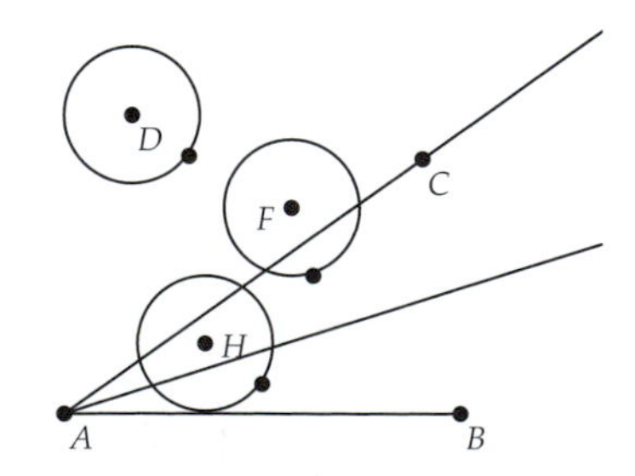

Move the original circle at D about the screen and note that the twice transformed circle at H is always the same distance and direction from the original circle. That is, the circle at H seems to be a *translation* of the circle at D.

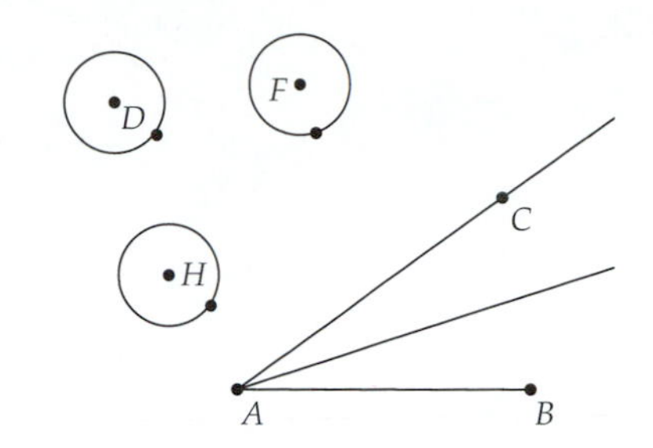

We proved in the last section that this composition must be a translation if the angles of rotation added to 360. But can we give a more geometric argument for this?

Undo your construction back to the original angle and bisector. Place a point L on the bisector to $\angle BAC$ and extend $\overline{AB}$ to M. Then, rotate point M about B by an angle equal to $\angle BAL$, yielding point N. Construct ray $\overrightarrow{BN}$.

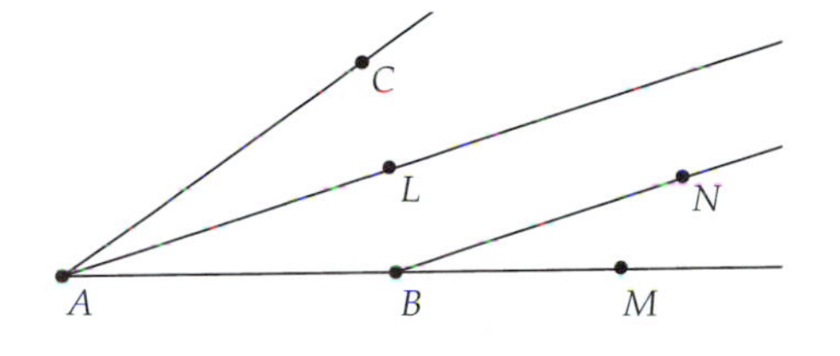

Exercise 5.8.4. Show that the original rotation about A can be written in terms of the reflections r_{AL}, r_{AB} and that the rotation about B can be written in terms of r_{BM}, r_{BN}. Use this to prove geometrically that the composition of the two rotations about A and B is a translation by twice the vector between AL and BN.

For the project report, provide detailed analysis of the constructions used in this project and complete answers to the exercises.

Chapter 6

Symmetry

Tyger, Tyger, burning bright,
In the forests of the night;
What immortal hand or eye,
Could frame thy fearful symmetry?

—"The Tyger" by William Blake

In Chapter 1 we saw how our axiomatic and abstract understanding of geometry had its origins with the Greeks' desire for perfection of reasoning. This goal of the "ideal" permeated all aspects of Greek culture, as evidenced by their love of the golden ratio—the perfect harmony of proportion. Western cultures have embraced this love of harmony and balance, as evidenced by a focus on *symmetry* in art, music, architecture, and design.

In the preceding chapter, isometries were used to define several types of *symmetry*. Symmetry is a common property of both natural and man-made objects.

In general, given a geometric figure F, we will call a transformation f a *symmetry* of F if f maps F back to itself. That is, $f(F) = F$. We say that F is *invariant* (or unchanged) under f. Note that this does not mean that every point of F remains unchanged, only that the total *set* of points making up F is unchanged.

Given a figure such as a flower, or snail shell, or triangle, the set of all symmetries of that figure is not just a random collection of functions. The set of symmetries of an object has a very nice *algebraic* structure.

Theorem 6.1. *The set of symmetries of a figure F forms a* group.

Proof: Recall that a group is a set of elements satisfying four properties (refer to the discussion preceding Exercise 1.4.6). For a set of functions, these properties would be:

1. Given any two functions in the set, the composition of the two functions is again in the set.

2. The composition of functions is an associative operation.

3. The identity function is a member of the set.

4. Given any function in the set, its inverse exists and is an element of the set.

To prove that the set S of symmetries of F forms a group, we need to verify that S has all four of these properties. Since the composition of isometries is associative, the second condition is automatically true.

The third condition is true since the identity is clearly a symmetry of any figure.

Let f, g be two symmetries of F. Since $f(F) = F$ and $g(F) = F$, then $g(f(F)) = g(F) = F$ and $g \circ f$ is a symmetry of F, and the first condition holds.

Since a symmetry f is a transformation, then it must have an inverse f^{-1}. Since $f(F) = F$, then $F = f^{-1}(F)$ and the inverse is a symmetry. $\square$

Why is it important that the symmetries of a figure form a group? Groups are a fundamental concept in abstract algebra and would seem to have little relation to geometry. In fact, there is a very deep connection between algebra and geometry. As M. A. Armstrong states in the preface to *Groups and Symmetry*, "groups measure symmetry" [3]. Groups reveal to us the algebraic structure of the symmetries of an object, whether those symmetries are the geometric transformations of a pentagon, or the permutations of the letters in a word, or the configurations of a molecule. By studying this algebraic structure, we can gain deeper insight into the geometry of the figures under consideration.

To completely delve into this beautiful connection between algebra and geometry would take us far afield of this brief survey of geometry. We will, however, try to classify some special types of symmetries that often appear in art and nature, symmetries which are also isometries. In the following sections, we will assume that the symmetries under study are Euclidean isometries—reflections, translations, rotations, and glide reflections.

6.1 Finite Plane Symmetry Groups

We will first look at those symmetry groups of an object that are *finite*, that is, those groups of symmetries that have a finite number of elements. What can be said about symmetries in a finite group? Suppose that f is a symmetry in a finite symmetry group and consider the repeated compositions of f with itself, $f, f^2 = f \circ f, f^3, \ldots$ This set cannot contain all different elements, as then the group would be infinite. Thus, for some $i \neq j$, we have $f^i = f^j$ and $f^i \circ f^{-j} = id$, where $f^{-j} = (f^{-1})^j$ and id is the identity function. We can assume that $i > j$.

Definition 6.1. We say that a symmetry f has *finite order* in a symmetry group G if for some positive integer n, we have $f^n = id$.

We then have the following:

Lemma 6.2. *All of the symmetries in a finite symmetry group are of finite order.*

What are the symmetries of finite order? Consider a translation given by $T(x, y) = (x + v_1, y + v_2)$, with $(v_1, v_2) \neq (0, 0)$. Clearly, $T^n(x, y) = (x + nv_1, y + nv_2)$, and thus $T^n(x, y) = (x, y)$ iff $(v_1, v_2) = (0, 0)$. So, the only translations of finite order are the identity translations. Similarly, we can show that a non-trivial glide reflection must have infinite order.

Thus, the symmetries of finite order consist solely of reflections and rotations. But, which reflections and rotations?

Before we answer that question, let's consider the symmetry group of a simple geometric figure, an equilateral triangle. Which rotations and reflections will preserve the triangle?

A rotation R by 120 degrees clockwise about the centroid of the triangle will map the triangle back to itself, permuting the labels on the vertices. R^2 will be a rotation of 240 degrees and will also leave the triangle invariant. The rotation R^3 will yield the identity isometry. The effects of applying R, R^2, and R^3 are shown at right, with $R(\Delta ABC)$ the right-most triangle and $R^2(\Delta ABC)$ the left-most triangle.

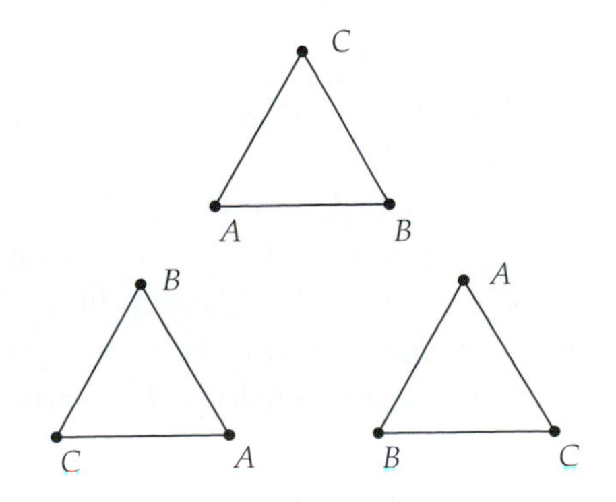

What about reflections? If we construct the perpendicular bisector for a side of the triangle, then that bisector will pass through the opposite vertex. Reflection across the bisector will just interchange the other two vertices and thus will preserve the triangle. There are three such reflections.

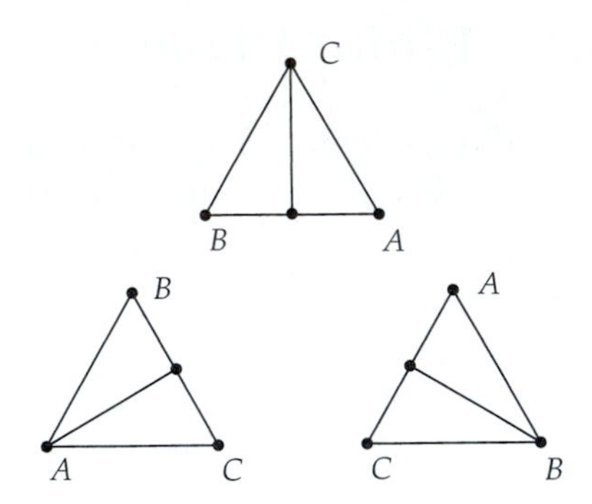

Are there any other possible isometries? To answer this question, let's consider the effect of an isometry on the labels A, B, C of the original triangle. An isometry must preserve the configuration of the triangle, so it must move the labels of the vertices around, leaving the shape of the triangle unchanged. How many labelings of the triangle are there? Pick one of the vertices to label. We have three possible labels for that vertex. After labeling this vertex, pick another to label. There are two possible labels for that vertex, leaving just one choice for the last vertex. Thus, there is a maximum of six labelings for the triangle, and therefore a maximum of six isometries. Since we have exhibited six different isometries (three rotations and three reflections), these must form the complete symmetry group for the triangle.

Let's note several interesting things about this symmetry group. First, the rotations by themselves form a group, $G = \{id = R^3, R_{120} = R, R_{240} = R^2\}$, and all the rotations in this group have the same center. Second, the reflections all have lines of symmetry passing through the center of the rotations, and there are exactly as many reflections as rotations. We will now show that these properties are shared by *all* finite symmetry groups.

Theorem 6.3. *Let R_1, R_2 be two rotations in a finite symmetry group G. Then R_1, R_2 must have the same center of rotation.*

Proof: Let a, b be the centers of rotation for R_1, R_2. Suppose that $a \neq b$. Then we know that $R_1 \circ R_2$ is either a rotation or translation (see Table 5.3 in Chapter 5). Clearly, it cannot be a non-identity translation, as a finite symmetry group has no non-trivial translations. If $R_1 \circ R_2 = id$, then $R_1 = R_2^{-1}$ and both have the same center.

Let α, β be the angles of rotation for R_1, R_2. Then we know that

$$R_1 \circ R_2 = R_{a,\alpha} \circ R_{b,\beta} = R_{c,\alpha+\beta}$$

for some point $c \neq a$ or b. Now $R_2^{-1} = R_{b,-\beta}$ is also in G and so $R_{b,-\beta} \circ R_{c,\alpha+\beta} = R_{d,\alpha}$ is in G for some point d.

We claim that $d \neq a$. For if $d = a$, then

$$R_{b,-\beta} \circ R_{c,\alpha+\beta} = R_{a,\alpha}$$

Or, equivalently,

$$R_{b,-\beta} \circ R_{a,\alpha} \circ R_{b,\beta} = R_{a,\alpha}$$

Then

$$R_{a,\alpha} \circ R_{b,\beta} = R_{b,\beta} \circ R_{a,\alpha}$$

Applying both sides to the point b, we get

$$R_{a,\alpha}(b) = R_{b,\beta} \circ R_{a,\alpha}(b)$$

Thus, $R_{a,\alpha}(b)$ would be a fixed point of $R_{b,\beta}$. But, the only fixed point of a rotation is its center, and so $R_{a,\alpha}(b) = b$. But, this implies that $a = b$, which is a contradiction.

Since $d \neq a$, then $R_{d,\alpha} \circ R_{a,-\alpha}$ is in G, but this composition is a translation.

Thus, it must be that both rotations have the same center. □

Theorem 6.4. *Among all rotations in a finite symmetry group G, let $R_{a,\alpha}$ be the one with smallest positive angle α. Then α divides* 360, *and if $R_{a,\beta}$ is another rotation in G, then $\beta = n\alpha$ for some integer n. That is, all rotations are multiples of some minimal angle.*

Proof: Since $R_{a,\alpha}$ is of finite order, then $R_{a,\alpha}^n = R_{a,n\alpha} = id$ for some n. Thus, $n\alpha = 360k$ for some integer k. If $n\alpha \neq 360$, then let i be the largest integer such that $i\alpha < 360$. Then $0 < 360 - i\alpha < \alpha$. But, then $R_{a,360} \circ R_{a,-i\alpha} = R_{a,360-i\alpha}$ must be in G, which would mean that there was a smaller rotation angle than α in G. Thus, $n\alpha = 360$.

Now suppose $R_{a,\beta}$ is another rotation in G, different from $R_{a,\alpha}$. We know that $\alpha < \beta$. Let j be the largest integer such that $j\alpha < \beta$. Then, $0 < \beta - j\alpha < \alpha$. Therefore, $\beta - j\alpha$ will be a rotation angle in G smaller than α and again we get a contradiction. □

We conclude that all rotations in a finite symmetry group are generated from a particular rotation having the smallest angle. Groups generated from a single element are called *cyclic groups*.

Definition 6.2. A group G is called *cyclic* if all of its members can be written as compositions of a single member with itself. That is, $G = \{id, f, f^2, f^3, \ldots\}$.

Corollary 6.5. *The set of rotations in a finite group form a cyclic group.*

Now, let's consider how rotations and reflections interact in a finite symmetry group.

Theorem 6.6. *If a finite symmetry group G contains a rotation and a reflection, then the line of symmetry for the reflection must pass through the center of the rotation.*

Proof: Let r_l and $R_{a,\alpha}$ be the reflection and rotation. We know that the composition of these is either a reflection or glide reflection. Since G cannot contain glide reflections, then $r_l \circ R_{a,\alpha}$ is a reflection, say r_m. Then $r_l \circ r_m = R_{a,\alpha}$ and l, m must pass through a. □

Theorem 6.7. *Let G be a finite symmetry group with n rotations (counting the identity as a rotation). If G has at least one reflection, say r_l, then it has exactly n reflections, which can be represented as $r_l \circ R_{a,i\alpha}$, $i = 0, \ldots, n-1$, where $R_{a,\alpha}$ generates the rotations of G.*

Proof: Consider the set $r_l \circ R_{a,i\alpha}$, with $i = 0, \ldots, n-1$. Clearly, all of the elements of this set are reflections. Also, no two elements of this set can be equivalent. Finally, all reflections in G are represented in this set, since if r_m is in G, then $r_l \circ r_m$ is a rotation and so must be some $R_{a,i\alpha}$. It follows that $r_m = r_l \circ R_{a,i\alpha}$. □

Definition 6.3. A finite symmetry group generated by a rotation and a reflection is called a *dihedral group.* If there are n rotations in the group, the group is called the *dihedral group of order n.*

Thus, we see that a finite symmetry group is either a cyclic group or a dihedral group. This result has been known historically as *Leonardo's Theorem* in honor of Leonardo Da Vinci (1452–1519). According to Hermann Weyl in his book *Symmetry* [40, page 66], Leonardo was perhaps the first person to systematically study the symmetries of a figure in his architectural design of central buildings with symmetric attachments.

Exercise 6.1.1. Find three examples in nature that have different finite symmetry groups. Sketch these and give the specific elements in their symmetry groups.

Exercise 6.1.2. Find the symmetry group for a square.

Exercise 6.1.3. Find the symmetry group for a regular pentagon.

Exercise 6.1.4. Show that the symmetry group for a regular n-gon must be finite.

Exercise 6.1.5. Show that the symmetry group for a regular n-gon must be D_n, the dihedral group of order n.

Exercise 6.1.6. Show that the dihedral group of order n can be generated by two reflections, that is, any element of the group can be expressed as a product of terms involving only these two reflections.

Exercise 6.1.7. Show that the number of symmetries of a regular n-gon is equal to the product of the number of symmetries fixing a side of the n-gon times the number of sides to which that particular side can be switched.

Exercise 6.1.8. Find a formula for the number of symmetries of a regular *polyhedron* by generalizing the result of the last exercise. Use this to find the number of symmetries for a regular tetrahedron (four faces and four vertices) and a cube.

6.2 Frieze Groups

Planar symmetry groups that are infinite must necessarily contain translations and/or glide reflections. In this section we will consider symmetry groups having translations in just one direction.

Definition 6.4. A *frieze group* G is a planar symmetry group with all translations in the same direction. Also, there exists a translation T_v in G such that v is of minimal (non-zero) length among all translations of G.

It turns out that all translations in a frieze group G will be generated from a single translation $T \in G$, which we will call the *fundamental translation* for G.

Lemma 6.8. *Let G be a frieze group. Then there exists a translation T in G such that if T' is any (non-identity) translation in G, then $T' = T^n$ for some non-zero integer n.*

Proof: The definition of frieze groups guarantees that there is a translation T in G, with translation vector v of minimal length. We claim that T is the fundamental translation of G. For if T' is any other (non-identity) translation in G, with translation vector v', then $v' \parallel v$ and so $v' = nv$ for some (non-zero) real number n.

If n is not an integer, let k be the nearest integer to n. Then $T^{-k} = (T^{-1})^k$ must be in G and since $T^{-k} \circ T' = T^{-k} \circ T^n = T^{n-k}$, then T^{n-k} is in G. But T^{n-k} has translation vector of length $|(n-k)|\,\|v\|$, which is less than the length of v, as $|(n-k)| < 1$.

Thus, n is an integer and all translations are integer powers of T. □

A frieze *pattern* is a pattern that is invariant under a frieze group. Such a pattern is generally composed of repetitions of a single pattern, or *motif*, in a horizontal direction as in Fig. 6.1.

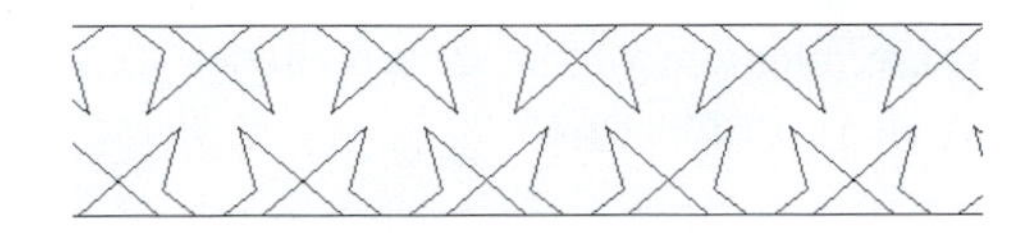

Fig. 6.1

Frieze groups have translations generated by a single translation T. By Exercise 5.3.8, frieze groups have invariant lines that are all pair-wise parallel, and also each is parallel to the translation vector of T. What else can be said of the invariant lines of a frieze group?

Theorem 6.9. *Let G be a frieze group with fundamental translation T. If l is invariant under T, then $S(l)$ is parallel (or equal) to l for all symmetries S in G.*

Proof: Consider the conjugation of T by S: $S \circ T \circ S^{-1}$. By Exercise 5.7.4, we know that $S \circ T \circ S^{-1}$ is a translation and so $S \circ T \circ S^{-1} = T^k$ for some positive integer k. In particular, $S \circ T = T^k \circ S$.

Let v be the translation vector for T. Then since $S \circ T = T^k \circ S$, the direction of the line through $S(P)$ and $S(T(P))$ will be the same as that of the line through $S(P)$ and $T^k(S(P))$. This direction is given by the vector kv. Also, the direction for the line through P and $T(P)$ is given by the vector v. Thus, for any point P, we have either $S(\overleftrightarrow{P\,T(P)}) \parallel \overleftrightarrow{P\,T(P)}$, or these two lines are coincident.

For a line l invariant under T, let P be a point on l. Using the results of the last paragraph, we see that $S(l) \parallel l$ or $S(l) = l$. □

Definition 6.5. If the frieze group G has exactly one line invariant under all the elements of G, we call that line the *midline* for G (or any pattern invariant under G).

Theorem 6.10. *For a frieze group G, the only symmetries possible in G are those generated by the fundamental translation T, a unique reflection about a midline m, reflections perpendicular to m, half-turns (180-degree rotations) about points on m, and glide reflections along m.*

Proof: Suppose G has a reflection r_l. If m is an invariant line for G, then $r_l(m) = m$. By Exercise 5.2.12, we have $l = m$ or $l \perp m$.

If $l = m$, then m is the only line that is invariant under r_l and so is the only line invariant under *all* elements of G. Thus, if G has a reflection parallel to one of the invariant lines of G, then G has exactly one invariant line, the midline, and has exactly one reflection parallel to this line.

If G has a rotation R and m is an invariant line, then $R(m) = m$. By Exercises 5.4.4 and 5.4.5 in section 5.4, R is a rotation of 180 degrees about a point on m, and among all parallel invariant lines of G, there is only one that is invariant under R. This will be the midline of G. Thus, any rotation in G is a half-turn about a point on the midline for G.

If G has a glide reflection $g = r_l \circ T'$ (with $T' \neq id$), then since $g^2 = T'^2$, which is a translation, $T'^2 = T^k$ for some positive integer k. If m is an invariant line for G, then $g(m) = m$. Since $g(m) = r_l(T^k(m)) = r_l(m)$, then $r_l(m) = m$ and so $l = m$ or $l \perp m$. If $l \perp m$, then l is perpendicular to the translation vector for T', and $g = r_l \circ T'$ is a reflection by Theorem 5.21. This contradicts our assumption that g is a glide reflection. Thus, $l = m$ and if G has a glide reflection, it must be along the midline, which is the unique invariant line for r_l.

Finally, we have to show that if G has a reflection r_m (or glide reflection $g = r_m \circ T^k$) about a line m that is invariant under all elements of G, and also has a half-turn H_O about a point O on another invariant line m' for G, then $m = m'$. This is immediate, since m must be invariant under all elements of G, in particular $H_O(m) = m$. But, H_O has a unique invariant line and so $m = m'$. □

We conclude that a frieze group G has a unique invariant line, the midline, if any of the following conditions are satisfied:

1. G contains a reflection parallel to an invariant line of the fundamental translation.

2. G contains a half-turn.

3. G contains a glide reflection.

The non-translational isometries in a frieze group can all be generated using the fundamental translation, as follows.

Theorem 6.11. *Let G be a frieze group. Then*

(i) All half-turns in G are generated by a single half-turn in G and powers (repeated compositions) of T.

(ii) All perpendicular reflections are generated from a single perpendicular reflection in G and powers of T.

(iii) All glide reflections are generated by a single reflection (or glide reflection) in G that is parallel to the direction of translation, together with powers of T. Also, if v is the translation vector for T, then a glide reflection must have a glide vector equal to kv or half of kv for some positive integer k.

Proof: For statement (i), let H_A, H_B be two half-turns in G. Then by Exercise 5.4.12, we know that $H_B \circ H_A$ is a translation, thus $H_B \circ H_A = T^k$ for some positive integer k. Since $H_A^{-1} = H_A$, then $H_B = T^k \circ H_A$.

The proofs of statements (ii) and (iii) are left as exercises. □

Thus, all frieze groups (of non-trivial translational symmetry) are generated by a translation $\tau = T_v$ and one or more of the following isometries: r_m (reflection across the midline m), r_u (with u perpendicular to m), H (half-turn rotation about O on m), and γ (glide reflection along m with glide vector equal to $\frac{v}{2}$). We omit the case of glides with glide vector v as these can be generated by τ and r_m.

In principle, this would give us a total of 2^4 possible frieze groups, each frieze group generated by τ and a subset of the four isometries r_m, r_u, H, and γ. However, many of these combinations will generate the same group. For example, we can choose u such that $H \circ \gamma = r_u$ and $r_u \circ \gamma = H$, and so the group generated by τ, γ, H must be the same as the group generated by τ, γ, r_u.

It turns out that there are only seven different frieze groups. We will list them by their generators. For example, the group listed as $< \tau, r_m >$ is the group generated by all possible compositions of these two isometries (compositions such as τ^3, r_m^5, $r_m^2 \circ \tau \circ r_m^4$, etc.).

1. $< \tau >$
2. $< \tau, r_m >$
3. $< \tau, r_u >$
4. $< \tau, \gamma > = < \gamma >$
5. $< \tau, H >$
6. $< \tau, r_m, H >$
7. $< \tau, \gamma, H > = < \gamma, H >$

(For a complete proof of this result see [31, page 392] or [38, page 190].)

Exercise 6.2.1. Show that the groups generated by τ, γ, H and γ, H are the same. [Hint: To show two sets equivalent, show that each can be a subset of the other.]

Exercise 6.2.2. Show that the groups generated by τ, r_m, H and τ, r_u, H are the same, assuming we choose r_u so that it intersects m at the center of H.

Exercise 6.2.3. Prove statement (ii) of Theorem 6.11. [Hint: Consider the composition of reflections.]

Exercise 6.2.4. Prove statement (iii) of Theorem 6.11.

Exercise 6.2.5. Show that if a frieze group has glide reflections, then the group must have a glide reflection g with glide vector of v or $\frac{v}{2}$, where v is the translation vector for the group.

Exercise 6.2.6. Show that if H_A and H_B are two half-turns of a frieze group G, then $AB = kv$ or $AB = kv + \frac{v}{2}$, where v is the translation vector for T and k is a positive integer. [Hint: Consider the action of $H_B \circ H_A$ on A.]

Exercise 6.2.7. Show that if r_u and r_v are two reflections of a frieze group G that are perpendicular to the translation vector v of T, then $AB = kv$ or $AB = kv + \frac{v}{2}$, where A, B are the intersection points of u, v, with an invariant line of G and k is a positive integer. [Hint: Consider the action of $r_v \circ r_u$ on A.]

Definition 6.6. A *subgroup* K of a group G is a non-empty subset of elements of G that is itself a group.

Exercise 6.2.8. Show that all elements of $< T, r_u >$ are either translations or reflections perpendicular to m. Use this to show that none of $< T, r_m >$, $< T, H >$, or $< T, \gamma >$ are subgroups of $< T, r_u >$. [Hint: Use Table 5.3.]

Exercise 6.2.9. Show that all elements of $< T, H >$ are either translations or half-turns. Use this to show that none of $< T, r_m >$, $< T, r_u >$, or $< T, \gamma >$ are subgroups of $< T, H >$.

Exercise 6.2.10. Show that all elements of $< T, \gamma >$ are either translations or glide reflections (with glide vector of $kv + \frac{v}{2}$). Use this to show that none of $< T, r_m >$, $< T, r_u >$, or $< T, H >$ are subgroups of $< T, \gamma >$.

Exercise 6.2.11. Show that all elements of $< T, r_m >$ are either translations, or r_m, or glide reflections (with glide vector of kv). Use this to show that none of $< T, \gamma >$, $< T, r_u >$, or $< T, H >$ are subgroups of $< T, r_m >$.

Exercise 6.2.12. Draw a diagram showing which of the seven frieze groups are subgroups of the others. [Hint: Use the preceding exercises.]

Exercise 6.2.13. In the figure below, there are seven frieze patterns, one for each of the seven frieze groups. Match each pattern to the frieze group that is its symmetry group.

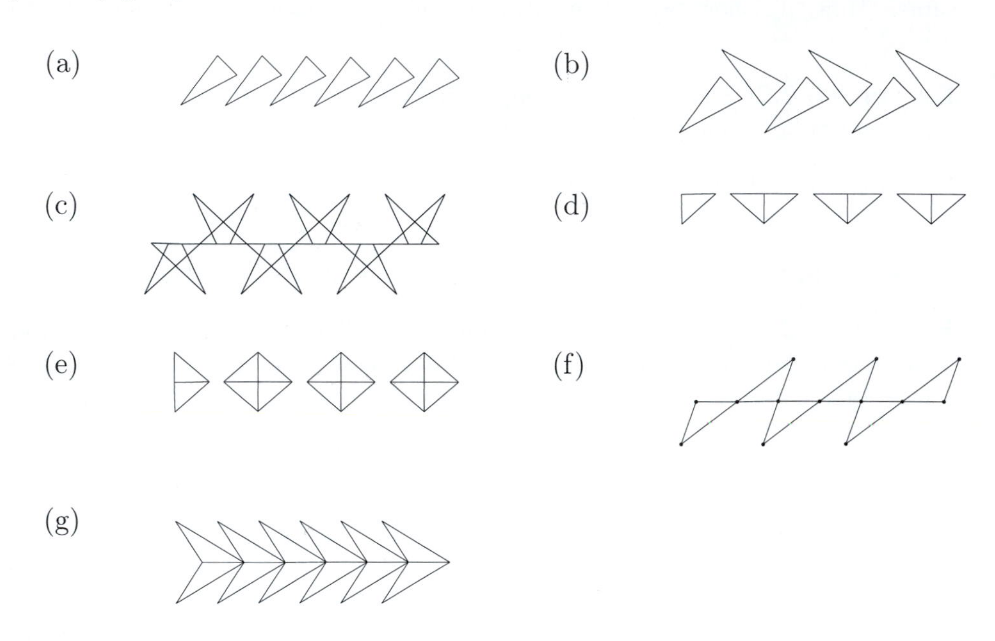

Exercise 6.2.14. Design two frieze patterns having different frieze groups and list the group for each of your patterns.

6.3 Wallpaper Groups

A second type of infinite planar symmetry group is a generalization of frieze groups into two dimensions.

Definition 6.7. A *wallpaper group* is a planar symmetry group containing translations in two different directions, that is, translations with non-parallel translation vectors. Also, there exists a translation T_v in G such that v is of minimal (non-zero) length among all translations of G.

The condition on minimal length translations will ensure that all rotations in G are of finite order.

Theorem 6.12. *Let G be a group having a translation T_v such that v has least length among all translation vectors in G. Then all rotations of G must have finite order.*

Proof: Suppose G contains a rotation $R_{a,\alpha}$ of infinite order. Then for all integers $k > 0$, we have that $R_{a,\alpha}^k \neq id$, and thus $k\alpha \pmod{360} \neq 0$.

Consider the set of numbers $S = \{k\alpha \pmod{360}\}$. Since S is an infinite set, then α cannot be rational, and if we split the interval from $[0, 360]$ into n equal sub-intervals of length $\delta = \frac{360}{n}$, then one of these sub-intervals must have an infinite number of elements of S. Suppose the interval $[i\delta, (i+1)\delta]$ has an infinite number of elements of S. In particular, the interval has two elements $k_1\alpha \pmod{360}$ and $k_2\alpha \pmod{360}$, with neither value equal to an endpoint of the interval, and $k_1\alpha \pmod{360} < k_2\alpha \pmod{360}$. Then $(k_2 - k_1)\alpha \pmod{360}$ will be an element of S in the interval $[0, \delta]$.

Thus, we can assume that for any δ close to 0, we can find a k such that $k\alpha \pmod{360} < \delta$. So if a symmetry group G contains a rotation $R_{a,\alpha}$ of infinite order, then it must have rotations of arbitrarily small angle $R_{a,\delta}$.

We know that T_v has translation vector v of smallest length among all possible translation vectors in G. Consider $R_{a,\delta} \circ T_v \circ R_{a,-\delta}$. This must be an element of G. It is left as an exercise to show that $R_{a,\delta} \circ T_v \circ R_{a,-\delta}$ is the translation $T_{R_{a,\delta}(v)}$. Then the translation $T_{R_{a,\delta}(v)} \circ T_{-v} = T_{R_{a,\delta}(v)-v}$ must be in G. However, for δ arbitrarily small, the length of the vector $R_{a,\delta}(v) - v$, which is the translation vector for $T_{R_{a,\delta}(v)-v}$, will be arbitrarily small, which contradicts the hypothesis of the theorem. □

Definition 6.8. A *discrete* symmetry group is one that has translations of minimal (non-zero) length.

Thus, frieze groups and wallpaper groups are by definition discrete groups of planar symmetries.

A wallpaper *pattern* is a pattern that is invariant under a wallpaper group. Such a pattern is generally composed of repetitions of a single pattern, or *motif*, in two different directions, as shown at right.

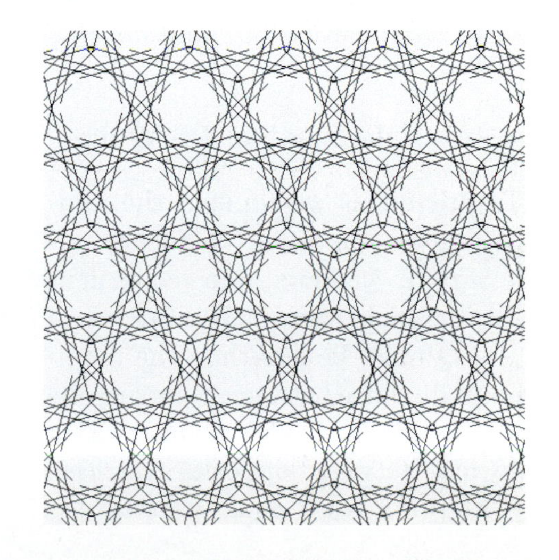

Wallpaper groups are certainly not trivial to classify. However, we can make a few important observations as to their structure.

First of all, every wallpaper group generates a *lattice* in the plane.

Definition 6.9. A *lattice* L spanned by two vectors v, w in the plane is the set of all integer combinations of v and w. That is, $L = \{sv + tw | s, t \in Z\}$, where Z is the set of integers.

Theorem 6.13. *Let $\mathcal{T}$ be the set of all translations of a wallpaper group G. Then, the set of all translations of the origin by elements of $\mathcal{T}$ forms a lattice that is spanned by v and w, where v is a vector of minimal length in $\mathcal{T}$ and w is a vector of minimal length in a different direction from v.*

Proof: First, we note that all translations in $\mathcal{T}$ must be generated by integer combinations of v and w. For suppose T_z was a translation with z not an integer combination of v and w. The two vectors v, w define a parallelogram in the plane and the set of all integer combinations of v, w will divide up the plane into congruent parallelograms.

Let $P = T_z(O) = T_z$. Then P must lie in one of these parallelograms. Let Q be the corner of the parallelogram containing P that is closest to P. Then the vector $P-Q$ cannot be congruent to any of the edges of the parallelogram, since if it were, then $P-Q = T_z(O) - T_{s_1 v + t_1 w}(0) = T_{s_2 v + t_2 w}(0)$, and z would be an integer combination of v and w.

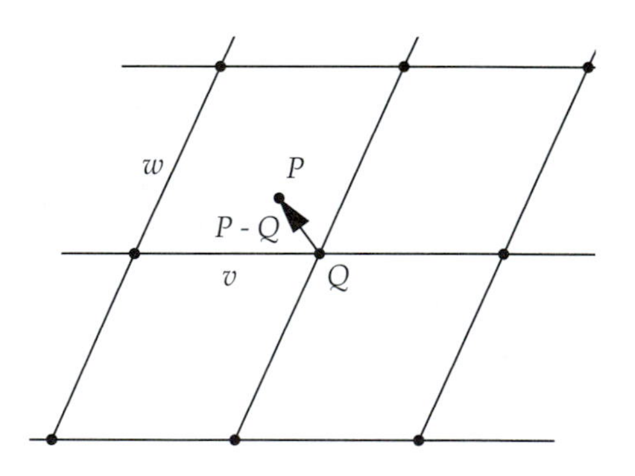

Also, the length of $P - Q$ must be less than the length of w.

To prove this we connect the midpoints of the sides of the parallelogram, yielding four congruent sub-parallelograms. The length of $P - Q$ must be less than the maximum distance between points in one of these sub-parallelograms, which occurs between opposite vertices. (The proof is left as an exercise.)

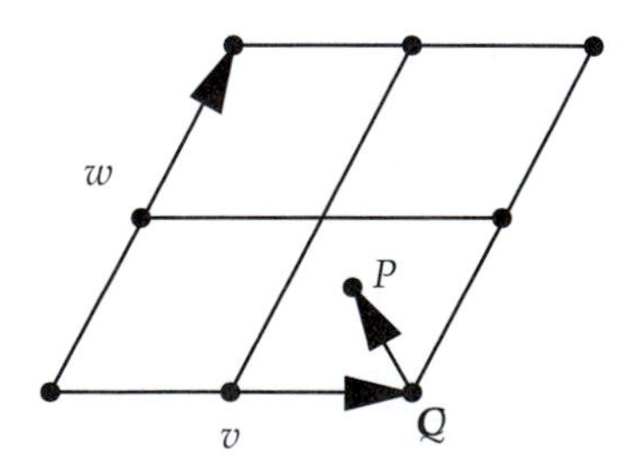

By the triangle inequality, the length of one side of a triangle is less than the sum of the lengths of the other sides. Thus, the distance between opposite vertices of one of the sub-parallelograms must be less than $\frac{\|v\|}{2} + \frac{\|w\|}{2} < 2\frac{\|w\|}{2} = \|w\|$, and so the length of $P - Q$ is also less than $\|w\|$.

Now, if P is inside the parallelogram, or on any side other than that parallel to v, we would get a translation vector that is skew to v and smaller than w, which would contradict the choice of w. Thus, $P - Q$ must lie along a side parallel to v. However, this contradicts the choice of v.

So, all translations have the form $T_z = T_{sv+tw}$ with s, t integers, and the proof is complete. □

What kinds of lattices are possible for a wallpaper group? In the parallelogram spanned by v and w, we have that $v-w$ and $v+w$ are the diagonals. By replacing w by $-w$ we could switch the order of these diagonals, which means we can assume $\|v - w\| \leq \|v + w\|$. Also, neither of these diagonals can be smaller than w because of how w is defined. Thus, we know that

$$\|v\| \leq \|w\|$$

And

$$\|w\| \leq \|v - w\|$$

And

$$\|v - w\| \leq \|v + w\|$$

Thus, $\|v\| \leq \|w\| \leq \|v - w\| \leq \|v + w\|$. Therefore, there are eight possible lattices:

1. **Oblique**
$\|v\| < \|w\| < \|v - w\| < \|v + w\|$.
The lattice is made of skew parallelograms.

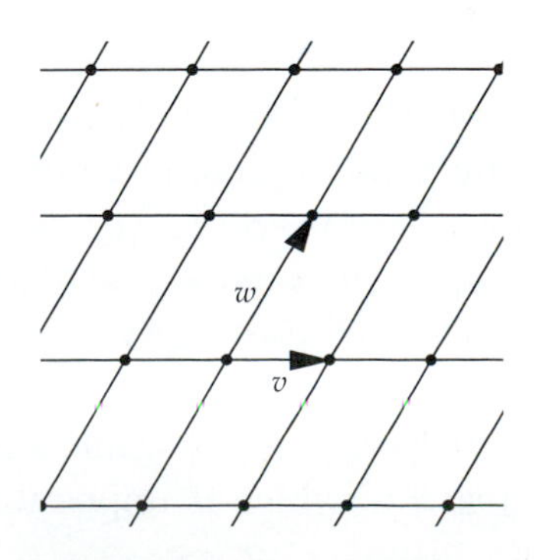

2. **Rectangular**
$\|v\| < \|w\| < \|v - w\| = \|v + w\|$. If the diagonals of the parallelogram are congruent, then the parallelogram must be a (non-square) rectangle.

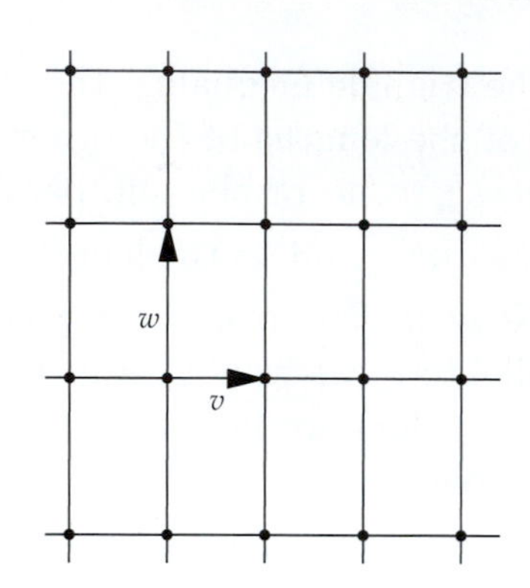

3. **Centered Rectangular**
$\|v\| < \|w\| = \|v - w\| < \|v + w\|$. If $\|w\| = \|v - w\|$, then these sides form an isosceles triangle and the head of w will be the center of a rectangle built on every other row of the lattice, as shown.

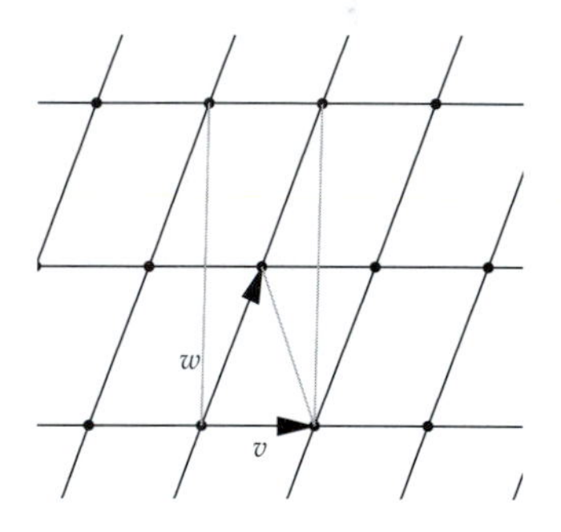

4. **Rhombal**
$\|v\| = \|w\| < \|v - w\| < \|v + w\|$. Again, we get an isosceles triangle, this time with sides being v and w. This case is essentially the same as the centered rectangle, with the rectangular sides built on the vectors $v - w$ and $v + w$, and the center of the rectangle at a lattice point.

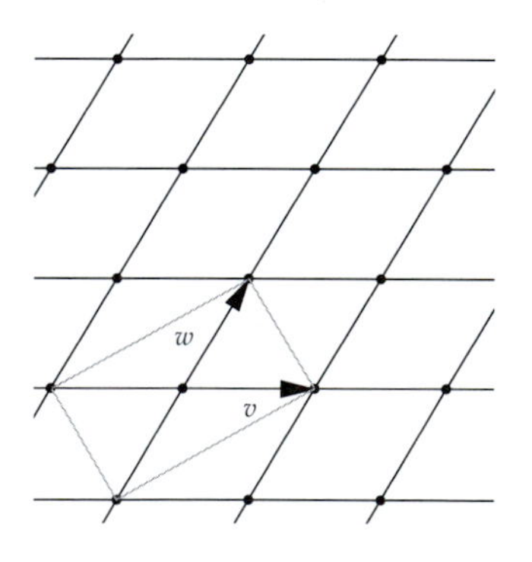

5. $\|v\| < \|w\| = \|v - w\| = \|v + w\|$. This is not a possible configuration for a lattice. Since $\|v - w\|^2 = (v - w)\bullet(v - w) = v\bullet v - 2v\bullet w + w\bullet w$, and $\|v + w\|^2 = v\bullet v + 2v\bullet w + w\bullet w$, then we would have that $v \bullet w = 0$. Then, $\|w\| = \|v - w\|$ would imply that $\|v\| = 0$, which is impossible.

6. **Square**
$\|v\| = \|w\| < \|v - w\| = \|v + w\|$.

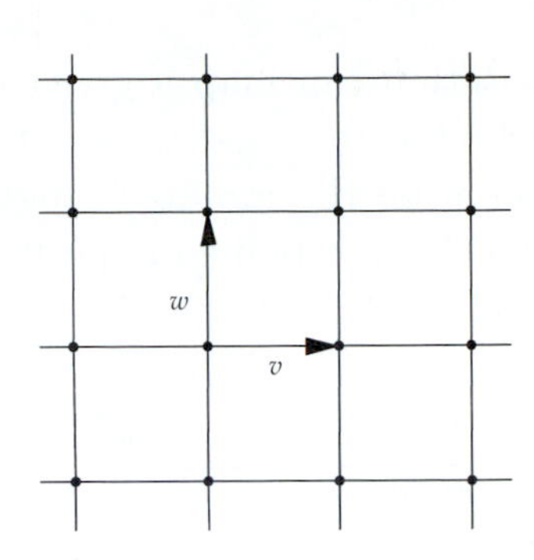

7. **Hexagonal**
$\|v\| = \|w\| = \|v - w\| < \|v + w\|$.

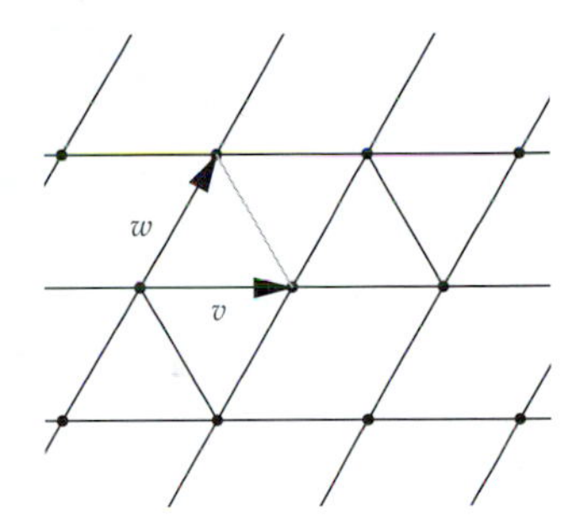

8. $\|v\| = \|w\| = \|v - w\| = \|v + w\|$. This is not a possible configuration for a lattice.

Thus, from eight possible lattice types, we see that there are really only five different lattices possible for a wallpaper group.

Theorem 6.14. *If L is a lattice in the plane, then it is either oblique, rectangular, centered rectangular (which includes rhombal), square, or hexagonal.*

Just as in the case of frieze groups, the possible rotational symmetries of a wallpaper group are restricted to specific values.

Theorem 6.15. *The rotations in a wallpaper group G must map elements of the lattice of the group back to the lattice. Furthermore, angles of rotation in a wallpaper group can only be* 60, 90, 120, *or* 180 *degrees.*

Proof: Let $sv + tw$ be a point in the lattice. Then T_{sv+tw} is in G. Let $R_{a,\alpha}$ be a rotation in G. Consider $R_{a,\alpha} \circ T_{sv+tw} \circ R_{a,-\alpha}$. In Exercise 5.7.6, we showed that this composition is the same as $T_{R_{a,\alpha}(sv+tw)}$. Since the composition of elements in G must yield an element of G, then $T_{R_{a,\alpha}(sv+tw)}$ is in G, and if O is the origin, then $T_{R_{a,\alpha}(sv+tw)}(O) = R_{a,\alpha}(sv + tw)$ is in the lattice. Thus, the rotation maps points of the lattice to other points of the lattice. Also, since there are only a finite number of lattice points to which the rotation can map a given lattice point, then the order of the rotation must be finite.

Let v be the minimal length translation vector among all translations and let P be a point in the lattice. Consider $R_{a,\alpha}(P)$ (labeled "$R(P)$" in Fig. 6.2).

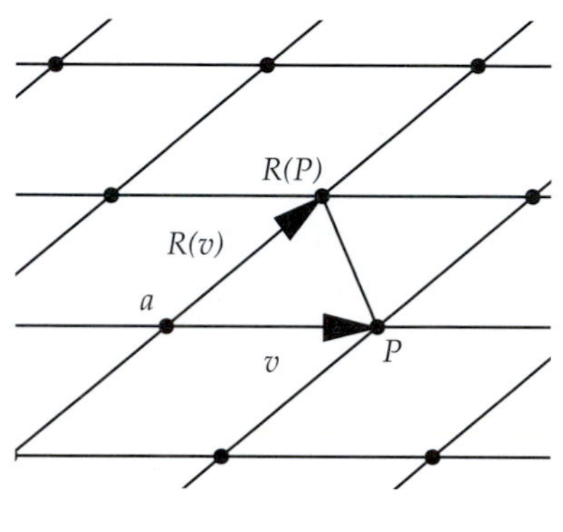

Fig. 6.2

The vector $P - R_{a,\alpha}(P)$ will be a translation vector in G (difference of two lattice vectors must be a translation vector). If the rotation angle α is less than 60 degrees, then this vector will have *shorter* length than v. This is impossible due to how v was chosen. The only possible rotation angles are then angles α with $60 \leq \alpha < 360$. (We will ignore the trivial case of rotation by 360.)

In our discussion of finite symmetry groups, we proved that if a rotation has finite order, then the angle of rotation must evenly divide 360. Then the only possible angles for a wallpaper rotation are $60 = \frac{360}{6}, 72 = \frac{360}{5}$, $90 = \frac{360}{4}$, $120 = \frac{360}{3}$, and $180 = \frac{360}{2}$.

Rotations by 60 and 120 are possible in hexagonal lattices. Rotations by 90 and 180 degrees are possible in square and rectangular lattices. It is left as an exercise to show that rotations of 72 degrees are impossible. □

This theorem has become known as the "Crystallographic Restriction." Many crystals have the property that if you slice them along a plane, the atoms of the crystal along that plane form a wallpaper pattern. Thus, the possible crystals having wallpaper symmetry is "restricted" to those generated by wallpaper groups.

How many wallpaper groups are there? It turns out that there are exactly 17 wallpaper groups. A careful proof of this result can be carried out by examining the five lattice types we have discussed and finding the symmetry groups that preserve each type. For details see [3, Chapter 26].

The 17 groups have traditionally been listed with a special notation consisting of the symbols p, c, m, and g, and the integers 1, 2, 3, 4, and 6. This is the crystallographic notation adopted by the International Union of Crystallography (IUC) in 1952.

In the IUC system the letter p stands for *primitive.* A lattice is generated from a polygonal cell that is translated to form the complete lattice. In the case of oblique, rectangular, square, and hexagonal lattices, the cell is precisely the original parallelogram formed by the vectors v and w and is, in this sense, a "primitive" cell. In the case of the centered-rectangle lattice, the cell is a rectangle, together with its center point, with the rectangle larger than the original parallelogram and not primitive. Thus, lattice types can be divided into two classes: primitive ones designated by the letter p and non-primitive ones designated by the letter c.

Other symmetries for a general wallpaper group will include reflections, rotations, and glide reflections. The letter m is used to symbolize a reflection and g symbolizes a glide reflection. The numbers 1, 2, 3, 4, and 6 are used to represent the orders of rotation for a group (e.g., 3 would signify a rotation of $\frac{360}{3} = 120$).

In the IUC system, a wallpaper group is designated by a string of letters and numbers. First, there is the letter p (for oblique, rectangular, square, or hexagonal lattices) or c (for centered-rectangular lattices). Then a set of non-translational generators for the group will be listed. These generators include reflections, glide reflections, and/or rotations. The two fundamental translations are not listed, but are understood to also be generators for the group.

For example, "p1" would symbolize the simplest wallpaper group, where the lattice is oblique and there are no non-translational symmetries.

What would "p2mm" symbolize? The lattice is primitive; there is a rotational symmetry of 180 degrees; and there are two reflection symmetries. Since the reflections are listed separately, neither can be generated from the other by combining with the fundamental translations. Thus, the reflection lines of symmetry cannot be parallel and the two reflections will generate a rotation, which must be of 180 degrees. The two reflection lines will then be in perpendicular directions and the lattice must be rectangular, centered-rectangular, or square. In fact, the lattice must be rectangular. (The proof is left as an exercise.)

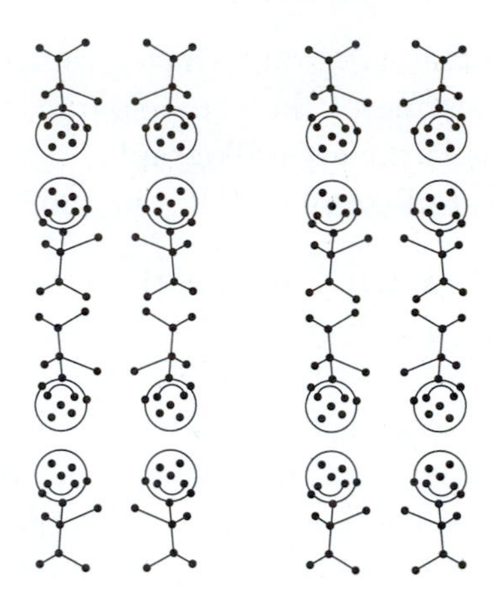

In Appendix E there is a complete listing of all 17 wallpaper groups.

Exercise 6.3.1. Sketch the lattices spanned by the following pairs of vectors, and specify which of the five types of lattices each pair will generate.

(a) $\vec{v} = (2, 0)$, $\vec{w} = (2, 4)$

(b) $\vec{v} = (1, \sqrt{3})$, $\vec{w} = (1, -\sqrt{3})$

(c) $\vec{v} = (-1, 1)$, $\vec{w} = (1, 1)$

Exercise 6.3.2. Show that the transformation $f(x, y) = (-x, y + 1)$ is a glide reflection and that a symmetry group generated by f and the translation $T(x, y) = (x + 1, y + 1)$ must be a wallpaper group. Which lattice type will this group have?

Exercise 6.3.3. Show that the symmetry group generated by the glide f, defined in the previous exercise, and the translation $T(x, y) = (x, y + 1)$ will not be a wallpaper group.

Exercise 6.3.4. Let G be a wallpaper group, and let H be the subset of all symmetries fixing a point of the lattice for G. If H is a subgroup of G of order 4 and all elements of H are of order 2, show that H must be the group generated by a half-turn and two reflections about perpendicular lines.

Exercise 6.3.5. With the same assumptions about G and H as in the preceding exercise, show that the lattice for G must be rectangular, centered-rectangular, or square.

Exercise 6.3.6. Show that the lattice for the wallpaper group p2mm must be rectangular. [Hint: Use the previous two exercises.]

Exercise 6.3.7. For the wallpaper group p2mm, with mm representing generating reflections along the two translation vectors, prove there are also reflections along lines through the midpoints of the lattice segments, and find formulas for these reflections in terms of the generators of p2mm.

Exercise 6.3.8. Show that the wallpaper group p2mm, with mm representing generating reflections along the two translation vectors, is the same group as pm'm', where m'm' represents two reflections through midpoints of the translation vectors. [Hint: Use the preceding exercise.]

Exercise 6.3.9. Show that a wallpaper pattern with translation symmetry given by two perpendicular vectors v, w, and wallpaper group p2mm, can be generated by applying the symmetries of the group to the shaded area of the rectangle spanned by v and w shown. Thus, this area can be thought of as a *generating area* for the pattern. [Hint: Use the previous exercise.]

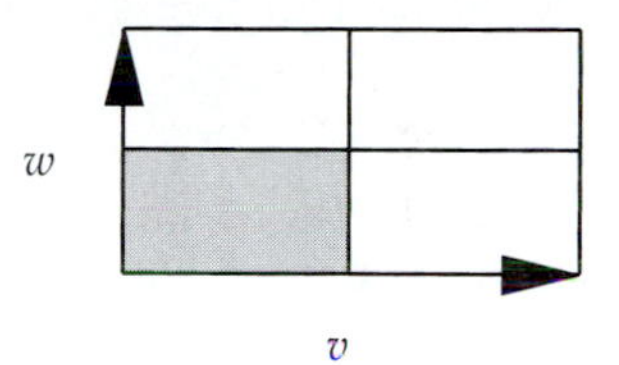

Exercise 6.3.10. (This problem assumes familiarity with the process of finding the matrix representation of a transformation with respect to a basis.) Let L be a lattice spanned by vectors $\vec{v}$ and $\vec{w}$. Let $R_{A,\alpha}$ be an element of a wallpaper group G for L. By Theorem 6.12, $R_{A,\alpha}$ has finite order, say n. Using the idea of conjugation, show that $R_{O,\alpha}$ is also in G. The matrix for $R_{O,\alpha}$, in the standard basis, is then

$$\begin{bmatrix} \cos(\frac{360}{n}) & -\sin(\frac{360}{n}) \\ \sin(\frac{360}{n}) & \cos(\frac{360}{n}) \end{bmatrix}$$

Show that the matrix for $R_{O,\alpha}$ with respect to the basis $\{\vec{v}, \vec{w}\}$ will have integer entries. Then, use the fact that the trace of a matrix is invariant under change of bases to give an alternate proof of the second part of Theorem 6.15.

Exercise 6.3.11. Show that in a parallelogram, the maximum distance between any two points on or inside the parallelogram occurs between opposite vertices of the parallelogram. [Hint: Let $A = lv + mw$ and $B = sv + tw$ be two points in the parallelogram. Show that these are farthest apart when s, t, l, and m are all 0 or 1.]

Exercise 6.3.12. Show that a wallpaper group cannot have rotations of 72 degrees about a point. [Hint: Use an argument similar to that used in Theorem 6.15 and consider the angle between the vector $-v$ and a double rotation of v by $R_{a,72}$.]

Exercise 6.3.13. Sketch a planar pattern that has a symmetry group that is not discrete. (Blank or completely filled sheets of paper are *not* allowed.)

6.4 Tiling the Plane

> A long time ago, I chanced upon this domain [of regular division of the plane] in one of my wanderings; I saw a high wall and as I had a premonition of an enigma, something that might be hidden behind the wall, I climbed over with some difficulty. However, on the other side I landed in a wilderness and had to cut my way through with great effort until—by a circuitous route—I came to the open gate, the open gate of mathematics.
>
> —Maurits Cornelis (M. C.) Escher (1898–1972)

6.4.1 Escher

Much of the renewed interest in geometric design and analysis in the modern era can be traced to the artistic creations of M. C. Escher. While he did not prove new theorems in geometry, he did use geometric insights to create fascinating periodic designs like the design in Fig. 6.3.

A beautiful book that describes Escher's artwork and the mathematics that underlies it is *M. C. Escher: Visions of Symmetry* [37], by Doris Schattschneider.

One of Escher's favorite themes was that of a tessellation of the plane by geometric shapes. A *tessellation* or tiling is a covering of the plane by repeated copies of a shape such that there are no gaps left uncovered and the copied shapes never overlap. In Fig. 6.3 we see the beginnings of a tessellation of the plane by a three-sided shape.

A tessellation is produced by repeating a basic figure, or set of figures, throughout the plane. Repetitions of the basic tile(s) are carried out by isometries of the plane. Thus, every tiling has associated with it a group of symmetries that map the tiling back to itself. For example, in Fig. 6.3 we see that this tiling is invariant under a rotation of 60 degrees and, thus, it must have a symmetry group equal to one of the two wallpaper groups p6 or p6m (see Appendix E). Since the tiling is not invariant under a reflection, then it must have a symmetry group of p6.

Escher received inspiration for his tiling designs from two sources: the intricate designs of Arab artists and the work of mathematicians on classifying planar symmetry groups.

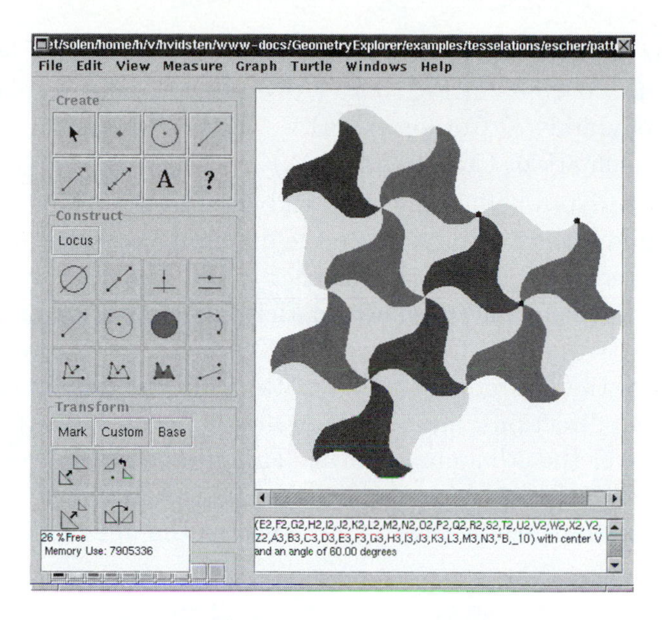

Fig. 6.3 A Tiling in the Spirit of Escher

At the age of 24, Escher made his first visit to the Alhambra, a fourteenth century palace in Grenada, Spain. He found almost every wall, floor, and ceiling surface covered with abstract geometric tilings. In Fig. 6.4 we see several sketches Escher made of the Alhambra tilings during a return trip to Spain in 1936.

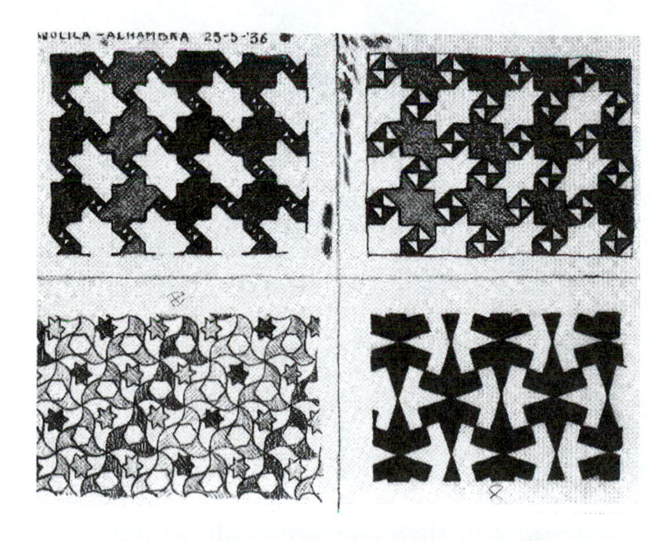

Fig. 6.4 Escher's Alhambra Sketches

Note that the pattern in the lower-left corner of Fig. 6.4 is essentially the same as the pattern in Fig. 6.3. (All Escher images displayed in this section are copyright (c) Cordon Art B.V., The Netherlands. M.C. Escher (TM) is a Trademark of Cordon Art B.V.)

The wide variety of tiling patterns and tile shapes exhibited at the Alhambra inspired Escher to investigate the different ways that one could tile the plane in a systematic fashion. Escher found the answer to this question through the work of the mathematician George Pólya. Pólya, in a 1924 article in the journal *Zeitschrift für Kristallographie*, gave a complete classification of the discrete symmetry patterns of the plane, with patterns repeated in two directions, namely, the wallpaper patterns. All 17 patterns were exhibited in the Alhambra tilings. This mathematical revelation was the "open gate" through which Escher was able to carefully design and produce his famous and imaginative tilings, such as the one in Fig. 6.5.

Fig. 6.5 An Escher Tiling

6.4.2 Regular Tessellations of the Plane

The simplest tessellations of the plane are made with a single tile that has the shape of a regular polygon. A regular polygon has the property that all sides have the same length and all interior angles created by adjacent sides are congruent. We will call a regular polygon with n sides a *regular n-gon.*

A regular 3-gon is an equilateral triangle, a regular 4-gon is a square, and so on. Which regular polygons tile the plane?

Let's consider the simplest n-gon tile, the equilateral triangle.

In the figure at the right, we have a tiling by equilateral triangles in which all triangles meet at common vertices.

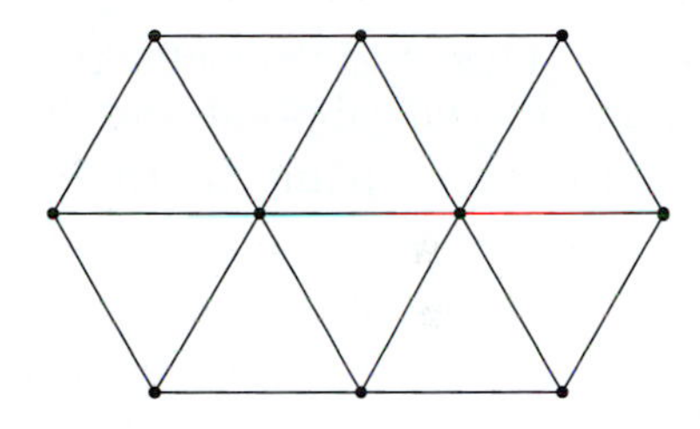

In the new tiling at the right, we have shifted the top row of triangles a bit. This configuration will still lead to a tiling of the plane, although triangles no longer share common vertices.

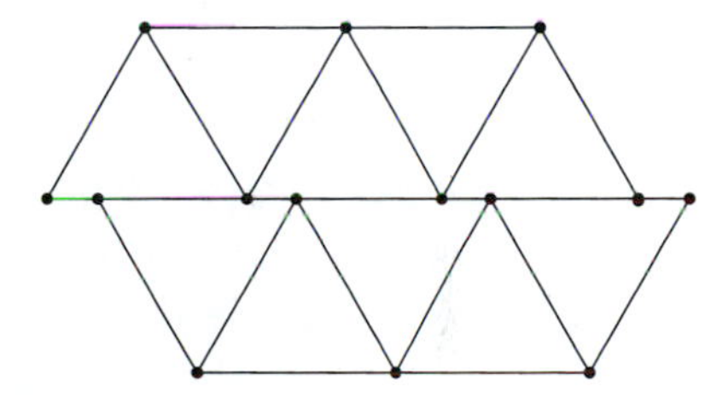

Definition 6.10. We will call a tessellation *regular* if it is made from copies of a single regular n-gon with all n-gons meeting at common vertices.

The second triangular tiling above is not a regular tiling. How many regular tilings are there? Clearly, the example above shows that there is a regular tiling with triangles. We can easily create a regular tiling with squares. The triangle tiling also shows that regular hexagonal tilings are possible. In fact, these three are the *only* regular tilings.

It is not hard to see why this is the case. At a common vertex of a regular tiling, suppose that there are k regular n-gons meeting at the vertex. Then, an angle of 360 degrees will be split into k parts by the edges coming out of this vertex. Thus, the interior angle of the n-gon must be $\frac{360}{k}$. On the other hand, suppose we take a regular n-gon, find its central point and draw edges from this point to the vertices of the n-gon. In the example at the right, we have done this for the regular hexagon. The point labeled "A" is the central point of the hexagon.

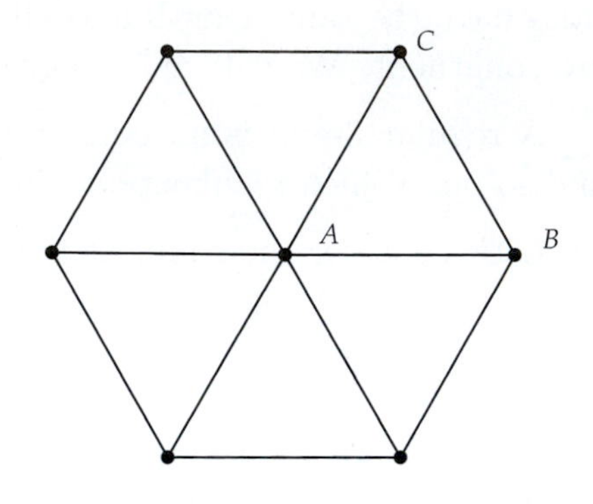

For a regular n-gon, this triangulation will yield n isosceles triangles, each with angle sum of 180 degrees. If we add up the sum of the angles in all of the triangles in the figure, we get a total angle sum of $180n$.

On the other hand, if we add up only those triangle angles defined at the central point, then the sum of these will have to be 360. For each of our isosceles triangles, the other two angles at the base of the triangle will be congruent (the angles at B and C in the hexagon example). Let's call these angles α. Then, equating the total triangle sum of $180n$ with the sum of the angles at the center and the sum of the base angles of each isosceles triangle, we get

$$180n = 360 + 2n\alpha$$

and thus,

$$2\alpha = 180 - \frac{360}{n}$$

Now, 2α is also the interior angle of each n-gon meeting at a vertex of a regular tessellation. We know that this interior angle must be $\frac{360}{k}$, for k n-gons meeting at a vertex of the tessellation. Thus, we have that

$$\frac{360}{k} = 180 - \frac{360}{n}$$

If we divide both sides by 180 and multiply by nk, we get

$$nk - 2k - 2n = 0$$

If we add 4 to both sides, we can factor this as

$$(n-2)(k-2) = 4$$

There are only three integer possibilities for n and k, namely 6, 4, and 3. These three possibilities directly correspond to the three regular tessellations with equilateral triangles, squares, and regular hexagons.

6.5 Project 9 - Constructing Tessellations

Escher modeled many of his tilings after those he saw at the Alhambra. In this project we will look at how to create one of these Moorish tilings, find its symmetry group, and construct other tilings with a specified symmetry group.

The tiling that we will construct is based on the dart-like shape found in the tiling at the lower right in Fig. 6.4.

Start *Geometry Explorer* and create a segment $\overline{AB}$ in the Canvas. Attach a point C to the segment by creating a point on top of $\overline{AB}$.

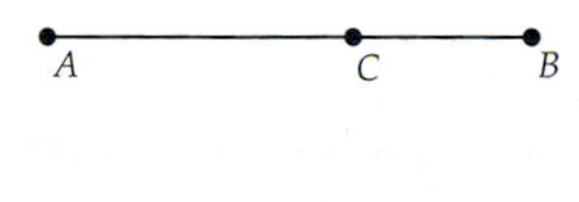

To construct a square on $\overline{CB}$, first select C and choose **Center** from the **Mark** pop-up menu in the Transform panel. Define a custom rotation of 90 degrees (choose **Rotation** from the **Custom** pop-up menu). Then select point B and hit the Rotate button in the Transform panel to get point D.

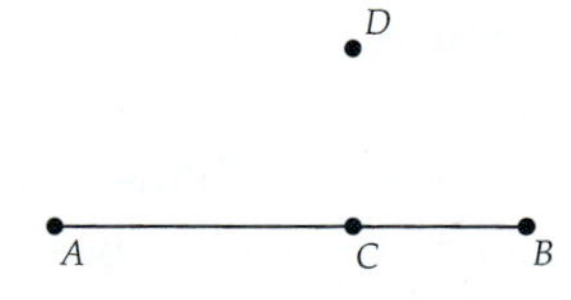

Set point D as a new center of rotation and rotate point C 90 degrees to get point E. Connect segments as shown.

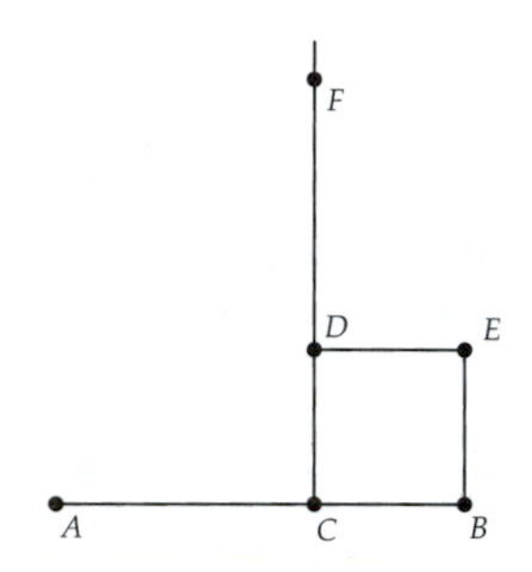

To make the point of our dart, create a ray from point C vertically through point D and attach a point F to this ray.

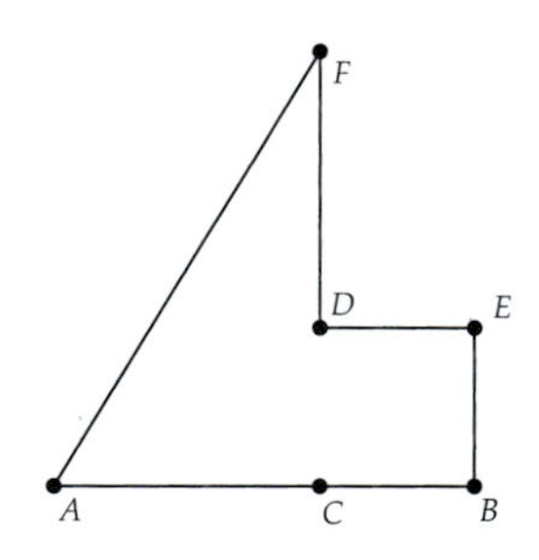

Next, hide the ray, hide $\overline{CD}$, and connect segments as shown.

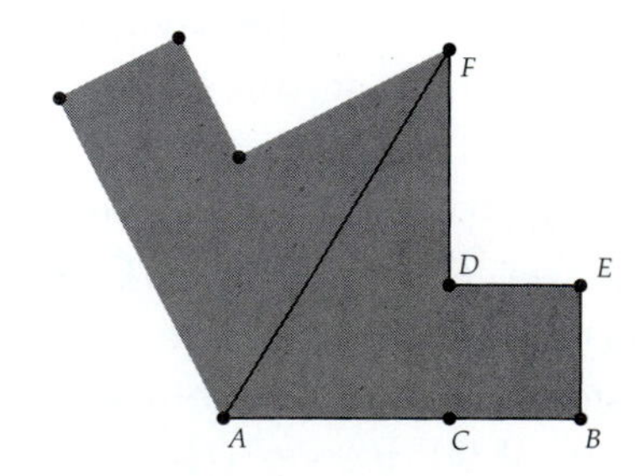

This completes the construction of half of the dart. Multi-select points A, B, E, D, and F and hit the Filled-Polygon button (third button in last row of Construct panel) to color our half-dart. Select segment $\overline{AF}$ and choose **Mirror** from the **Mark** pop-up menu. Select the half-dart by clicking inside the filled area and click on the Reflect button to reflect it across $\overline{AF}$ to get the entire dart.

Select point E and set it as a center of rotation. Select the entire dart by clicking and dragging to enclose the dart in a selection box. Then, rotate the dart three times (hit the Rotate button three times). In the figure at the right, we have changed the color of each component dart so that we can see the pieces better. Also, we have rescaled the Canvas, since the image grew too large. (To rescale the Canvas, choose **Rescale Geometry in Canvas** (**View** menu).

To make this a valid tiling (no gaps), we move point F (the "point" of the dart) to a position where it directly matches the base of the dart to its right, as shown in the figure.

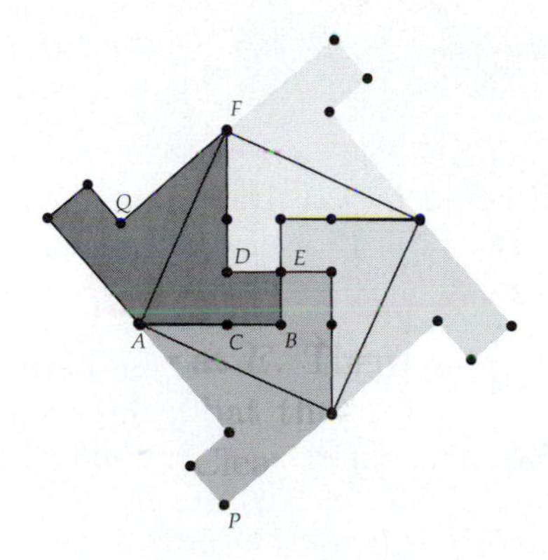

We have now constructed a basic "tile" that can be translated to completely cover the plane. The points labeled "P" and "Q" will be used to define a translation vector for our 4-dart region.

To translate this basic tile we multi-select P and Q and set these as a rectangular vector of translation (choose **Vector** from the **Mark** pop-up menu). Then we select the whole 4-dart region and hit the Translate button. Fig. 6.6 shows this two-tile configuration. It is clear that the 4-dart region will tile the plane if we continue to translate it vertically and horizontally.

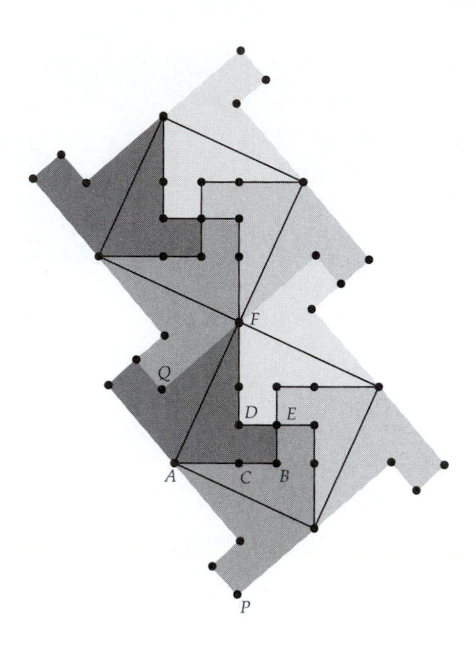

Fig. 6.6

Exercise 6.5.1. Determine the symmetry group for this figure. [Hint: It must be one of the 17 wallpaper groups. Why? Use Fig. 6.4 for a more complete tiling picture.]

Exercise 6.5.2. Choose one of the wallpaper groups and construct a tiling (other than a regular tiling) with this symmetry group. You can do this on the computer or by using paper cutouts if you wish.

For the project report give a thorough answer to the first question, illustrating all symmetries on the dart tiling. For the second exercise, describe your chosen symmetry group, illustrate the basic lattice cell for this group, and describe the steps you took to construct your tiling.

Chapter 7

Non-Euclidean Geometry

> I have discovered such wonderful things that I was amazed . . . out of nothing I have created a strange new universe.
>
> —János Bolyai (1802–1860), from a letter to his father, 1823

7.1 Background and History

Euclid's development of planar geometry was based on five postulates (or axioms):

1. Between any two distinct points, a segment can be constructed.

2. Segments can be extended indefinitely.

3. Given a point and a distance, a circle can be constructed with the point as center and the distance as radius.

4. All right angles are congruent.

5. Given two lines in the plane, if a third line l crosses the given lines such that the two interior angles on one side of l are less than two right angles, then the two lines if continued indefinitely will meet on that side of l where the angles are less than two right angles.

Postulates 1–4 seem very intuitive and self-evident. To construct geometric figures, one needs to construct segments, extend them, and construct circles. Also, geometry should be uniform so that angles do not change as objects are moved about.

The fifth postulate, the so-called *parallel postulate*, seems overly complex for an axiom. It is not at all self-evident or obvious and reads more like a theorem.

In fact, many mathematicians tried to find simpler postulates, ones that were more intuitively believable, to replace Euclid's fifth, with the hope that the fifth postulate could then be proved from the first four postulates and the new postulate.

We have already considered one of these substitutes, *Playfair's Postulate*:

> Given a line and a point not on the line, it is possible to construct exactly one line through the given point parallel to the line.

This postulate is certainly simpler to state and easier to understand when compared to Euclid's fifth postulate. However, in Chapter 2 we saw that Playfair's Postulate is logically equivalent to Euclid's fifth postulate, and so replacing Euclid's fifth postulate with Playfair's Postulate does not really simplify Euclid's axiomatic system.

Other mathematicians attempted to prove Euclid's fifth postulate as a *theorem* solely on the basis of the first four postulates. One popular method of proof was to assume the logical opposite of Euclid's fifth postulate (or Playfair's Postulate). If one could show the logical opposite to be false, or if one could obtain a contradiction to a known result by using the opposite, then Euclid's fifth postulate would be true as a *theorem* based on the first four postulates.

Giovanni Girolamo Saccheri (1667–1773) and Johann Lambert (1728–1777) both used this method of attack to prove Euclid's fifth postulate. Saccheri's work focused on quadrilaterals whose base angles are right angles and whose base-adjacent sides are congruent. In Euclidean geometry, such quadrilaterals must be rectangles, that is, the top (or summit) angles must be right angles. The proof of this result depends on Euclid's fifth postulate. Saccheri supposed that the top angles were either greater than or less than a right angle. He was able to show that the hypothesis of the obtuse angle resulted in a contradiction to theorems based on the first four Euclidean postulates. However, he was unable to derive any contradictions using the hypothesis of the acute angle. Lambert, likewise, studied the hypothesis of the acute angle. In fact, Lambert spent a good fraction of his life working on the problem of the parallel postulate. Even though both men discovered many important results of what has become known as *non-Euclidean geometry*, geometry based on a negation of Euclid's fifth postulate, neither could accept the possibility of non-Euclidean geometry. Saccheri eventually resorted to simply asserting that such a geometry was impossible:

> The hypothesis of the acute angle is absolutely false, because [it is] repugnant to the nature of the straight line! [17, page 125]

In the 1800s several mathematicians experimented with negating Playfair's Postulate, assuming that a non-Euclidean fifth postulate could be consistent with the other four. This was a revolutionary idea in the history of mathematics. János Bolyai (1802–1860), Carl Friedrich Gauss (1777–1855), and Nikolai Lobachevsky (1792–1856) explored an axiomatic geometry based on the first four Euclidean postulates plus the following postulate:

> Given a line and a point not on the line, it is possible to construct *more than one* line through the given point parallel to the line.

This postulate has become known as the *Bolyai-Lobachevskian Postulate.* Felix Klein, who was instrumental in classifying non-Euclidean geometries based on a surface with a particular conic section, called this the *hyperbolic postulate.* A geometry constructed from the first four Euclidean postulates, plus the Bolyai-Lobachevskian Postulate, is known as *Bolyai-Lobachevskian geometry*, or *hyperbolic geometry.*

Gauss, one of the greatest mathematicians of all time, was perhaps the first to believe that hyperbolic geometry could be consistent. Harold Wolfe in *Non-Euclidean Geometry* describes how Gauss wrote a letter to a friend about his work.

> The theorems of this geometry appear to be paradoxical and, to the uninitiated, absurd; but calm, steady reflection reveals that they contain nothing at all impossible. For example, the three angles of a triangle become as small as one wishes, if only the sides are taken large enough; yet the area of the triangle can never exceed a definite limit, regardless of how great the sides are taken, nor indeed can it ever reach it. All my efforts to discover a contradiction, an inconsistency, in this Non-Euclidean Geometry have been without success. [41, page 47]

In fact Gauss, Bolyai, and Lobachevsky developed the basic results of hyperbolic geometry at approximately the same time, though Gauss developed his results before the other two, but refused to publish them.

What these three mathematicians constructed was a set of theorems based on the first four Euclidean postulates plus the hyperbolic postulate. This did not mean, however, that they showed hyperbolic geometry to be a *consistent* system. There was still the possibility that one of the three just missed finding a theorem in hyperbolic geometry that would lead to a contradiction of a result based on the first four Euclidean postulates.

The consistency of hyperbolic geometry was demonstrated by Eugenio Beltrami (1835–1900), Felix Klein (1849–1925), and Henri Poincaré (1854–1912) in the late 1800s to early 1900s. They created *models* of hyperbolic geometry inside Euclidean geometry, with perhaps strange definitions of points, lines, circles, and angles, but models nonetheless. In each of their models, they showed that Euclid's first four postulates were true and that the hyperbolic postulate was true as well. Since each model was created within Euclidean geometry, if hyperbolic geometry had an internal contradictory statement, then that statement, when translated into its Euclidean environment, would be an internal contradiction in Euclidean geometry! Thus, if one believed Euclidean geometry was consistent, then hyperbolic geometry was equally as consistent.

In the next section we will introduce two models of hyperbolic geometry, the Poincaré model and the Klein model. Then, in the following sections of the chapter, we will look at results that are true in *any* model satisfying the five axioms of hyperbolic geometry.

7.2 Models of Hyperbolic Geometry

7.2.1 Poincaré Model

In the Poincaré model for 2-dimensional hyperbolic geometry, a point is defined to be any point interior to the unit disk. That is, any point $P = (x, y)$, with $x^2 + y^2 < 1$. The collection of all such points will be called the *Poincaré disk*.

Definition 7.1. A *hyperbolic line* (or *Poincaré line*) is a Euclidean arc, or Euclidean line segment, within the Poincaré disk that meets the boundary circle at right angles.

At right are two "lines" in the Poincaré model.

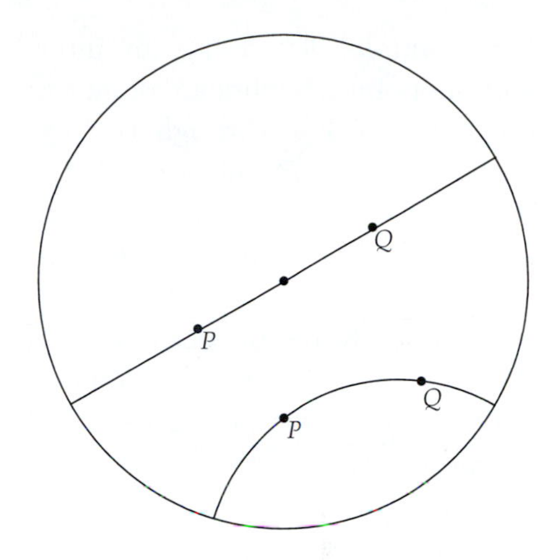

To verify that this actually is a model for hyperbolic geometry, we need to show that Euclid's first four postulates, plus the hyperbolic postulate, are satisfied in this model.

We will start with the first two postulates of Euclid—that unique segments can always be constructed through two points and that segments can always be extended. By segments we will mean subsets of the Poincaré lines as defined thus far.

Given two points P and Q, suppose they lie on a diameter of the boundary circle in the Poincaré disk. Then, as shown in the figure above, we can construct the Euclidean segment $\overline{PQ}$ along the diameter. Since this diameter meets the boundary circle at right angles, then $\overline{PQ}$ will lie on a Poincaré (hyperbolic) line and so will be a hyperbolic segment.

Now suppose that P and Q do not lie on a diameter. Then by the work we did on orthogonal circles at the end of Chapter 2, there is a unique circle through P and Q that meets the boundary circle at right angles. We find this circle by constructing the inverse point P' to P with respect to the boundary circle. The circle through P, P', and Q will then meet the boundary circle at right angles. We conclude that if P and Q do not lie on a diameter, then we can find a hyperbolic segment through P and Q.

To verify Euclid's second postulate, that lines (hyperbolic) can always be extended, we first note that the points of our geometry are not allowed to be on the boundary circle, by the definition of the Poincaré model. This allows us to extend any hyperbolic segment.

For example, let X be an intersection point (Euclidean point) of the Poincaré line through two hyperbolic points P and Q with the boundary circle. Then, since P cannot be on the boundary, the distance along the circle arc from P to X will always be positive, and thus we can find another point Y between these two points with $\overline{YQ}$ extending $\overline{PQ}$.

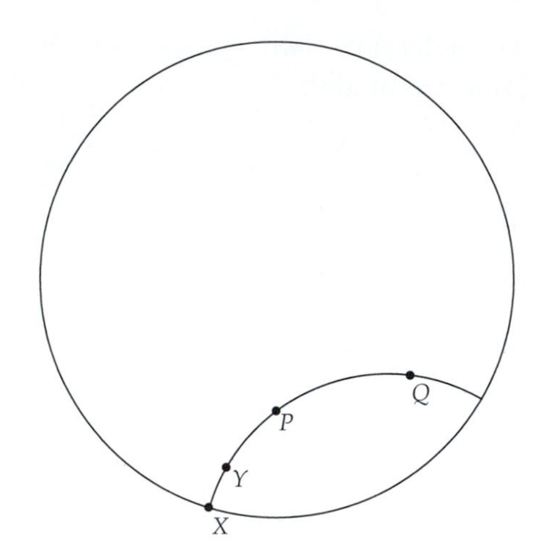

Hyperbolic Distance

To define circles for the third postulate, we need a notion of distance. Since the boundary of the Poincaré disk is not reachable in hyperbolic geometry, we want a definition of distance such that the distance goes to infinity as we approach the boundary of the Poincaré disk.

In the figure at the right, we have two points P and Q in the Poincaré disk. There is a unique hyperbolic line (Euclidean arc $RPQS$) on which P and Q lie that meets the boundary of the disk at points R and S.

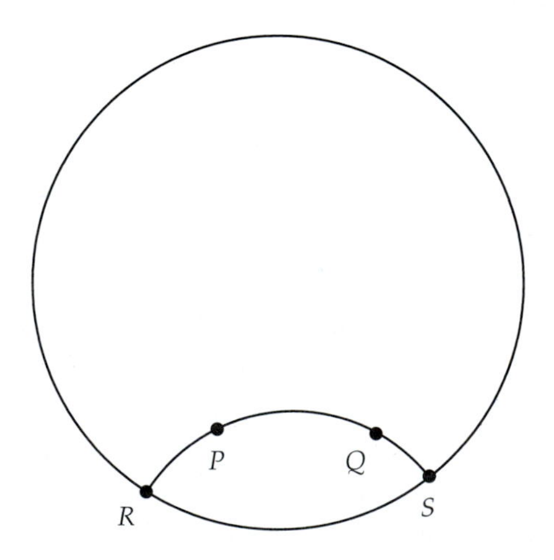

Definition 7.2. The *hyperbolic distance* from P to Q is

$$d_P(P,Q) \quad = \quad \left| \ln\left(\frac{(PS)}{(PR)} \frac{(QR)}{(QS)}\right) \right| \tag{7.1}$$

where R and S are the points where the hyperbolic line through P and Q meets the boundary circle, PS is the Euclidean distance between P and S, and likewise for PR, QR, and QS.

This function satsifies the critical defining properties of a distance function. It is non-negative and equal to zero only when $P = Q$. It is additive along lines, and it satisfies the triangle inequality (for triangles constructed of hyperbolic segments). These properties will be proved in detail in the last section of this chapter.

One thing that is clear from looking at the form of the distance function is that as P or Q approach the points on the boundary (R or S), the fraction inside the log function goes to ∞ or 0, and thus the distance function itself goes to infinity.

We can now define hyperbolic circles.

Definition 7.3. A hyperbolic circle c of radius r centered at a point O in the Poincaré disk is the set of points in the Poincaré disk whose hyperbolic distance to O is r.

Here are some hyperbolic circles with their associated hyperbolic centers.

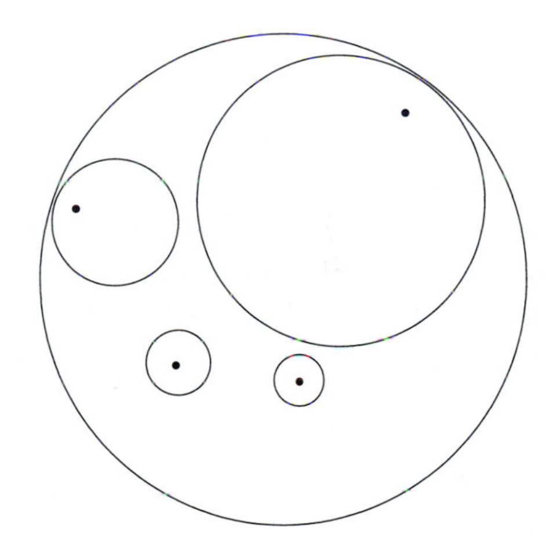

We must now verify that circles always exist. To construct the circle of radius r at O, we note that through any line passing through O, we can find points that are r units away (measured in the hyperbolic distance function). This is because no matter how close O might be to the boundary points R or S, we can always find points between O and those boundary points whose distance to O will grow without bound.

For the fourth postulate, we will define angles just as they are defined in Euclidean geometry. We use the Euclidean tangent lines to Poincaré lines (i.e., Euclidean arcs) in the Poincaré model to determine angles. That is, the angle determined by two hyperbolic lines will be the angle made by their Euclidean tangents. Since angles inherit their Euclidean meaning, the fourth postulate is automatically true.

For the fifth postulate (the hyperbolic postulate), consider a line l and a point P not on l as shown at right.

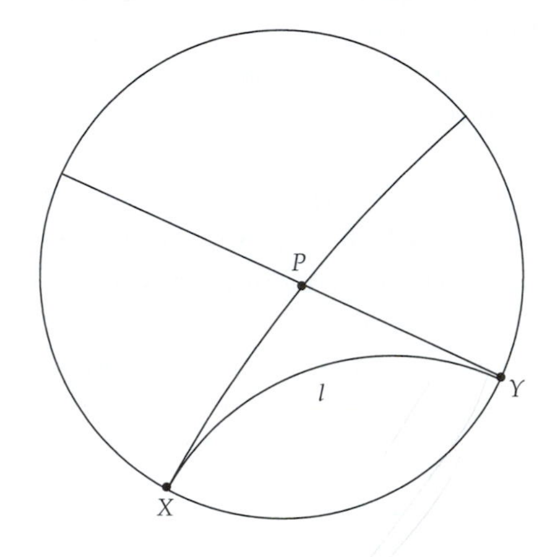

Let X and Y be the intersection points of l with the boundary circle. Then by Theorem 2.38 we know that there are two circular arcs, one through P and X and one through P and Y, that are orthogonal to the circle boundary. Also, neither of these can intersect l at a point other than P inside the boundary circle. For, suppose that the arc through P and X intersected l at a point Q inside the circle. Then Q and X would be on l and also on the arc through P and X. By the uniqueness part of Theorem 2.38, these two arcs must be coincident. But, this is impossible as P was assumed not to lie on l. Thus, the two arcs through P will be two hyperbolic lines that do not intersect l inside the boundary circle and by definition are parallel lines to l.

We see that all of the first four Euclidean postulates hold in this geometry and the hyperbolic parallel postulate holds as well. We conclude that this strange geometry in the Poincaré disk is just as logically consistent as Euclidean geometry. If there were contradictory results about lines, circles, and points in this new geometry, they would have to be equally contradictory in the Euclidean context in which this geometry is embedded.

We note here that there is really nothing special about using the unit disk in the Poincaré model. We could just as well have used any circle in the plane and defined lines as diameters or circular arcs that meet the boundary circle at right angles.

7.2.2 Mini-Project - The Klein Model

The Poincaré model preserves the Euclidean notion of angle, but at the expense of defining lines in a fairly strange manner. Is there a model of hyperbolic geometry, built within Euclidean geometry, that preserves *both* the Euclidean definition of lines *and* the Euclidean notion of angle? Unfortunately, this is impossible. If we had such a model, and ΔABC was any triangle, then the angle sum of the triangle would be 180 degrees, which is a property that is equivalent to the parallel postulate of Euclidean geometry [Exercise 2.1.8 in Chapter 2].

A natural question to ask is whether it is possible to find a model of hyperbolic geometry, built within Euclidean geometry, that preserves *just* the Euclidean notion of lines.

In this project we will investigate a model first put forward by Felix Klein, where hyperbolic lines are segments of Euclidean lines. Klein's model starts out with the same set of points we used for the Poincaré model, the set of points inside the unit disk.

However, lines will be defined differently. A hyperbolic line (or Klein line) in this model will be any chord of the boundary circle (minus its points on the boundary circle).

Here is a collection of Klein lines.

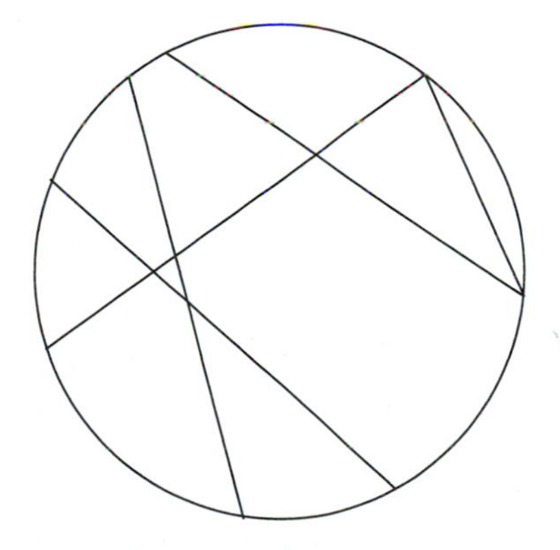

Exercise 7.2.1. Show that Euclid's first two postulates are satisfied in this model.

In order to verify Euclid's third postulate, we will need to define a distance function.

Definition 7.4. The *hyperbolic distance* from P to Q in the Klein model is

$$d_K(P,Q) = \frac{1}{2}\left|\ln\left(\frac{(PS)}{(PR)}\frac{(QR)}{(QS)}\right)\right| \tag{7.2}$$

where R and S are the points where the hyperbolic line (chord of the circle) through P and Q meets the boundary circle.

Note the similarity of this definition to the definition of distance in the Poincaré model. We will show at the end of this chapter that the Klein and Poincaré models are *isomorphic*. That is, there is a one-to-one map between the models that preserves lines and angles and also preserves the distance functions.

Just as we did in the Poincaré model, we now define a circle as the set of points a given (hyperbolic) distance from a center point.

Exercise 7.2.2. Show that Euclid's third postulate is satisifed with this definition of circles. [Hint: Use the continuity of the logarithm function, as well as the fact that the logarithm is an unbounded function.]

Euclid's fourth postulate deals with right angles. Let's skip this postulate for now and consider the hyperbolic postulate. It is clear that given a line and a point not on the line, there are many parallels (non-intersecting lines) to the given line through the point. Draw some pictures on a piece of paper to convince yourself of this fact.

Now, let's return to the question of angles and, in particular, right angles. What we need is a notion of *perpendicularity* of lines meeting at a point. Let's start with the simplest case, where one of the lines, say l, is a diameter of the Klein disk. Suppose we define another line m to be (hyperbolically) perpendicular to l at a point P if it is perpendicular to l in the Euclidean sense.

Shown here are several Klein lines perpendicular to the Klein line l, which is a diameter of the boundary circle.

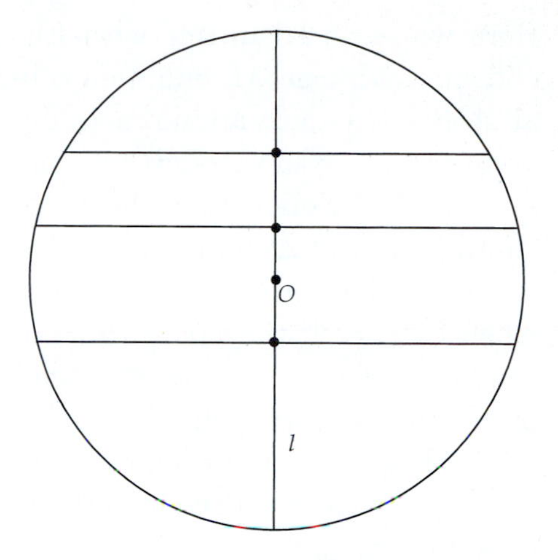

It is clear that we cannot extend this definition to non-diameter chords directly. If we did so, then right angles would have the same meaning in the Klein model as they do in the Euclidean plane, which would mean that parallels would have to satisfy the Euclidean parallel postulate.

The best we can hope for is an extension of *some* property of perpendiculars to a diameter. If we consider the extended plane, with the point at infinity attached, then all the perpendicular lines to the diameter l in the figure above meet at the point at infinity. The point at infinity is the *inverse* point to the origin O, with respect to the unit circle (as was discussed at the end of Chapter 2). Also, O has a unique position on l—it is the Euclidean midpoint of the chord defining l.

If we move l to a new position, say to line l', so that l' is no longer a diameter, then it makes sense that the perpendiculars to l would also move to new perpendiculars to l', but in such a way that they still intersected at the inverse point of the midpoint of the chord for l'. This inverse point is called the *pole* of the chord.

Definition 7.5. The *pole* of chord $\overline{AB}$ in a circle c is the inverse point of the midpoint of $\overline{AB}$ with respect to the circle.

From our work in Chapter 2, we know that the pole of chord AB is also the intersection of the tangents at A and B to the circle.

Here we see a Klein line with Euclidean midpoint M and tangents at A and B (which are not actually points in the Klein model) meeting at pole P. Notice that the three chords ($m1$, $m2$, and $m3$) inside the circle have the property that, when extended, they pass through the pole. It makes sense to extend our definition of perpendicularity to state that these three chords will be perpendicular (in the hyperbolic sense) to the given line ($\overline{AB}$) at the points of intersection.

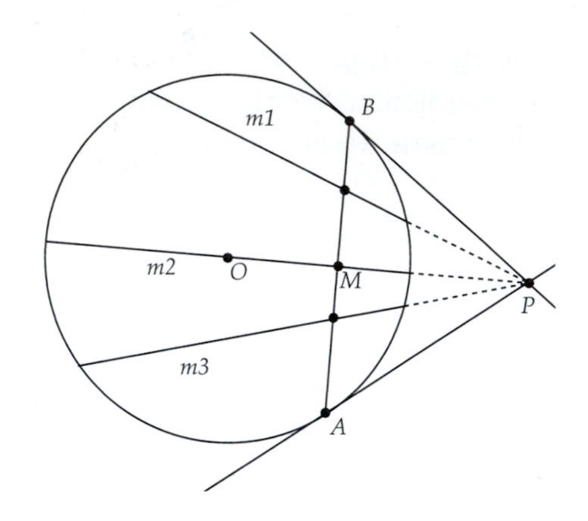

Definition 7.6. A line m is perpendicular to a line l (in the Klein model) if the Euclidean line for m passes through the pole P of l.

Exercise 7.2.3. Use this definition of perpendicularity to sketch some perpendicular lines in the Klein model. Then, use this definition to show that the common perpendicular to two parallel Klein lines exists in *most* cases. That is, show that there is a Klein line that meets two given parallel Klein lines at right angles, *except* in one special case of parallels. Describe this special case.

Given a Klein line l (defined by chord $\overline{AB}$) and a point P not on l, there are two chords $\overline{BC}$ and $\overline{AD}$, both passing through P and parallel to l (only intersections are on the boundary). These two parallels possess the interesting property of dividing the set of all lines through P into two subsets: those that intersect l and those that are parallel to l. These special parallels ($\overline{AD}$ and $\overline{BC}$) will be called *limiting parallels* to l at P. (A precise definition of this property will come later.)

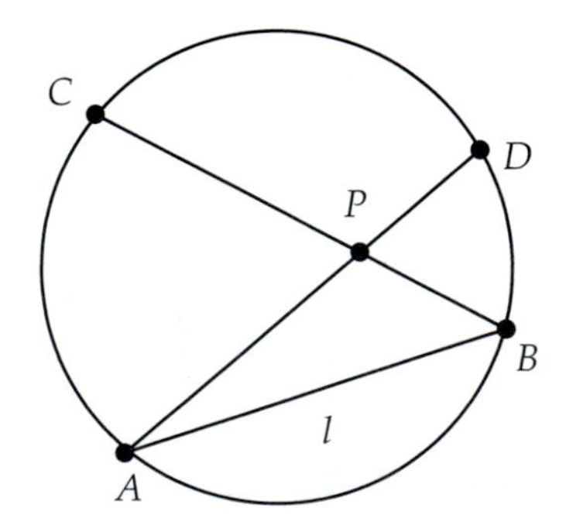

From P drop a perpendicular to l at Q as shown. Consider the hyperbolic angle $\angle QPT$, where T is a point on the hyperbolic ray from P to B. This angle will be called the *angle of parallelism* for l at P.

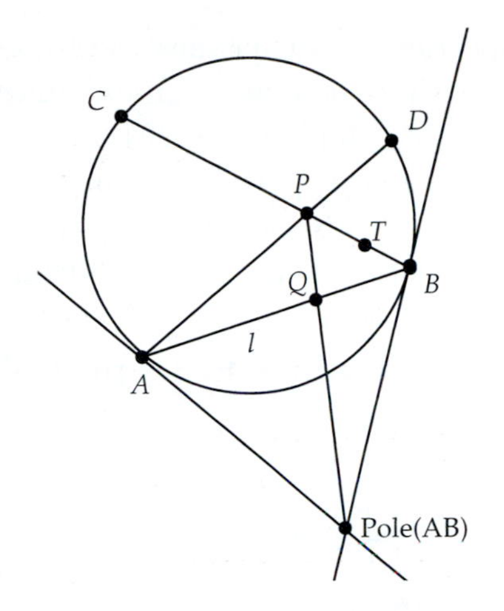

Exercise 7.2.4. Considering the figure above, explain why the angle made by QPT (the angle of parallelism) cannot be a right angle. Then use this result, and the fact that $\overline{AD}$ and $\overline{BC}$ are limiting parallels, to show that the angle of parallelism cannot be greater than a right angle.

7.3 Basic Results in Hyperbolic Geometry

We will now look at some basic results concerning lines, triangles, circles, and the like, that hold in all models of hyperbolic geometry.

Just as we use diagrams to aid in understanding the proofs of results in Euclidean geometry, we will use the Poincaré model (or the Klein model) to draw diagrams to aid in understanding hyperbolic geometry. However, we must be careful. Too much reliance on figures and diagrams can lead to hidden assumptions. We must be careful to argue solely from the postulates or from theorems based on the postulates.

One aid in our study of hyperbolic geometry will be the fact that all results in Euclidean geometry that do not depend on the parallel postulate (those of *neutral geometry*) can be assumed in hyperbolic geometry immediately. For example, we can assume that results about congruence of triangles, such as SAS, will hold in hyperbolic geometry. In fact, we can assume the first 28 propositions in Book I of Euclid (see Appendix A).

Also, we can assume results on isometries, including reflections and rotations, found in sections 5.1, 5.2, and 5.4, as they do not depend on the parallel postulate. These results will hold in hyperbolic geometry, assuming that we have a distance function that is well defined. We also assume basic properties of betweenness and continuity of distance and angle. We saw earlier that these assumptions must be added to Euclid's axiomatic system to ensure completeness of that system, so it is reasonable to assume these properties in hyperbolic geometry as well.

We will start our study with the main area in which hyperbolic geometry distinguishes itself from Euclidean geometry—the area of parallels.

7.3.1 Parallels in Hyperbolic Geometry

As we saw in the Klein model in the last project, if we have a hyperbolic line l and a point P not on l, there are always two parallel lines m, n through P with special properties. This is true in any model of hyperbolic geometry.

Theorem 7.1. *(Fundamental Theorem of Parallels in Hyperbolic Geometry) Given a hyperbolic line l and a point P not on l, there are exactly two parallel lines m, n through P that have the following properties:*

1. *Every line through P lying within the angle made by one of the parallels m, n and the perpendicular from P to l must intersect l while all other lines through P are parallel to l.*

2. *m, n make equal acute angles with the perpendicular from P to l.*

Proof: Drop a perpendicular to l through P, intersecting l at Q.

Consider all angles with side $\overline{PQ}$. The set of these angles will be divided into those angles $\angle QPA$ where $\overrightarrow{PA}$ intersects l and those where it does not. By continuity of angle measure, there must be an angle that separates the angles where $\overrightarrow{PA}$ intersects l from those where it does not. Let $\angle QPC$ be this angle.

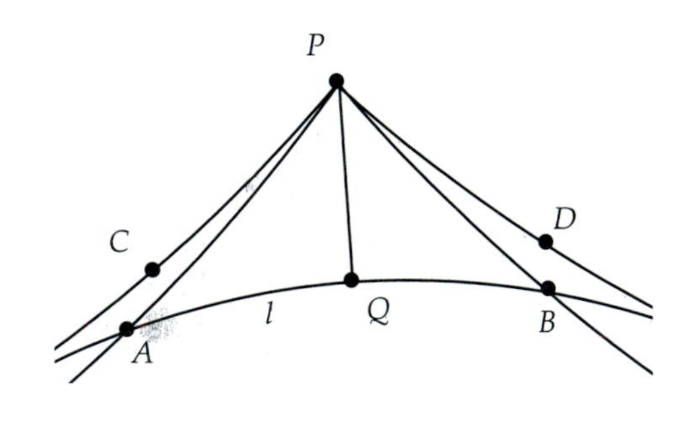

Let $\overrightarrow{PD}$ be the reflection of $\overrightarrow{PC}$ across $\overleftrightarrow{PQ}$. Since reflections preserve parallelism, we have that $\overrightarrow{PD}$ must separate intersecting lines from parallels. Also, reflections preserve angle, so $\angle QPD$ must be congruent to $\angle QPC$. This finishes the proof of part (1) of the theorem.

We now show that both angles are less than 90 degrees. It suffices to show that neither can be a right angle. Suppose that $\angle QPC$ is a right angle. Then by the preceding paragraph, we know that $\angle QPD$ must be a right angle. Points C, P, D are then collinear and make up one parallel to l through P by Euclid's Proposition 27, which does not depend on Euclid's fifth postulate. By the hyperbolic parallel postulate, there must be another line m' through P parallel to l. But, m' has to lie within one of the two right angles $\angle QPC$ or $\angle QPD$, which would be a contradiction to what was just proved about angles $\angle QPC$ and $\angle QPD$ separating intersecting and non-intersecting lines. □

Definition 7.7. The two special parallels defined in the previous theorem are called *limiting parallels* (also called *asymptotic parallels* or *sensed parallels*) to l through P. These will be lines through P that separate intersecting and non-intersecting lines to l. There will be a *right* and a *left* limiting parallel to l through P. Other lines through P that do not intersect l will be called *ultraparallels* (or *divergent parallels*) to l. The angle made by a limiting parallel with the perpendicular from P to l is called the *angle of parallelism at P.*

The properties of limiting parallels have no counterpart in Euclidean geometry and thus it is hard to develop an intuition for them. In the project on the Klein model, we saw how to construct limiting parallels to a given Klein line. It may be helpful to review that construction to have a concrete mental picture of how limiting parallels work in hyperbolic geometry.

In the next few theorems, we will review some of the basic results on limiting parallels. We will prove these results in a *synthetic* fashion, independent of any model. The proofs will focus on the case of right-limiting parallels. The proofs for left-limiting parallels follow by symmetry of the arguments used.

Theorem 7.2. *Let $\overleftrightarrow{PP'}$ be the right-limiting parallel to a line l through P. Then $\overleftrightarrow{PP'}$ is also the right-limiting parallel to l through P'.*

Proof: There are two cases to consider.

In the first case point P' is to the right of P on $\overleftrightarrow{PP'}$ (Fig. 7.1). Drop perpendiculars from P and P' to l at Q and Q'. Then by Euclid's Prop. 27, $\overleftrightarrow{P'Q'} \parallel \overleftrightarrow{PQ}$, and we know that all points on $\overleftrightarrow{P'Q'}$ will be on the same side of $\overleftrightarrow{PQ}$.

Let R be a point to the right of P' on $\overleftrightarrow{PP'}$. We need to show that every ray $\overrightarrow{P'S}$ lying within $\angle Q'P'R$ will intersect l.

Since S is interior to $\angle QPR$, we know that ray $\overrightarrow{PS}$ will intersect l at some point T. $\overrightarrow{PS}$ will also intersect $\overline{P'Q}$ at a point U, since this ray cannot intersect either of the other two sides of triangle $\Delta PQP'$.

Now, consider ΔQUT. $\overrightarrow{P'S}$ intersects side $\overline{UT}$ and does not intersect side $\overline{QU}$ (S and U must be on opposite sides of $\overline{P'Q'}$). Thus, $\overrightarrow{P'S}$ must intersect side $\overline{QT}$ and thus intersects l.

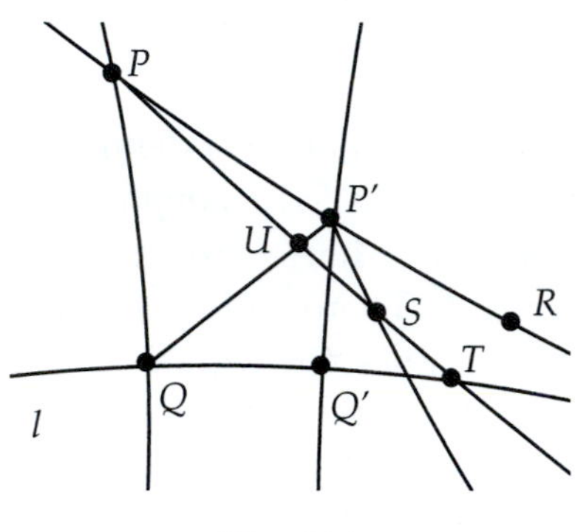

Fig. 7.1

In the second case, point P' is to the left of P on $\overleftrightarrow{PP'}$. A similar proof can be given in this case. □

Note that in this proof we are using the fact that a line intersecting a triangle at a side must intersect one of the other sides. This fact is known as *Pasch's axiom* and is independent of Euclid's original set of five postulates. Since we normally assume this axiom in Euclidean geometry, and it does not depend on any parallel properties, we will likewise assume it in hyperbolic geometry.

The next theorem tells us that the property of being a limiting parallel is a *symmetric* property. That is, we can talk of a pair of lines being limiting parallels to each other without any ambiguity.

Theorem 7.3. *If line m is a right-limiting parallel to l, then l is conversely a right-limiting parallel to m.*

Proof: Let $m = \overleftrightarrow{PD}$ be the right-limiting parallel to l through P. Drop a perpendicular from P to l at Q (Fig. 7.2). Also, drop a perpendicular from Q to m at R. We know that R must be to the right of P, since if it were to the left, then we would get a contradiction to the exterior angle theorem for ΔQRP. Let B be a point on l that is on the same side of $\overline{PQ}$ as D. To show that l is limiting parallel to m, we need to show that any ray interior to $\angle BQR$ must intersect m.

Let $\overrightarrow{QE}$ be interior to $\angle BQR$. We will show that $\overrightarrow{QE}$ must intersect m. Drop a perpendicular from P to $\overleftrightarrow{QE}$ at F. By an exterior angle theorem argument, we know that F will lie on the same side of Q on $\overleftrightarrow{QE}$ as E.

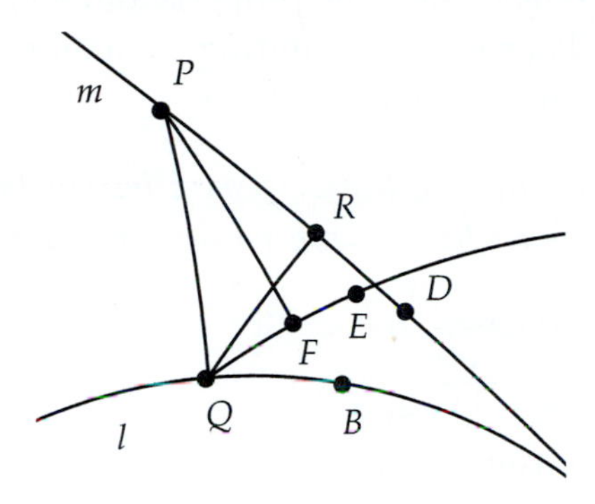

Fig. 7.2

Also, since in ΔPQF the greater side lies opposite the greater angle, we know that $PQ > PF$.

Now, we can use the rotation results from Chapter 5 to rotate $\overline{PF}$, $\overleftrightarrow{PD}$, and $\overleftrightarrow{FE}$ about P by the angle $\theta = \angle FPQ$, as shown in Fig. 7.3. Since $PQ > PF$, F will rotate to a point F' on $\overline{PQ}$ and $\overleftrightarrow{FE}$ will rotate to a line $\overleftrightarrow{F'E'}$ that is ultraparallel to l, because of the right angles at Q and F'. Also, $\overleftrightarrow{PD}$ will rotate to a line $\overleftrightarrow{PD'}$ that is interior to $\angle QPR$ and so will intersect l at some point G.

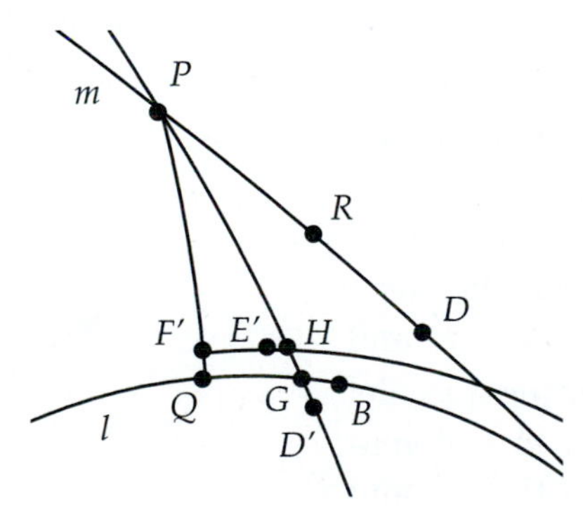

Fig. 7.3

Since $\overleftrightarrow{F'E'}$ intersects ΔPQG and does not intersect $\overline{QG}$, it must intersect $\overline{PG}$ at some point H. Rotating $\overleftrightarrow{F'E'}$ and $\overleftrightarrow{PD'}$ about P by $-\theta$ shows that $\overrightarrow{QE}$ will intersect m at the rotated value of H. □

The next theorem tells us that the property of being a limiting parallel is a *transitive* property. That is, if l is limiting parallel to m and m is limiting parallel to n, then l is limiting parallel to n.

Theorem 7.4. *If two lines are limiting parallel (in the same direction) to a third line, then they must be limiting parallel to each other.*

Proof: There are two cases to consider.

In the first case suppose $\overleftrightarrow{AB}$ and $\overleftrightarrow{CD}$ are both right-limiting parallel to $\overleftrightarrow{EF}$ with points A and C on opposite sides of $\overleftrightarrow{EF}$. Then, $\overline{AC}$ will intersect $\overleftrightarrow{EF}$ at some point G. Let $\overrightarrow{CH}$ be any ray interior to $\angle ACD$. Then, since $\overleftrightarrow{CD}$ is right-limiting parallel to $\overleftrightarrow{EF}$, we have that $\overrightarrow{CH}$ will intersect $\overleftrightarrow{EF}$ at some point J.

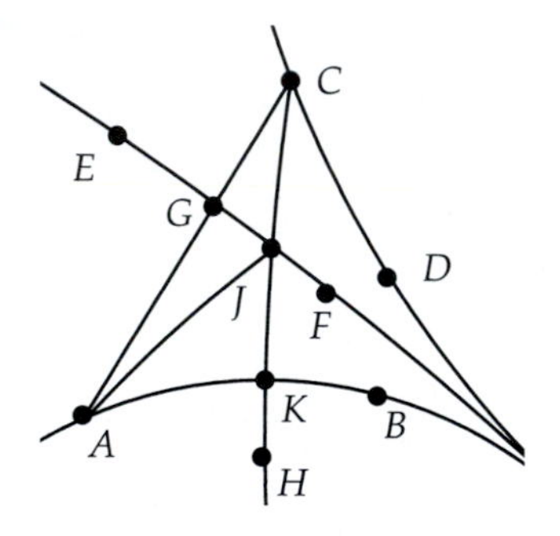

Connect A and J. Since limiting parallelism is symmetric and $\overleftrightarrow{AB}$ is right-limiting parallel to $\overleftrightarrow{EF}$, then $\overleftrightarrow{EF}$ is right-limiting parallel to $\overleftrightarrow{AB}$ and is limiting parallel at all points, including J. Thus, $\overrightarrow{CJ}$ will intersect $\overleftrightarrow{AB}$ at some point K.

Since $\overleftrightarrow{AB}$ and $\overleftrightarrow{CD}$ do not intersect (they are on opposite sides of $\overleftrightarrow{EF}$), and for all rays $\overrightarrow{CH}$ interior to $\angle ACD$ we have that $\overrightarrow{CH}$ intersects $\overleftrightarrow{AB}$, then $\overleftrightarrow{CD}$ is right-limiting parallel to $\overleftrightarrow{AB}$.

For the second case we assume that $\overleftrightarrow{AB}$ and $\overleftrightarrow{CD}$ are both on the same side of $\overleftrightarrow{EF}$. Suppose that $\overleftrightarrow{AB}$ and $\overleftrightarrow{CD}$ are not right-limiting parallel to each other. Without loss of generality, we may assume that C and E are on opposite sides of $\overleftrightarrow{AB}$. Let $\overleftrightarrow{CG}$ be the right-limiting parallel to $\overleftrightarrow{AB}$.

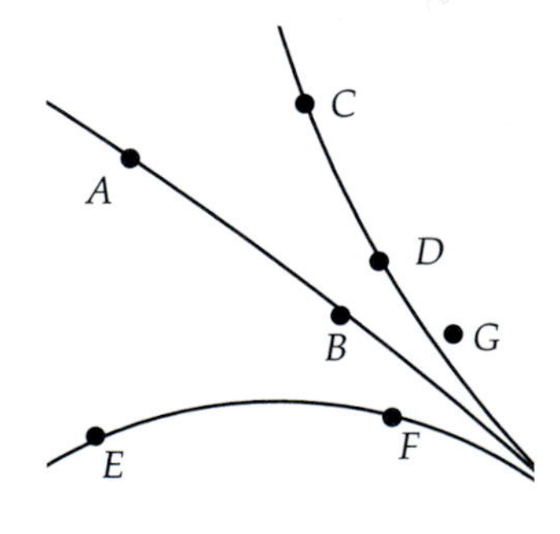

By the first part of this proof, $\overleftrightarrow{CG}$ and $\overleftrightarrow{EF}$ must be right-limiting parallel. But, $\overleftrightarrow{CD}$ is already the right-limiting parallel to $\overleftrightarrow{EF}$ at C, and thus G must be on $\overleftrightarrow{CD}$ and $\overleftrightarrow{CD}$ is right-limiting parallel to $\overleftrightarrow{AB}$. □

7.3.2 Omega Points and Triangles

In the previous section we saw that given a line l and a point P not on l, there were two special lines called the right- and left-limiting parallels to l through P. These separated the set of lines through P that intersect l from those that did not intersect l.

In the Poincaré and Klein models, these limiting parallels actually meet at points on the boundary circle. While these boundary points are not actually valid points in the geometry, it is still useful to consider the boundary points as representing the special relationship that limiting parallels have with the given line.

We will call these special points *omega points*, or *ideal points*. Thus, a given line will have in addition to its set of "ordinary" points a special pair of omega points. All limiting parallels to the given line will pass through these omega points.

Definition 7.8. Given a line l, the *right [left] omega point to* l represents the set of all right- [left-] limiting parallels to l. We say two lines intersect at an omega point if one is right- [left-] limiting parallel to the other. If Ω is an omega point of l and P is an ordinary point, then by the *line* through P and Ω we will mean the line through P that is right [left-] limiting parallel to l.

Definition 7.9. Given two omega lines $\overleftrightarrow{P\Omega}$ and $\overleftrightarrow{Q\Omega}$, the *omega triangle* defined by P, Q, and Ω is the set of points on $\overline{PQ}$ and the two rays defined by the limiting parallels $\overleftrightarrow{P\Omega}$ and $\overleftrightarrow{Q\Omega}$.

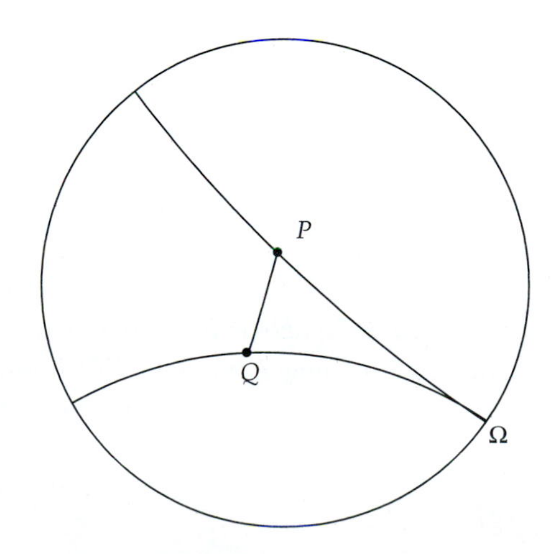

It is interesting to note that these omega triangles, while not true triangles, still share many important properties of ordinary triangles.

Theorem 7.5. *(Pasch's Axiom for Omega Triangles, Part I) If a line passes through a vertex P or Q of an omega triangle $PQ\Omega$ and passes through an interior point of the triangle, then it must intersect the opposite side. If a line passes through Ω, and an interior point, it must intersect side $\overline{PQ}$.*

Proof: The proof of the first part of the theorem is a direct consequence of the fact that $\overleftrightarrow{P\Omega}$ and $\overleftrightarrow{Q\Omega}$ are limiting parallels and will be left as an exercise.

For the proof of the second part, suppose that a line passes through Ω and through a point X within the omega triangle.

By *passing through* Ω we mean that the line through X is a limiting parallel to $\overleftrightarrow{P\Omega}$ (or $\overleftrightarrow{Q\Omega}$). By the first part of this proof, we know that $\overleftrightarrow{PX}$ will intersect $\overleftrightarrow{Q\Omega}$ at some point Y. By Pasch's axiom, we know that the line $\overleftrightarrow{X\Omega}$ will intersect side $\overline{PQ}$ of ΔPQY.

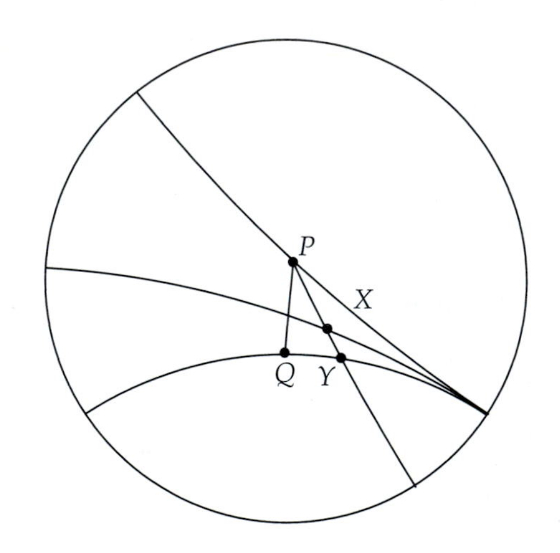

□

Theorem 7.6. *(Pasch's Axiom for Omega Triangles, Part II) If a line intersects one of the sides of an omega triangle $PQ\Omega$ but does not pass through a vertex, then it will intersect exactly one of the other two sides.*

Proof: The proof of this theorem is left as a series of exercises below. □

Theorem 7.7. *(Exterior Angle Theorem for Omega Triangles) The exterior angles of an omega triangle $PQ\Omega$ made by extending $\overline{PQ}$ are greater than their respective opposite interior angles.*

Proof: Extend $\overline{PQ}$ to R. It suffices to show that $\angle RQ\Omega$ is greater than $\angle QP\Omega$.

We can find a point X on the right side of $\overline{PQ}$ such that $\angle RQX \cong \angle QP\Omega$. Suppose $\overrightarrow{QX}$ intersected $\overleftrightarrow{P\Omega}$ at S. Then $\angle RQX$ would be an exterior angle to ΔPQS and would equal the opposite interior angle in this triangle, which contradicts the exterior angle theorem for ordinary triangles.

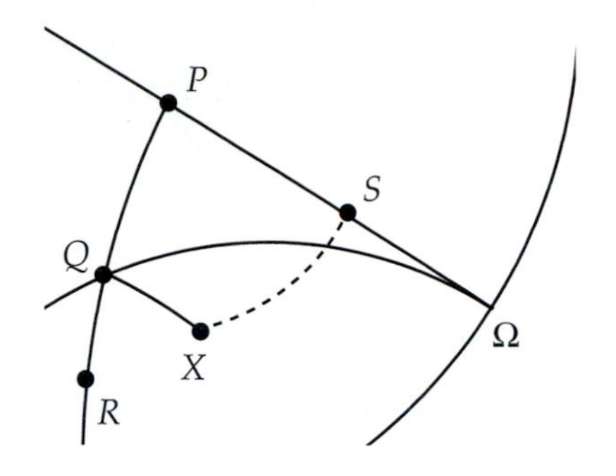

Could $\overline{QX}$ lie on $\overleftrightarrow{Q\Omega}$? Suppose it does.

Let M be the midpoint of $\overline{PQ}$ and drop a perpendicular from M to $\overleftrightarrow{Q\Omega}$ at N. Now, on the line $\overleftrightarrow{P\Omega}$, we can find a point L, on the other side of $\overline{QP}$ from N, with $\overline{QN} \cong \overline{PL}$. If $\overline{QX}$ lies on $\overleftrightarrow{Q\Omega}$, then $\angle NQM \cong \angle LPM$ (both are supplementary to congruent angles). Thus, by SAS, ΔNQM and ΔLPM are congruent and $\angle MLP$ is a right angle. Since the angles at M are congruent, L, M, and N lie on a line. But, this would imply that $\overline{LN}$ is perpendicular to both $\overleftrightarrow{P\Omega}$ and $\overleftrightarrow{Q\Omega}$ and that the angle of parallelism is 90 degrees, which is impossible.

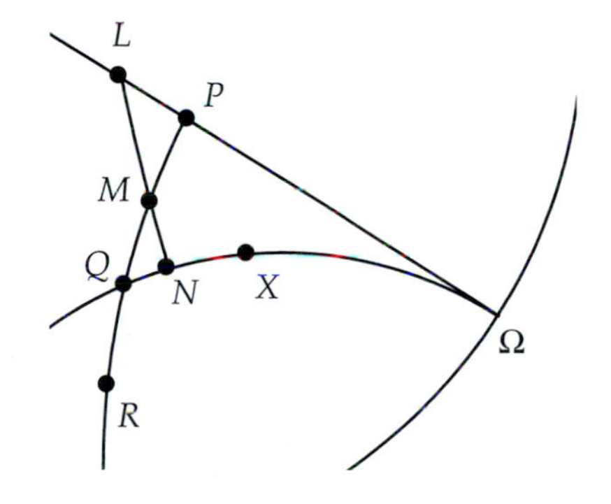

Thus, $\angle RQX$ is less than angle $\angle RQ\Omega$ and since $\angle RQX \cong \angle QP\Omega$, we are finished with the proof. □

Theorem 7.8. *(Omega Triangle Congruence) If $\overline{PQ}$ and $\overline{P'Q'}$ are congruent and $\angle PQ\Omega$ is congruent to $\angle P'Q'\Omega'$, then $\angle QP\Omega$ is congruent to $\angle Q'P'\Omega'$ (Fig. 7.4).*

Proof: Suppose one of the angles is greater, say $\angle QP\Omega$. We can find R interior to $\angle QP\Omega$ such that $\angle QPR \cong \angle Q'P'\Omega'$. Then $\overrightarrow{PR}$ will intersect $\overleftrightarrow{Q\Omega}$ at some point S. On $\overleftrightarrow{Q'\Omega'}$ we can find S' with $\overline{QS} \cong \overline{Q'S'}$. Triangles ΔPQS and $\Delta P'Q'S'$ are then congruent by SAS and $\angle Q'P'S' \cong \angle QPS$. Also, $\angle QPS \cong \angle Q'P'\Omega'$. Thus, $\angle Q'P'S' \cong \angle Q'P'\Omega'$, which is impossible.

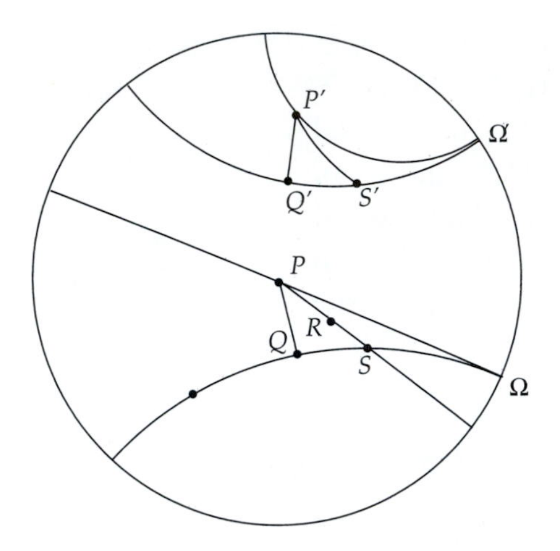

Fig. 7.4

□

Exercise 7.3.1. Illustrate Theorem 7.3 in the Klein model and explain why it must be true using the properties of that model.

Exercise 7.3.2. Illustrate Theorem 7.4 in the Klein model and explain why it must be true using the properties of that model.

Exercise 7.3.3. Let l be a hyperbolic line and let r_l be reflection across that line. Use the parallelism properties of reflections to show that r_l maps limiting parallels of l to other limiting parallels of l. Use this to show that r_l fixes omega points of l.

Exercise 7.3.4. Show that the omega points of a hyperbolic line l are the only omega points fixed by reflection across l. [Hint: If a reflection r_l fixes an omega point of a line $l' \neq l$, show that r_l must fix l' and get a contradiction.]

Exercise 7.3.5. Show that if a rotation R fixes an omega point, then it must be the identity rotation. [Hint: Use the preceding exercise.]

Exercise 7.3.6. Show that for an omega triangle $PQ\Omega$, the sum of the angles $\angle PQ\Omega$ and $\angle QP\Omega$ is always less than 180 degrees.

Exercise 7.3.7. Show that if a line intersects an omega triangle $PQ\Omega$ at one of the vertices P or Q, then it must intersect the opposite side.

Exercise 7.3.8. (Partial Proof of Theorem 7.6)

Suppose a line intersects side $\overleftrightarrow{P\Omega}$ of an omega triangle $PQ\Omega$ but does not pass through a vertex. Let R be the point of intersection and connect R to Q. Use Pasch's axiom and Theorem 7.5 to show that a line through R that does not pass through P or Q must intersect $\overline{PQ}$ or $\overleftrightarrow{Q\Omega}$.

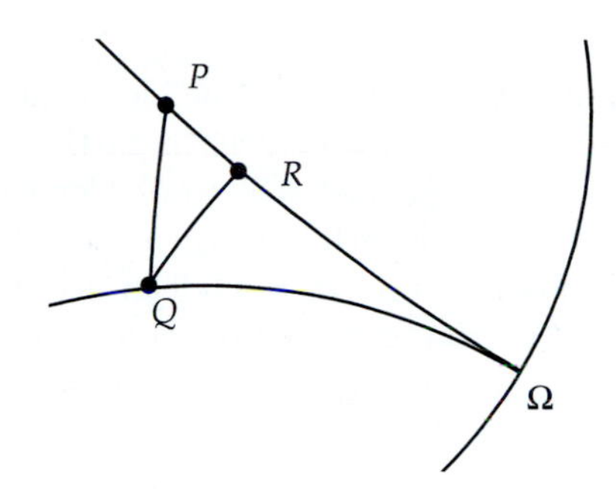

Exercise 7.3.9. (Partial Proof of Theorem 7.6)

Suppose a line intersects side $\overline{PQ}$ of an omega triangle $PQ\Omega$ but does not pass through a vertex. Let R be the point of intersection. We can find the limiting parallel $\overleftrightarrow{R\Omega}$. Show that a line through R must intersect either $\overleftrightarrow{P\Omega}$ or $\overleftrightarrow{Q\Omega}$. [Hint: Use Theorem 7.5.]

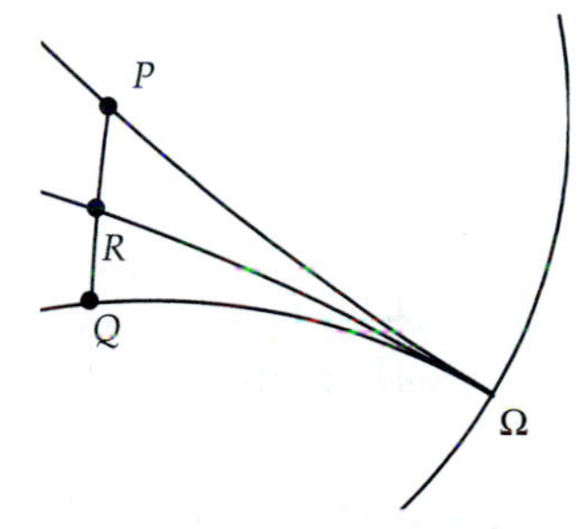

Exercise 7.3.10. In this exercise we will prove Angle-Angle Congruence for omega triangles. Suppose $PQ\Omega$ and $P'Q'\Omega'$ are two omega triangles with $\angle PQ\Omega \cong \angle P'Q'\Omega'$ and $\angle QP\Omega \cong \angle Q'P'\Omega'$. Show that $\overline{PQ} \cong \overline{P'Q'}$. [Hint: Suppose the segments are not congruent. See if you can derive a contradiction to Exercise 7.3.6.]

Exercise 7.3.11. Let $\overline{PQ}$ be a segment of length h. Let l be a perpendicular to $\overline{PQ}$ at Q and $\overleftrightarrow{PR}$ the limiting parallel to l at P. Define the *angle of parallelism* to be $a(h) = \angle QPR$, where the angle is measured in degrees. Use a theorem on omega triangles to show this definition is well defined, that it depends only on h.

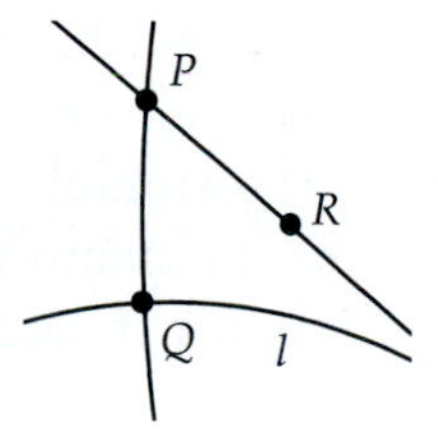

Exercise 7.3.12. Show that if $h < h'$, then $a(h) > a(h')$. That is, the function $a(h)$ is *order-reversing.*

Exercise 7.3.13. Show that $a(h)$ is a one-to-one function on the set of positive real numbers h.

Exercise 7.3.14. In Euclidean and hyperbolic geometry, angles are said to be *absolute*. One can *construct* a particular right angle and from there all other angles. In Euclidean geometry, however, length is not absolute. One must *choose* a segment to be of unit length, and then other lengths can be measured. Measures of length are thus *arbitrary* in Euclidean geometry. Assuming that it is always possible to carry out the construction described in the definition of $a(h)$, use the preceding exercise to show that length is *absolute* in hyperbolic geometry.

7.4 Project 10 - The Saccheri Quadrilateral

Girolamo Saccheri was a Jesuit priest who, like Gauss and others mentioned at the beginning of this chapter, tried to negate Playfair's Postulate and find a contradiction to known results based on the first four Euclidean postulates. His goal in this work was not to study hyperbolic geometry, but rather to prove Euclid's fifth postulate as a *theorem.* Just before he died in 1733, he published *Euclides ab Omni Naevo Vindicatus* ("Euclid Freed of Every Flaw"), in which he summarized his work on negating the parallel postulate by the "hypothesis of the acute angle."

Saccheri's idea was to study quadrilaterals whose base angles are right angles and whose base-adjacent sides are congruent. Of course, in Euclidean geometry, such quadrilaterals must be rectangles; that is, the top (or summit) angles must be right angles. Saccheri negated the parallel postulate by assuming the summit angles were less than 90 degrees. This was the "hypothesis of the acute angle." Saccheri's attempt to prove the parallel postulate ultimately failed because he could not find a contradiction to the acute angle hypothesis.

To see why Saccheri was unable to find a contradiction, let's consider his quadrilaterals in hyperbolic geometry.

Start *Geometry Explorer* and create a segment $\overline{AB}$ on the screen. This will be the base of our quadrilateral.

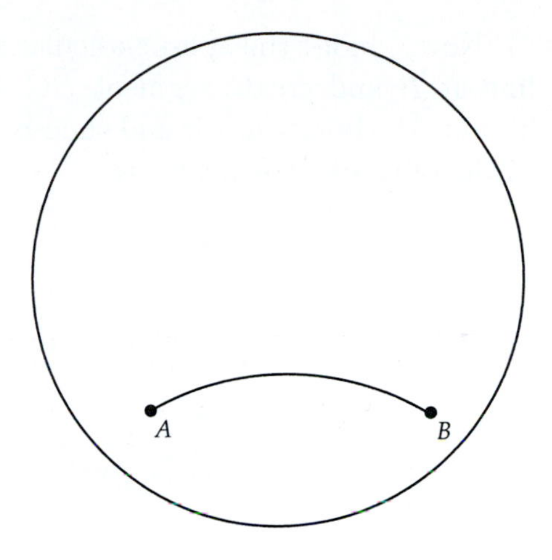

Multi-select $\overline{AB}$ and A and click on the Perpendicular button in the Construct panel to construct the perpendicular to $\overline{AB}$ at A. Likewise, construct the perpendicular to $\overline{AB}$ at B.

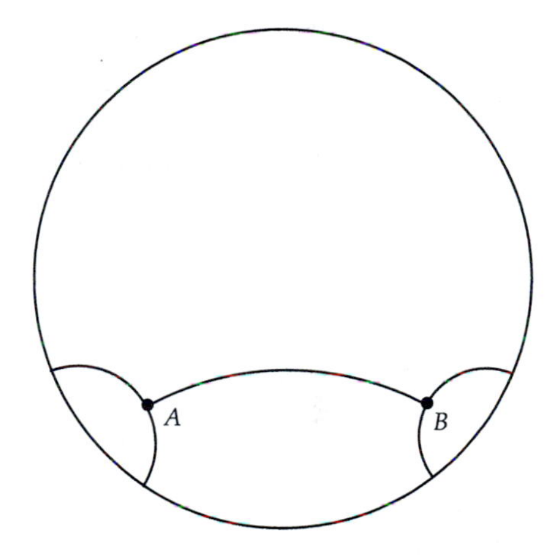

Now, attach a point C along the perpendicular at B as shown. To attach a point to a line, create a point on top of the line. This will cause the point to move only along the line.

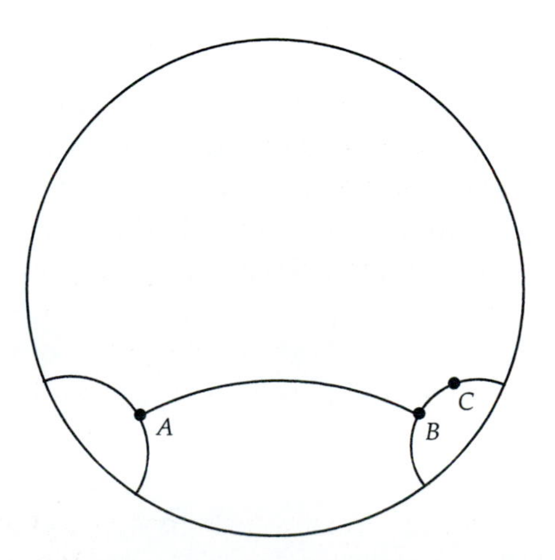

Now, hide the perpendicular line at B and create segment $\overline{BC}$. To hide the line, select it and choose **Hide Object** (**View** menu).

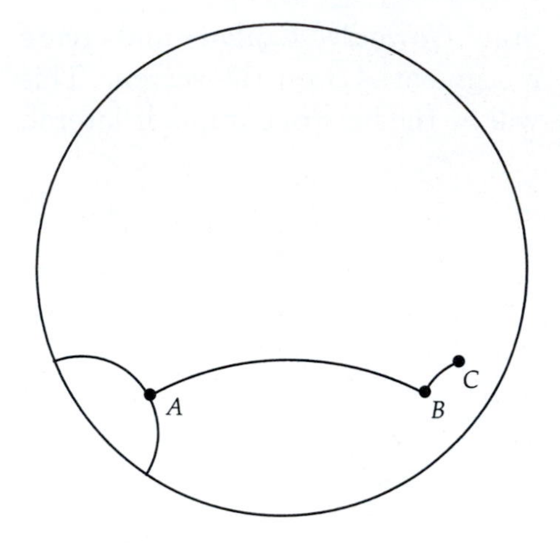

Multi-select point A and segment $\overline{BC}$ and click on the Circle constructor tool in the Construct panel, yielding a circle centered at A of *hyperbolic* radius the *hyperbolic* length of $\overline{BC}$. Multi-select the circle and the perpendicular at A and click the Intersection button (Construct panel) to find their intersection.

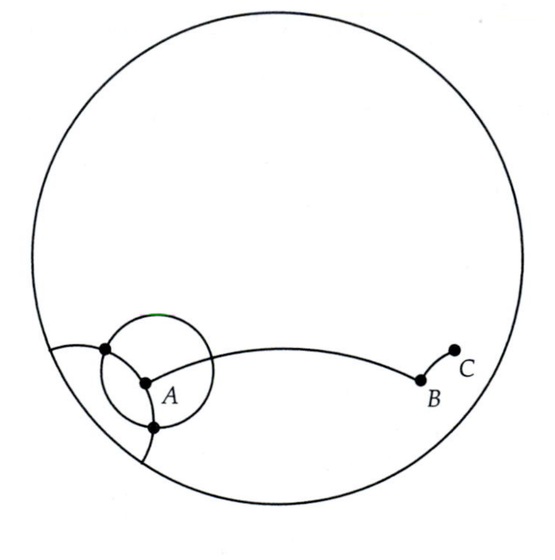

The upper intersection point, D, is all we need, so hide the circle, the perpendicular, and the lower intersection point. Then, connect A to D and D to C to finish the construction of a Saccheri Quadrilateral.

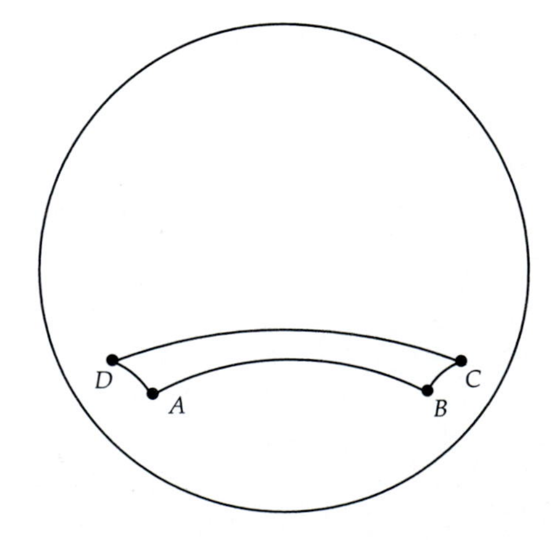

At this point we can use our Saccheri Quadrilateral to study many fascinating properties in hyperbolic geometry. Let's look at a couple.

Measure the two angles at D and C in the quadrilateral. For this configuration they are equal and less than 90 degrees, which is what they would be in Euclidean geometry. It appears that Saccheri's acute angle hypothesis holds in hyperbolic geometry, at least for this one case. Move points A and B and check the summit angles.

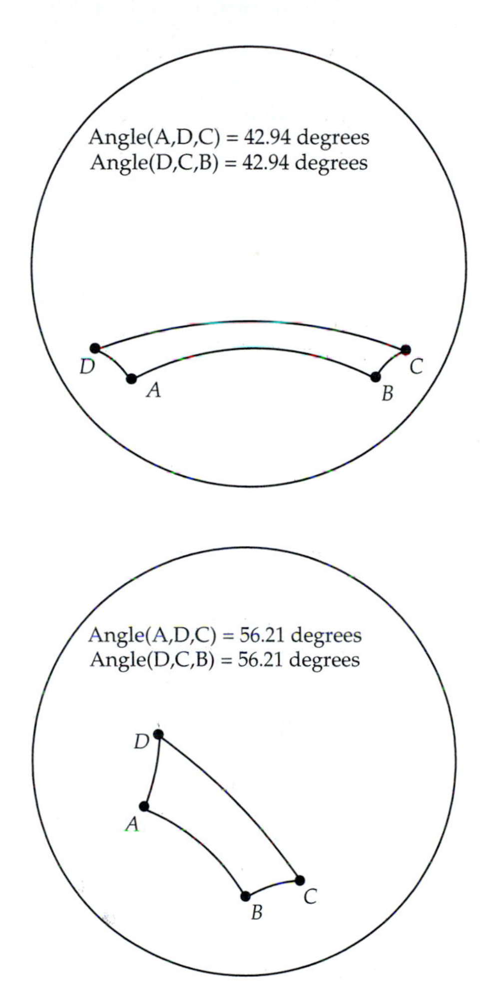

It appears that this result holds for all configurations of our Saccheri Quadrilateral (except perhaps when orientations switch, because our intersection point switches from above the rectangle to below it and the angles become greater than 180 degrees). In fact, this is a theorem in hyperbolic geometry. The summit angles of a Saccheri Quadrilateral are always equal and less than 90 degrees (i.e., are acute).

Saccheri could find no contradiction in assuming that the summit angles of this quadrilateral were acute because in hyperbolic geometry (which can be modeled within Euclidean geometry), the summit angles are always acute. If he had been able to find a contradiction, then that would also be a contradiction in Euclidean geometry. However, Saccheri could not believe what his own work was telling him. As mentioned at the beginning of this chapter, he found the hypothesis of the acute angle "repugnant to the nature of the straight line."

Exercise 7.4.1. Prove that the summit angles of a Saccheri Quadrilateral are always congruent. [Hint: Start by showing the diagonals are congruent.]

Exercise 7.4.2. Fill in the "Why?" parts of the following proof that the summit angles of a Saccheri Quad are always acute. Refer to the figure at right.

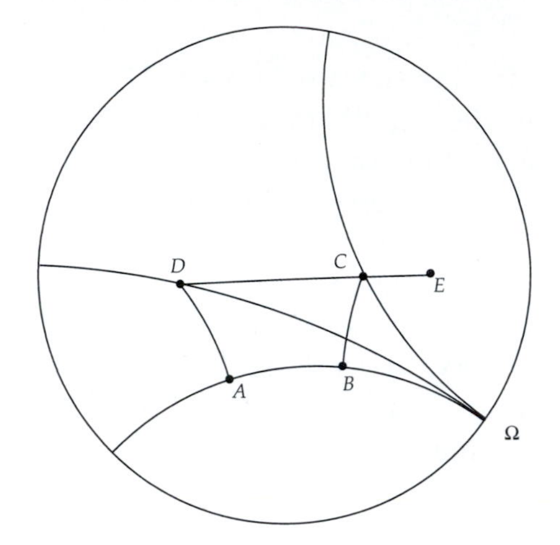

At C we can construct the right-limiting parallel $\overleftrightarrow{C\Omega}$ to $\overline{AB}$. At D we can also construct the right-limiting parallel $\overleftrightarrow{D\Omega}$ to $\overline{AB}$. These two limiting parallels will also be limiting parallel to each other. (Why?)

$\overleftrightarrow{DC}$ must be an ultraparallel (non-intersecting line) to $\overleftrightarrow{AB}$. (Why?) [Hint: Construct midpoints to $\overline{AB}$ and $\overline{CD}$ and use triangles to show that the two lines share a common perpendicular.]

Thus, the right-limiting parallels through D and C, respectively, must lie within the angles $\angle ADC$ and $\angle BCE$, where E is a point on $\overrightarrow{DC}$ to the right of C. Now $\angle AD\Omega$ and $\angle BC\Omega$ are equal in measure. (Why?)

Furthermore, in omega triangle $CD\Omega$, $\angle EC\Omega$ must be greater than interior angle $\angle CD\Omega$. (Why?)

Finally, this proves that $\angle ADC$ and $\angle BCD$ are both less than 90 degrees. (Why?)

For your report give a careful and complete summary of your work done on this project.

7.5 Lambert Quadrilaterals and Triangles

7.5.1 Lambert Quadrilaterals

In the last section we defined a Saccheri Quadrilateral as a quadrilateral $ABCD$, where the base angles $\angle BAD$ and $\angle CBA$ are right angles and the side lengths AD and BC are congruent (Fig. 7.5).

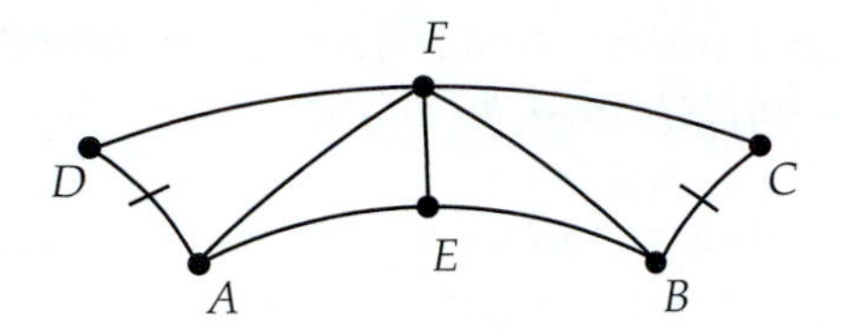

Fig. 7.5

Let E and F be the midpoints of the base and summit of a Saccheri quadrilateral. By the results of the last section, we know that ΔADF and ΔBCF are congruent by SAS and thus $\overline{AF} \cong \overline{BF}$. Thus, ΔAEF and ΔBEF are congruent by SSS and the angle at E must be a right angle. A similar argument will show that $\angle DFE$ and $\angle CFE$ are also right angles. Thus,

Theorem 7.9. *The segment joining the midpoints of the base and summit of a Saccheri quadrilateral makes right angles with the base and summit.*

If we look at the two quadrilaterals $AEFD$ and $BEFC$, they both share the property of having three right angles.

Definition 7.10. A *Lambert Quadrilateral* is a quadrilateral having three right angles.

The midpoint construction just described gives a natural way to associate a Lambert quadrilateral with a given Saccheri quadrilateral. We can also create a Saccheri quadrilateral from a given Lambert quadrilateral.

Theorem 7.10. *Let $ABCD$ be a Lambert quadrilateral with right angles at A, B, and C. If we extend $\overline{AB}$ and $\overline{CD}$, we can find points E and F such that $\overline{AB} \cong \overline{AE}$ and $\overline{CD} \cong \overline{CF}$ (Fig. 7.6). Then $EBDF$ is a Saccheri quadrilateral.*

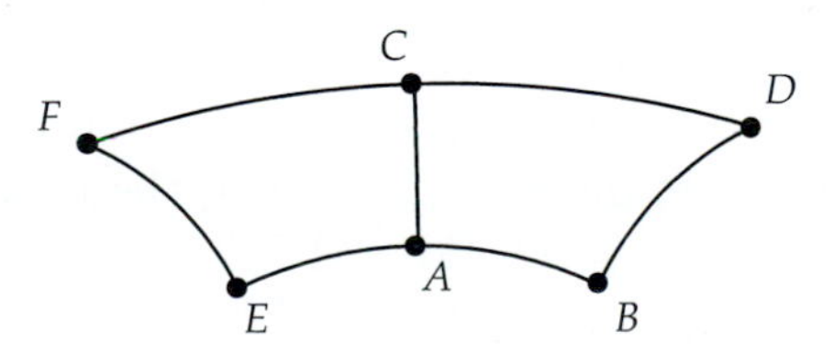

Fig. 7.6

Proof: The proof is left as an exercise. □

Corollary 7.11. *In a Lambert Quadrilateral the fourth angle (the one not specified to be a right angle) must be acute.*

Proof: By the preceding theorem, we can embed a given Lambert quadrilateral in a Saccheri quadrilateral and we know the summit angles of a Saccheri quadrilateral are acute. □

Corollary 7.12. *Rectangles do not exist in hyperbolic geometry.*

Proof: This is an immediate consequence of the preceding corollary. □

Here is another interesting fact concerning Lambert Quads.

Theorem 7.13. *In a Lambert quadrilateral the sides adjoining the acute angle are greater than the opposite sides.*

Proof: Given a Lambert quadrilateral $ABDC$ with right angles at A, B, and C, suppose that $DB < AC$ (Fig. 7.7). Then there is a point E on the line through B, D with D between B and E such that $\overline{BE} \cong \overline{AC}$. It follows that $ABEC$ is a Saccheri quadrilateral, and $\angle ACE \cong \angle BEC$, and both angles are acute.

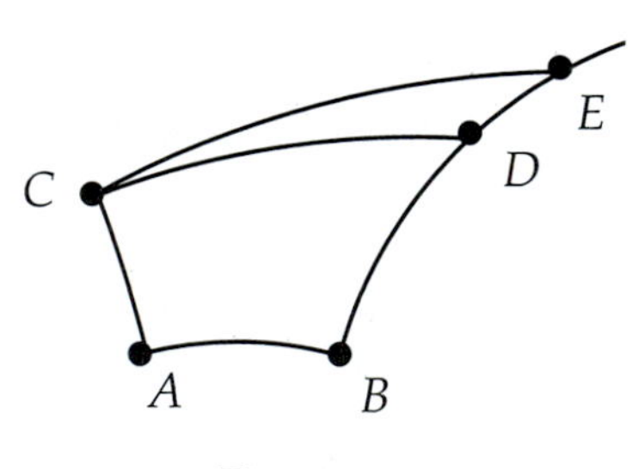

Fig. 7.7

However, since A and E are on opposite sides of $\overleftrightarrow{CD}$ (A and B are on the same side and B and E are on opposite sides), then $\overline{CD}$ lies within $\angle ACE$ and thus $\angle ACE$ contains $\angle ACD$. So, $\angle ACE$ must be greater than a right angle. This contradicts the fact that $\angle ACE$ must be acute.

Thus, $DB \geq AC$. If $DB = AC$, then we would have a Saccheri quadrilateral $ABDC$ and the summit angles would be congruent, which is impossible. We must have, then, that $DB > AC$. A similar argument shows $CD > AB$. □

7.5.2 Triangles in Hyperbolic Geometry

Here are two basic (but still amazing) facts about triangles in hyperbolic geometry.

Theorem 7.14. *The angle sum for any hyperbolic right triangle is always less than 180 degrees.*

Proof: Let ΔABC be a right triangle with a right angle at A. Let D be the midpoint of $\overline{BC}$ (Fig. 7.8).

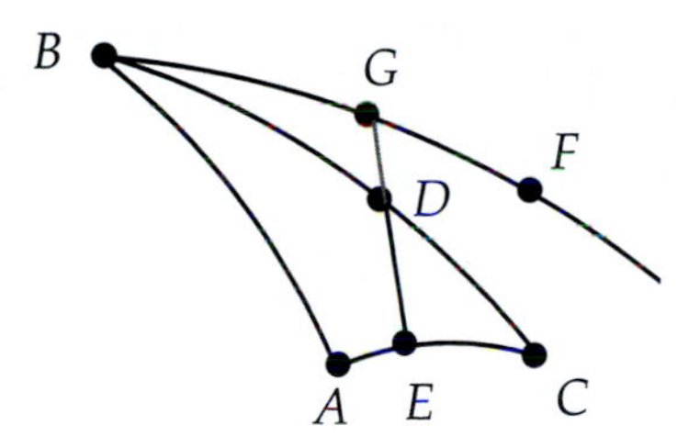

Fig. 7.8

Drop a perpendicular from D to $\overleftrightarrow{AC}$ intersecting at E. Then E must lie between A and C. For by Pasch's Axiom we know that the perpendicular must pass through one of the two sides $\overline{AB}$ or $\overline{AC}$. Clearly, it cannot pass through A, B, or C. Suppose it passed through $\overline{AB}$ at X. Then ΔXEA would have two right angles, contradicting the absolute geometry result that the sum of two angles in a triangle must be less than two right angles (Euclid, Book I, Prop. 17).

Now, we can find a point F such that $\angle DCE$ is congruent to $\angle DBF$. On $\overrightarrow{BF}$ we can find a point G such that $\overline{BG} \cong \overline{EC}$. By SAS we have that $\Delta ECD \cong \Delta GBD$. Thus, E, D, and G are collinear and $\angle DGB$ is a right angle.

Quadrilateral $ABGE$ is thus a Lambert Quadrilateral and $\angle ABG$ must be acute. Since $\angle DCE$ is congruent to $\angle DBF$, then $m\angle DCE + m\angle ABD = m\angle DBF + m\angle ABD < 90$. Thus, the sum of the angles in ΔABC is less than two right angles. □

Theorem 7.15. *The angle sum for any hyperbolic triangle is less than 180 degrees.*

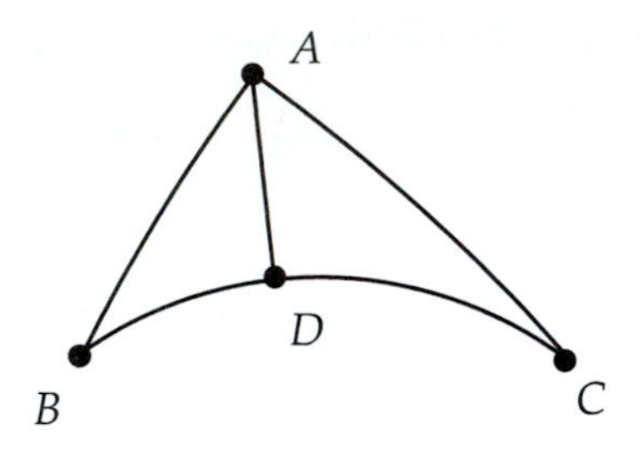

Fig. 7.9

Proof: Let ABC be a triangle (Fig. 7.9). We have proved the result if one of the angles is right. Thus, suppose none of the three angles are right angles. Then two must be acute, as otherwise we would have the sum of two angles being greater than or equal to 180 degrees.

Let the angles at B and C be acute. Drop a perpendicular from A down to $\overleftrightarrow{BC}$, intersecting at D. Then D is between B and C. (If it intersected elsewhere on the line, we could construct a triangle with two angles more than 180 degrees.)

Now, ΔABC can be looked at as two right triangles ΔABD and ΔADC. Using the previous result, we know that the sum of the angles in ΔABD is less than 180, as is the sum of the angles in ΔADC. So the sum of the angles in the two triangles together is less than 360. But clearly, this sum is the same as the sum of the angles in ΔABC plus the two right angles at D. Thus, the sum of the angles in ΔABC is less than $360 - 180 = 180$. □

Definition 7.11. Given ΔABC we call the difference between 180 degrees and the angle sum of ΔABC the *defect* of ΔABC.

Corollary 7.16. *The sum of the angles of any quadrilateral is less than 360 degrees.*

Proof: The proof is left as an exercise. □

Definition 7.12. Given quadrilateral $ABCD$ the *defect* is equal to 360 degrees minus the sum of the angles in the quadrilateral.

Theorem 7.17. *Given ΔABC let line l intersect sides $\overline{AB}$ and $\overline{AC}$ at points D and E, respectively. Then the defect of ΔABC is equal to the sum of the defects of ΔAED and quadrilateral $EDBC$.*

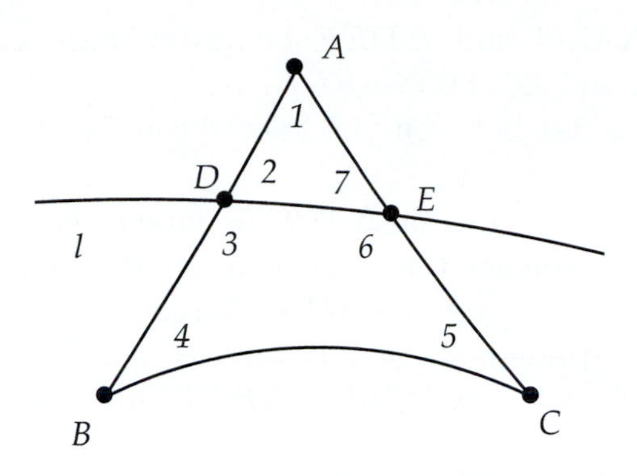

Fig. 7.10

Proof: If l intersects at any of A, B, or C, the result is clear. Suppose that D and E are interior to $\overline{AB}$ and $\overline{AC}$. Label all interior angles as shown in Fig. 7.10. Then

$$defect(\Delta ABC) = 180 - (\angle 1 + \angle 4 + \angle 5)$$

And,

$$\begin{aligned} defect(\Delta ADE) + defect(DECB) = 180 - (\angle 1 + \angle 2 + \angle 7) \\ + 360 - (\angle 3 + \angle 4 + \angle 5 + \angle 6) \end{aligned}$$

Since $\angle 2 + \angle 3 = 180$ and $\angle 6 + \angle 7 = 180$, the result follows. □

We can use this result to prove one of the most amazing facts about triangles in hyperbolic geometry—similar triangles are congruent!

Theorem 7.18. *(AAA Congruence) If two triangles have corresponding angles congruent, then the triangles are congruent (Fig. 7.11).*

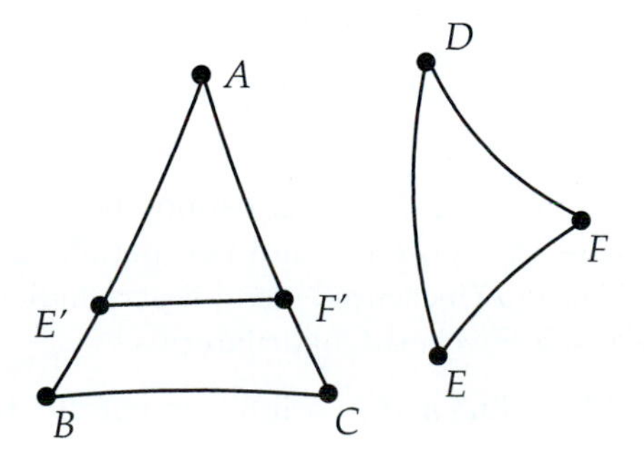

Fig. 7.11

Proof: Let ΔABC and ΔDEF be given with $\angle ABC \cong \angle DEF$, $\angle BCA \cong \angle EFD$, and $\angle CAB \cong \angle FDE$.

If any pair of sides between the two is congruent, then by ASA the triangles would be congruent.

So, either a pair of sides in ΔABC is larger than the corresponding pair in ΔDEF or is smaller than the corresponding pair. Without loss of generality we can assume $\overline{AB}$ and $\overline{AC}$ are larger than $\overline{DE}$ and $\overline{DF}$.

Then we can find points E' and F' on $\overline{AB}$ and $\overline{AC}$ so that $\overline{AE'} \cong \overline{DE}$ and $\overline{AF'} \cong \overline{DF}$. By SAS, $\Delta AE'F' \cong \Delta DEF$, and the two triangles have the same defect.

But, since ΔDEF and ΔABC have the same defect, we have that $\Delta AE'F'$ and ΔABC have the same defect. From the previous theorem, the defect of quadrilateral $E'F'CB$ is then zero, which is impossible.

Thus, all pairs of sides are congruent. □

Exercise 7.5.1. Prove Theorem 7.10.

Exercise 7.5.2. Prove Corollary 7.16.

Exercise 7.5.3. Prove that the summit is always larger than the base in a Saccheri Quadrilateral.

Exercise 7.5.4. Show that two Saccheri Quadrilaterals with congruent summits and congruent summit angles must be congruent quadrilaterals; that is, the bases must be congruent and the sides must be congruent. [Hint: Suppose they were not congruent. Show that you can then construct a rectangle using the quadrilateral with the longer sides.]

Exercise 7.5.5. Let l and m intersect at O at an acute angle. Let $A, B \neq O$ be points on l and drop perpendiculars to m from A and B, intersecting m at A', B'. If $OA < OB$, show that $AA' < BB'$.

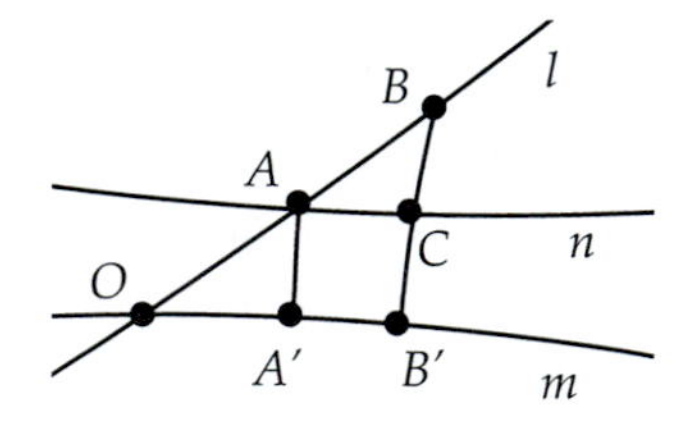

[Hint: Argue that the perpendicular to $\overline{AA'}$ at A must intersect $\overline{BB'}$ and use Lambert Quadrilateral results.]

Exercise 7.5.6. Show that parallel lines cannot be everywhere equidistant in hyperbolic geometry. [Hint: Suppose l, m are two parallel lines and that at points A, B, C on l, the distance to m (measured along a perpendicular to m) is the same. Derive a contradiction using Saccheri Quadrilaterals.]

Exercise 7.5.7. Let l be a line and m a limiting parallel to l through a point P. Show that the perpendicular distance from P to l decreases as you move P along m in the direction of the omega point of l. [Hint: Use the angle of parallelism results from Exercises 7.3.11 and 7.3.12.]

Exercise 7.5.8. Show that if l and m are limiting parallel, then they cannot have a common perpendicular.

Exercise 7.5.9. Show that two hyperbolic lines cannot have more than one common perpendicular.

Exercise 7.5.10. In the Poincaré model, show that two parallel lines that are not limiting parallel must have a common perpendicular. [Hint: Argue that you can assume one line is the x-axis, and then show that you can find the point on the other line (using Euclidean circle geometry) to form a common perpendicular.] Note: This result is true for any model of hyperbolic geometry. For the proof, see [17, page 158].

Exercise 7.5.11. Prove that two Saccheri Quadrilaterals with equal bases and equal summit angles must be congruent. [Hint: Suppose they were not congruent. Show that you can then construct a quadrilateral having angle sum equal to 360.]

Exercise 7.5.12. Given ΔABC let l be a cevian line (a line through a vertex and an opposite side). The cevian line divides the triangle into two sub-triangles. Show that the defect of ΔABC is equal to the sum of the defects of the component sub-triangles. (This result suggests that the defect works much like the concept of area for a hyperbolic triangle.)

Exercise 7.5.13. Is it possible to construct scale models of a figure in hyperbolic geometry? Briefly explain your answer.

7.6 Area in Hyperbolic Geometry

In Chapter 2 we saw that areas in Euclidean geometry were defined in terms of figures that were *equivalent*. Two figures are equivalent if the figures can be split up (or subdivided) into a finite number of pieces so that pairs of corresponding pieces are congruent. By using the notion of equivalence, we were able to base all Euclidean area calculations on the simple figure of a rectangle.

In hyperbolic geometry this is unfortunately not possible, as rectangles do not exist! Thus, we need to be a bit more careful in building up a notion of area. We will start with some defining axioms of how area works.

Area Axiom I If A, B, C are distinct and not collinear, then the area of triangle ABC is positive.

Area Axiom II The area of equivalent sets must be the same.

Area Axiom III The area of the union of disjoint sets is the sum of the separate areas.

These are reasonable axioms for area in hyperbolic geometry. Note that Axiom II automatically implies that the area of congruent sets is the same.

Since area is axiomatically based on triangles, we will need the following result.

Theorem 7.19. *Two triangles ABC and $A'B'C'$ that have two sides congruent, and the same defect, are equivalent and thus have the same area.*

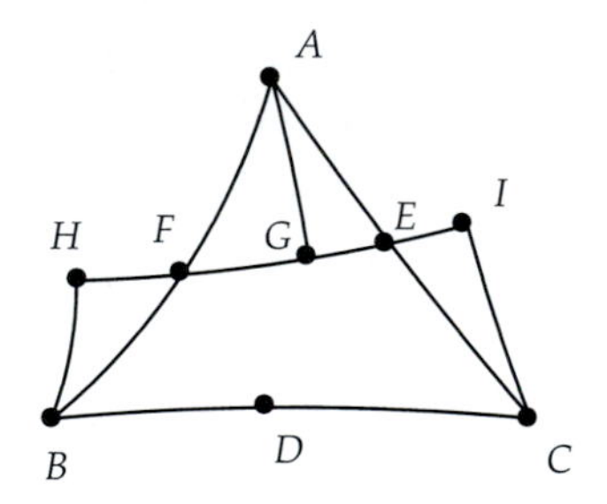

Fig. 7.12

Proof: Suppose that $\overline{BC} \cong \overline{B'C'}$. Let D, E, and F be the midpoints of sides $\overline{BC}$, $\overline{AC}$, and $\overline{AB}$, as shown in Fig. 7.12. Construct the line through F and E and drop perpendiculars to this line from A, B, and C meeting the line at G, H, and I, respectively.

Now, right triangles ΔBHF and ΔAGF will be congruent by AAS. Similarly, ΔAGE and ΔCIE will be congruent. Thus, $\overline{BH} \cong \overline{AG} \cong \overline{CI}$ and $BHIC$ is a Saccheri Quadrilateral. Also, it is clear that this quadrilateral is equivalent to the original triangle ΔABC (move appropriate pieces around).

Now, it might be the case that the positions of G, H, and I are switched, as shown in Fig. 7.13. However, one can easily check that in all configurations, $BHIC$ is a Saccheri Quadrilateral (with summit $\overline{BC}$) that is equivalent to ΔABC.

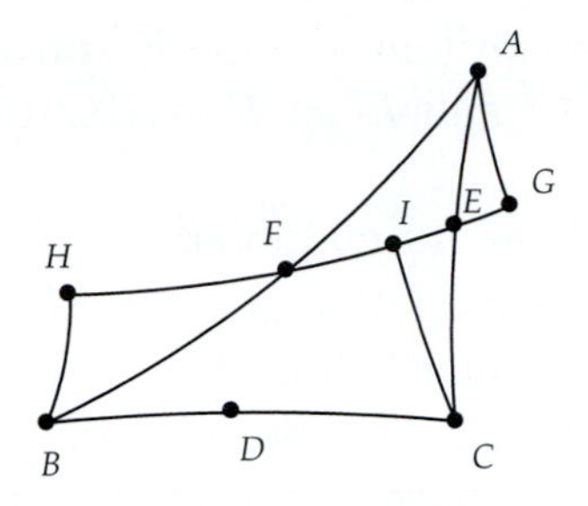

Fig. 7.13

Consider the summit angles of $BHIC$. The sum of these two summit angles must equal the angle sum of ΔABC, and thus the summit angles are each half of the angle sum of ΔABC.

Similarly, we can construct a Saccheri Quadrilateral for $\Delta A'B'C'$ that is equivalent to $\Delta A'B'C'$ and with summit $\overline{B'C'}$.

Since $\overline{BC} \cong \overline{B'C}'$ and the summit angles for both quadrilaterals are congruent (equal defects implies equal angle sums), then by Exercise 7.5.4 we know that the two quadrilaterals are congruent. Thus, the triangles are equivalent. □

We can prove an even more general result.

Theorem 7.20. *Any two triangles with the same defect are equivalent and thus have the same area.*

Proof: If one side of ΔABC is congruent to a side of $\Delta A'B'C'$, then we can use the previous theorem to show the result. So, suppose no side of one matches a side of the other. We can assume side $\overline{A'C'}$ is greater than $\overline{AC}$ (Fig. 7.14). As in the last theorem, let E, F be midpoints of $\overline{AC}$ and $\overline{AB}$. Construct the perpendiculars from B and C to the line through E and F, and construct the Saccheri Quadrilateral $BHIC$.

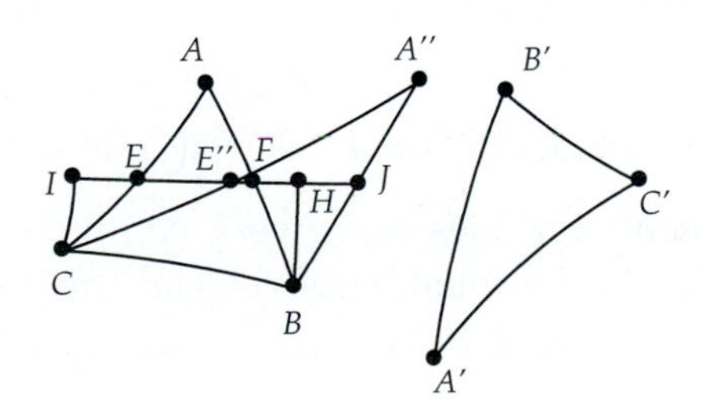

Fig. 7.14

Since $A'C' > AC$, we can find a point E'' on $\overleftrightarrow{EF}$ such that the length of $\overline{CE''}$ is half that of $\overline{A'C'}$ and $E'' \neq E$. Extend $\overline{CE''}$ to a point A'' such that $\overline{A''C} \cong \overline{A'C'}$.

Clearly, $\overleftrightarrow{EF}$ cuts $\overline{A''C}$ at its midpoint E''. It is left as an exercise to show that $\overleftrightarrow{EF}$ cuts $\overline{A''B}$ at its midpoint, point J. Then, as in the previous theorem, $\Delta A''BC$ will be equivalent to the Saccheri Quadrilateral $BHIC$ and thus also equivalent to ΔABC. Since triangles $\Delta A''BC$ and $\Delta A'B'C'$ share a congruent side, they must be equivalent, and hence the original two triangles are equivalent. □

We have shown that any two triangles having the same defect can be split into pieces that can be made congruent. Is the converse also true? Suppose we have two equivalent triangles. That is, the two triangles can be subdivided into sub-triangles with corresponding pairs congruent. Then for each pair, the defect will be the same. Also, if a set of sub-triangles forms a polygonal shape, it is easy to see that the defect of the polygon is equal to the sum of the defects of the triangles of which it is comprised. Thus, the defects of the original triangles will be the sum of the defects of all sub-triangles, which match pair-wise. We summarize this as follows.

Theorem 7.21. *Any two triangles that are equivalent (and thus have the same area) must have the same defect.*

Our conclusion from this exploration of hyperbolic area is that the defect and area share exactly the same properties. This tells us that the area must be a function of the defect. Since the area and defect are both linear functions in terms of triangle subdivisions, the area should be a linear function of the defect, and it must be positive. Therefore, $area = k^2 defect + c$, for some constants k and c. But, the area of the empty set must be zero, and thus $area = k^2 defect$. We can fix k by choosing some triangle to have unit area. We summarize this discussion in the following theorem.

Theorem 7.22. *If we have defined an area function for hyperbolic geometry satisfying Axioms I–III, then there is a positive constant k such that for any triangle ΔABC, we have*

$$area(\Delta ABC) = k^2 \; defect(\Delta ABC)$$

Note that our discussion does not give a complete proof of this result, only an argument for the reasonableness of the theorem. For a complete, rigorous proof see [32, pages 351–352].

Since the defect measures how much the angle sum of a triangle is *below* 180 and the lowest the angle sum can be is zero, we have the following corollary.

Corollary 7.23. *In hyperbolic geometry the area of a triangle is* at most $180k^2$ *(or πk^2, if we use radian measure for angles).*

Exercise 7.6.1. Show that in Theorem 7.20 $\overleftrightarrow{EF}$ cuts $\overline{A''B}$ at its midpoint J. [Hint: Suppose that the intersection of $\overleftrightarrow{EF}$ with $\overline{A''B}$ is not the midpoint. Connect E'' to the midpoint of $\overline{A''B}$; construct a second Saccheri quadrilateral; and then show that you can use the perpendicular bisector of $\overline{BC}$ to construct a triangle with more than 180 degrees.]

Exercise 7.6.2. Show that there is no finite triangle in hyperbolic geometry that achieves the maximum area bound.

Exercise 7.6.3. It is possible that our universe is hyperbolic in its geometry. Could you use the measurements of triangle areas on earth to determine if the universe were hyperbolic? Why or why not?

7.7 Project 11 - Tiling the Hyperbolic Plane

In Euclidean geometry there are just three different regular tessellations of the Euclidean plane—the ones generated by equilateral triangles, by squares, and by regular hexagons. How many regular tilings are there in hyperbolic geometry?

We can argue in a similar fashion as we did in the Euclidean tilings of Chapter 6. If we have k hyperbolic regular n-gons meeting at a common vertex of a tiling, then the interior angles of the n-gons would be $\frac{360}{k}$. Also, we can find a central point and triangulate each n-gon. The triangles in the triangulation will consist of isosceles triangles with base angles of $\alpha = \frac{360}{2k}$. The total angle sum of all the triangles in the triangulation will be *less* than $180n$. At the same time, this angle sum can be split into the angles around the center point (which add to 360) plus the base angles of the isosceles triangles (which add to $2n\alpha$). Thus, we get that

$$360 + 2n\alpha < 180n$$

Or,

$$2\alpha < 180 - \frac{360}{n}$$

Since the angle formed by adjacent edges of the n-gon is $2\alpha = \frac{360}{k}$, we get

$$\frac{360}{k} < 180 - \frac{360}{n}$$

Dividing both sides by 360 and then rearranging, we get

$$\frac{1}{n} + \frac{1}{k} < \frac{1}{2}$$

If there is a regular tessellation by n-gons meeting k at a vertex in hyperbolic geometry, then $\frac{1}{n} + \frac{1}{k} < \frac{1}{2}$. On the other hand, if this inequality is true, then a tiling with n-gons meeting k at a vertex must exist. Thus, this inequality completely characterizes regular hyperbolic tilings. We will call a regular hyperbolic tessellation of n-gons meeting k at a vertex a *(n,k)* tiling.

As an example, let's see how to generate a $(5, 4)$ tiling. In a (5,4) tiling, we have regular pentagons meeting four at a vertex. How do we construct regular pentagons of this kind? First of all, it is clear that the interior angle of the pentagon must be 90 degrees ($\frac{360}{4}$). If we take such a pentagon and triangulate it via triangles constructed to a central point, the angles about the central point will be 72 degrees and the base angles of the isosceles triangles will be 45 degrees (half the interior angle). Thus, to build the pentagon we need to construct a hyperbolic triangle with angles of 72, 45, and 45.

In Euclidean geometry, there are an infinite number of triangles that have a specified set of three angles, and these triangles are all similar to each other. In hyperbolic geometry, two triangles with congruent pairs of angles must be congruent themselves!

Thus, we know that a hyperbolic triangle with angles of 72, 45, and 45 degrees must be unique. *Geometry Explorer* has a built-in tool for building hyperbolic triangles with specified angles.

To build the interior triangle for our pentagon, we will first need a point at the origin. *Geometry Explorer* has a special menu in the hyperbolic main window titled **Misc**. We will use two options under this menu—one to create a point at the origin and the other to assist in creating a 72,45,45 triangle.

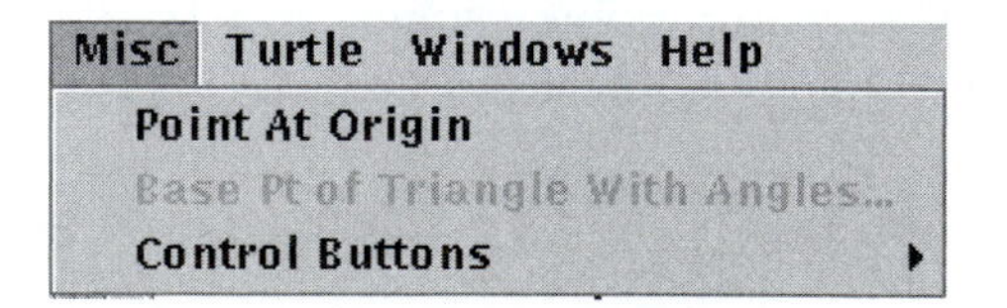

Choose **Point At Origin** (**Misc** menu) to create a point at the origin in the Poincaré plane.

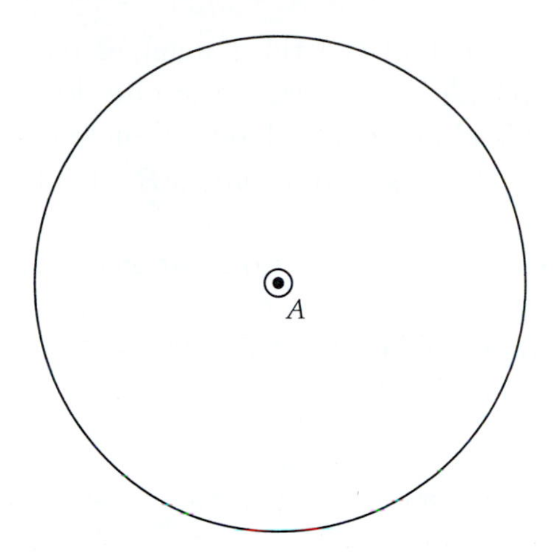

To create our triangle, we will first create the base segment of a triangle with angles of 72, 45, and 45. Select the point at the origin and choose **Base Pt of Triangle with Angles...**(**Misc** menu). A dialog box will pop up as shown. Note how the angles α, β, and γ are designated. In our example we want $\alpha = 72$ and the other two angles to be 45. Type these values in and hit "Okay."

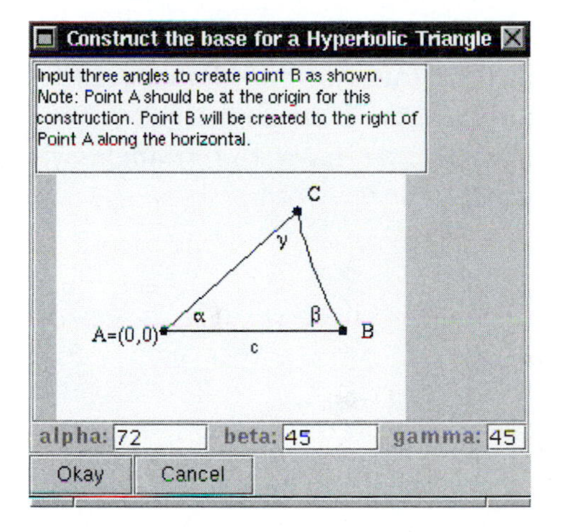

In the Canvas, a new point will be created that corresponds to point B in the dialog box just discussed. Create a line through points A and B as shown.

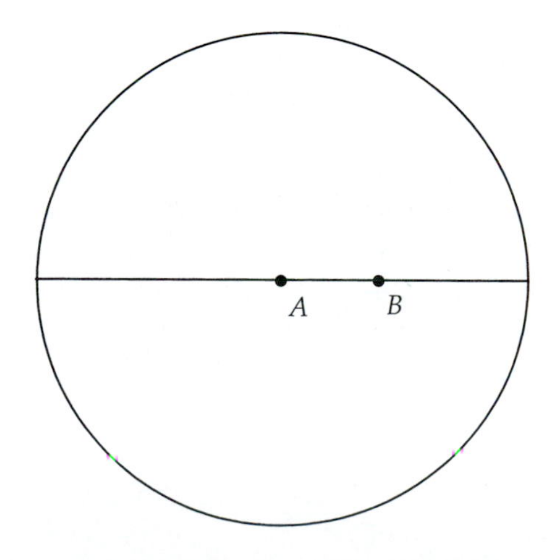

To find the third point of our triangle, we need to rotate line $\overleftrightarrow{AB}$ about point A by an angle of 72 degrees and rotate $\overleftrightarrow{AB}$ about B by an angle of -45 degrees. Carry out these two rotations and then multi-select the two rotated lines and construct the intersection point, point C.

Now, hide the three lines and any extraneous line points. Let's measure the three angles just to verify that we have the triangle we want.

Angle(B,A,C) = 72.00 degrees
Angle(A,C,B) = 45.00 degrees
Angle(C,B,A) = 45.00 degrees

Looks good. Next, hide the angle measurements and connect C and B with a segment. Then, select A as a center of rotation and define a custom rotation of 72 degrees. Rotate segment $\overline{BC}$ four times to get a regular pentagon.

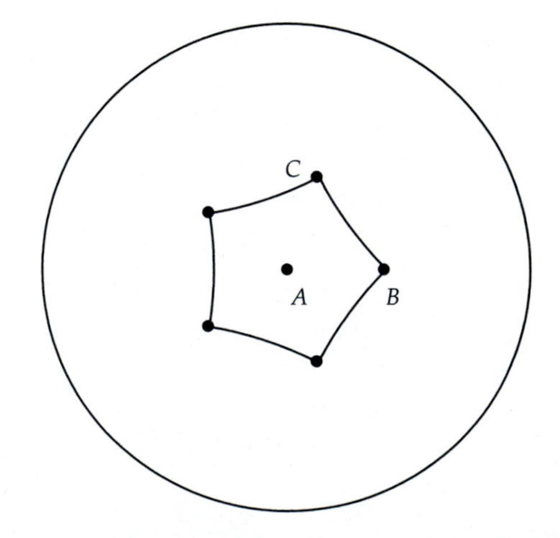

Finally, select point C as a center of rotation and define a new custom rotation of 90 degrees. Then, rotate the pentagon three times, yielding four pentagons meeting at right angles!

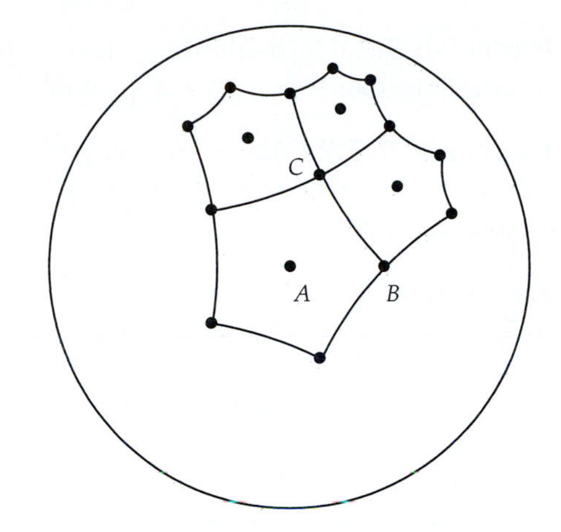

If we continue to rotate the pentagon about exterior points in this figure, we see that a tiling of the hyperbolic plane is indeed possible with regular pentagons meeting at right angles.

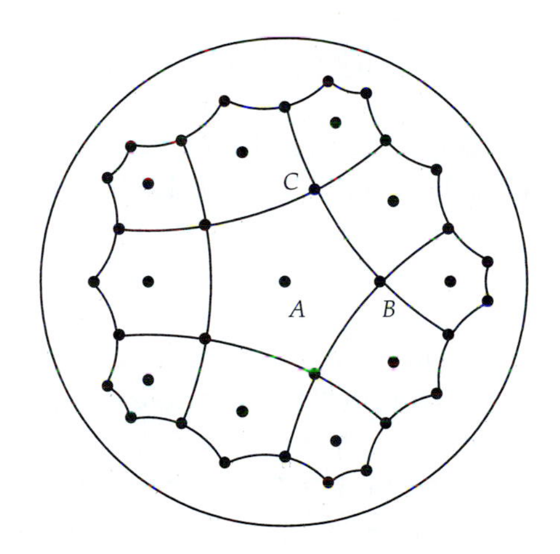

However, if we move the point at the origin, we see that the tiling breaks up in a quite nasty way. Why is this the case? The problem here is that by translating the origin point, we have created compound translations and rotations for other parts of the figure. In hyperbolic geometry compositions of translations are not necessarily translations again as they are in Euclidean geometry.

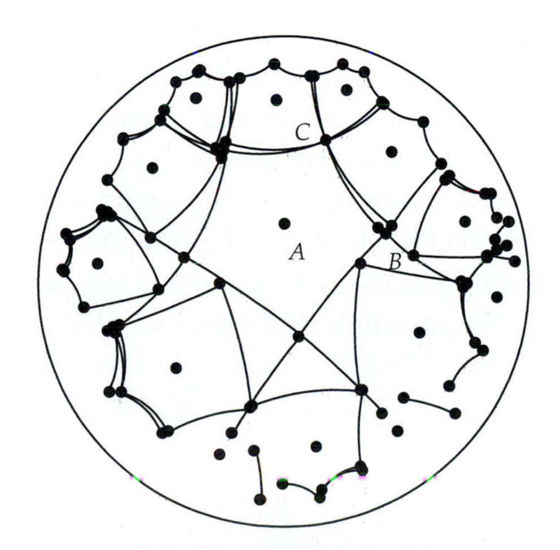

Exercise 7.7.1. In elliptic geometry the sum of the angles in a triangle is more than 180 degrees. Show that it is possible to have a $(3, 3)$ tiling in Elliptic geometry.

Exercise 7.7.2. There are only three regular tilings (up to scaling) in Euclidean planar geometry. How many regular tilings are there in hyperbolic (two-dimensional) geometry?

Exercise 7.7.3. Describe and illustrate the steps you would take to construct a (6,5) tiling in the Poincaré Plane.

For your report give a careful and complete summary of your work done in this project.

7.8 Models and Isomorphism

The two models of hyperbolic geometry covered so far in this chapter, the Poincaré model and the Klein model, are very similar. In fact, there is a conformal map taking one to the other that maps lines to lines, angles to angles, and the distance function of one model to the distance function of the other. The existence of such a function implies that the two models are *isomorphic*—they have identical geometric properties.

We will construct a one-to-one map from the Klein model to the Poincaré model as follows.

First, construct the sphere of radius 1 given by $x^2+y^2+z^2 = 1$. Consider the unit disk $(x^2 + y^2 = 1)$ within this sphere to be the Klein disk.

Let N be the north pole of the sphere and let P be a point on the Klein line l as shown. Project P orthogonally downward to the bottom of the sphere, yielding point Q. Then, stereographically project from N, using the line from N to Q, yielding point P' in the unit disk.

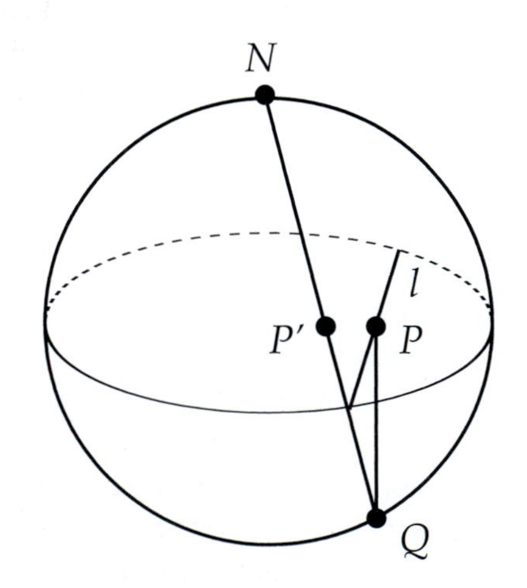

Define a function F from the Klein disk to the Poincaré disk by $F(P) = P'$, where P is a point in the Klein disk and P' is the unique point defined by the construction above. Since projection is one-to-one from the disk onto the lower hemisphere, and since stereographic projection is also one-to-one and onto from the lower hemisphere back to the unit disk, then the map F is a one-to-one, onto function from the Klein disk to the Poincaré disk.

The inverse to F, which we will denote by F', is the map that takes P' to P. That is, it projects P' onto the sphere and then projects this spherical point up to the disk, to point P. From our work in Chapter 3 on stereographic projection, we know that the equation for F' will be

$$F'(x, y) = (\frac{2x}{1 + x^2 + y^2}, \frac{2y}{1 + x^2 + y^2})$$

How do the maps F and F' act on lines in their respective domains?

Let l be a Klein line. Projecting l orthogonally downward will result in a circular arc c on the sphere that meets the unit circle (equator) at right angles. Since stereographic projection preserves angles and maps circles to circles, stereographic projection of c back to the unit disk will result in a circular arc that meets the unit circle at right angles—a Poincaré line.

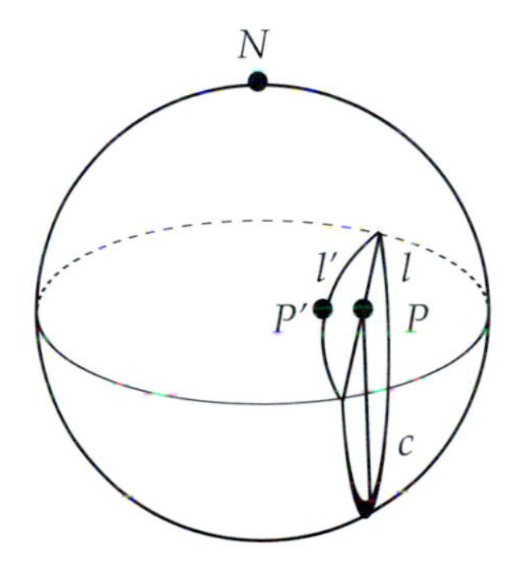

We see, then, that F maps Klein lines to Poincaré lines, *and* it preserves the ideal points of such lines, as the points where such lines meet the unit circle are not moved by F. Thus, F preserves the notions of *point* and *line* between the two models.

What about the notion of angle? Let's review the construction of a perpendicular in the Klein model. Recall that the *pole* of chord $\overline{AB}$ in a circle c is the inverse point of the midpoint of $\overline{AB}$ with respect to the circle. We defined a Klein line m to be perpendicular to a Klein line l based on whether l was a diameter of the Klein disk. If it is a diameter, then m is perpendicular to l if it is perpendicular to l in the Euclidean sense. If l is not a diameter, then m is perpendicular to l if the Euclidean line for m passes through the pole P of l.

Here are two lines $l = \overline{AB}$ and m that are perpendicular in the Klein model.

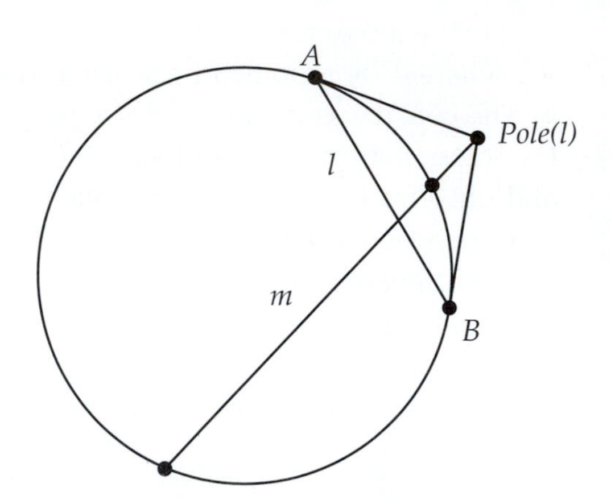

Since the pole of chord $\overline{AB}$ is also the intersection of the tangents at A and B to the circle, we see that the pole of $\overline{AB}$ will also be the *center* of the circle passing through A and B that is orthogonal to the unit circle. That is,

Lemma 7.24. *The pole of the Klein line $l = \overline{AB}$ will be the center of the orthogonal circle defining the Poincaré line $F(l)$.*

We can now prove that F preserves orthogonality between the models.

Theorem 7.25. *Two Klein lines l and m are perpendicular if and only if the corresponding Poincaré lines $F(l)$ and $F(m)$ are perpendicular.*

Proof: There are three cases to consider. First, suppose l and m are both diameters. Then, $F(l)$ and $F(m)$ are both diameters, and perpendicularity has the same (Euclidean) definition in both models.

Second, suppose l is a diameter and m is not and suppose that l and m are perpendicular. Then, the diameter l bisects chord m and thus passes through the pole of m. Now, $F(l) = l$ and since the center of $F(m)$ is the pole of l, $F(l)$ must pass through the center of $F(m)$. This implies that l is orthogonal to the tangent line to $F(m)$ at the point where it intersects $l = F(l)$ (Fig. 7.15), and so $F(l)$ is orthogonal to $F(m)$. Reversing this argument shows that if $F(l)$ and $F(m)$ are orthogonal, then so are l and m.

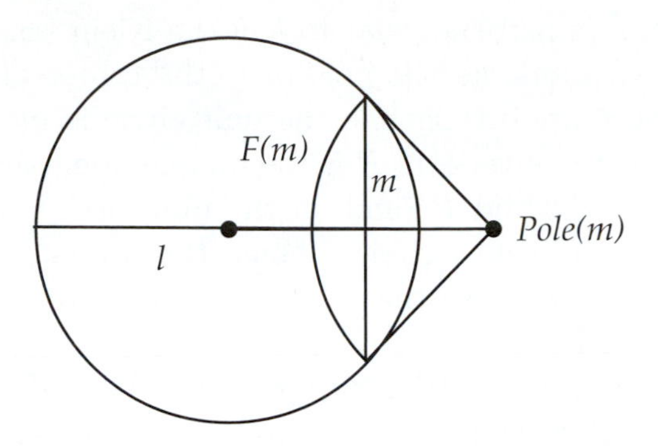

Fig. 7.15

Last, suppose both l and m are not diameters of the unit circle. Then, l passes through the pole of m and vice versa. Also, these poles are the centers of $F(l)$ and $F(m)$ (Fig. 7.16).

Suppose $F(m)$ is perpendicular to $F(l)$ (in the Poincaré sense). Let P and Q be the points where $F(m)$ meets the unit circle. Let c be the circle on which $F(l)$ lies and c' be the circle on which $F(m)$ lies. By Corollary 2.40, we know that inversion of c', and the unit circle, through c will switch P and Q. This is due to the fact that both c' and the unit circle are orthogonal to c, and so both will be mapped to themselves by inversion in c. Thus, P and Q are inverse points with respect to c, and the line through P and Q must go through the center of c (the pole of l). Then, m is perpendicular to l in the Klein sense.

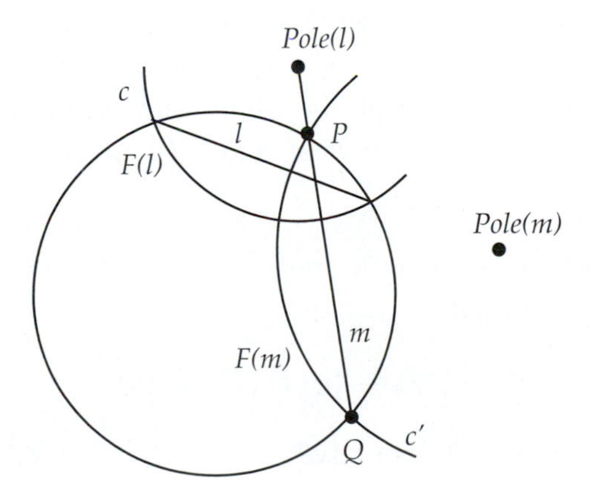

Fig. 7.16

Conversely, if m is perpendicular to l in the Klein sense, then the (Euclidean) line for m intersects the pole of l (the center of circle c). Since the unit circle and c are orthogonal, the unit circle is mapped to itself by inversion in c. But, the inverse to P is the unique point that lies on the ray through the center of c and P (and on the unit circle) that gets mapped to another point on the unit circle. Thus, the inverse to P must be Q, and by Theorem 2.39 we have that c' is orthogonal to c, and the lines are perpendicular in the Poincaré sense. □

Since F preserves the definition of right angles between the Poincaré and Klein models, this gives us a natural way to define *all* angles in the Klein model: we will define the measure of a Klein angle to be the value of the Poincaré angle it corresponds to.

Definition 7.13. Given three points A', B', C' in the Klein disk, the measure of the Klein angle defined by these points is the value of $\angle ABC$ in the Poincaré model, where $F(A') = A$, $F(B') = B$, and $F(C') = C$.

With this definition, F is a one-to-one map from the Klein model to the Poincaré model that preserves points and lines. F will be an isomorphism between these models if we can show that F preserves the distance functions of the models.

To show this, we will borrow a couple of results from the next chapter on isometries in hyperbolic geometry. Just as there are Euclidean isometries that will take any Euclidean line to the x-axis, so there are transformations in the Klein and Poincaré models that preseve the Klein and Poincaré distance functions and that map any line in these models to the diameter of the unit disk.

If we assume this property of the two models, then to show F preserves the distance functions of the models, it is enough to show that F preserves distances for hyperbolic lines that lie along the x-axis in the unit circle. Equivalently, we can show the inverse map F' preserves distances along such lines.

Theorem 7.26. *Let $P = (x_1, 0)$ and $Q = (x_2, 0)$ be two points in the Poincaré disk. Then*

$$d_P(P,Q) = d_K(F'(P), F'(Q))$$

Proof: By definition of d_P we have

$$d_P(P,Q) = \left| \ln\left(\frac{1+x_1}{1-x_1}\frac{1-x_2}{1+x_2}\right) \right|$$

Also, since stereographic projection maps any diameter of the unit circle to itself, we have

$$d_K(F'(P), F'(Q)) = \frac{1}{2}\left|\ln(\frac{1+u_1}{1-u_1}\frac{1-u_2}{1+u_2})\right|$$

where $F'(P) = (u_1, 0)$ and $F'(Q) = (u_2, 0)$.

We know that $F'(x, 0) = (\frac{2x}{1+x^2}, 0)$. So,

$$\begin{aligned} 1 \pm u_1 &= 1 \pm \frac{2x_1}{1+x_1^2} \\ &= \frac{(1 \pm x_1)^2}{1+x_1^2} \end{aligned}$$

and similarly for u_2.

Thus, we get

$$\begin{aligned} d_K(F'(P), F'(Q)) &= \frac{1}{2}\left|\ln(\frac{\frac{(1+x_1)^2}{1+x_1^2}}{\frac{(1-x_1)^2}{1+x_1^2}}\frac{\frac{(1-x_2)^2}{1+x_2^2}}{\frac{(1+x_2)^2}{1+x_2^2}})\right| \\ &= \frac{1}{2}\left|\ln((\frac{1+x_1}{1-x_1}\frac{1-x_2}{1+x_2})^2)\right| \\ &= \frac{1}{2}2\left|\ln(\frac{1+x_1}{1-x_1}\frac{1-x_2}{1+x_2})\right| \\ &= \left|\ln(\frac{1+x_1}{1-x_1}\frac{1-x_2}{1+x_2})\right| \\ &= d_P(P, Q) \end{aligned}$$

□

We conclude that the map F is an isomorphism of the Klein and Poincaré models. That is, any geometric property valid in one of these models must be valid in the other model and vice versa.

Chapter 8

Non-Euclidean Transformations

> Though the text of your article on 'Crystal Symmetry and Its Generalizations' is much too learned for a simple, self-made pattern man like me, some of the text-illustrations and especially Figure 7, page 11, gave me quite a shock . . .
>
> If you could give me a simple explanation how to construct the following circles, whose centres approach gradually from the outside till they reach the limit, I should be immensely pleased and very thankful to you! Are there other systems besides this one to reach a circle limit?
>
> Nevertheless I used your model for a large woodcut (of which I executed only a sector of 120 degrees in wood, which I printed 3 times). I am sending you a copy of it.
>
> —M. C. Escher (1898–1972), from a letter to H. S. M. Coxeter, as reported in [10]

The illustration that gave Escher "quite a shock" was a drawing that Coxeter had produced of a regular tiling of the Poincaré disk by triangles. This tiling inspired Escher to create his "Circle Limit I" woodcut. A version of this image created by Doug Dunham is shown in Fig. 8.1, with a few of the basic triangle tiles outlined in bold arcs. For more on Dunham's work on Escher-like hyperbolic tilings, see [13].

Fig. 8.1 Dunham's Version of Circle Limit I

The tiling illustrated in this figure is a *regular* tiling—all triangles used to build the tiling are congruent via *hyperbolic* transformations; that is, one-to-one and onto functions of the Poincaré disk to itself that preserve the Poincaré distance function.

In Chapter 5 we saw that the group of distance-preserving functions of the Euclidean plane consisted of Euclidean reflections, rotations, translations, and glide reflections. Such isometries not only preserved lengths of segments, but preserved angles as well. We also saw that the set of all Euclidean isometries formed an algebraic structure called a *group*. Before we study the nature of hyperbolic isometries, we will look at an alternate way to represent Euclidean isometries that will be useful in defining hyperbolic isometries.

Each element of the group of Euclidean isometries can be represented by a complex function. For example, the function $f(z) = z + (v_1 + iv_2)$ represents translation by the vector $v = (v_1, v_2)$, and $g(z) = e^{i\phi}z$ represents a rotation about the origin by an angle ϕ. Rotations about other points in the plane, say rotation about a by an angle of ϕ, can be defined by a sequence of isometries: first, translation by the vector $-a$, then rotation by ϕ about the origin, and then translation by the vector a. Thus, the desired rotation can be represented as $h(z) = (e^{i\phi}(z-a)) + a = e^{i\phi}z + (a - e^{i\phi}a)$.

Any translation or rotation can be represented as a complex function of the form

$$f(z) = e^{i\phi}z + b \tag{8.1}$$

with b complex and ϕ real. These are the orientation-preserving isometries of the Euclidean plane. The set of all such functions, which are sometimes called *rigid motions* of the plane, forms a group called the *Euclidean group.*

What are the analogous orientation- and distance-preserving functions in hyperbolic geometry? In particular, what are the orientation- and distance-preserving functions in the Poincaré model? Since all rigid Euclidean isometries can be realized as certain one-to-one and onto complex functions, a good place to look for hyperbolic transformations might be in the *entire* class of one-to-one and onto complex functions.

But, which functions should we consider? Since we are concentrating on the Poincaré model, we need to find one-to-one and onto orientation-preserving functions that preserve the Euclidean notion of angle, but do not preserve Euclidean length. In section 3.5, we studied functions that preserved angles and preserved the *scale* of Euclidean lengths *locally.* Such functions were called *conformal* maps. Euclidean rigid motions such as rotations and translations preserve angles and length *globally* and so comprise a subset of all conformal maps.

If we consider the entire set of conformal maps of the plane onto itself, then by Theorem 3.17 such maps must have the form $f(z) = az + b$, where $a \neq 0$ and b is a complex constant. Since $a = |a|e^{i\phi}$, then f is the composition of a translation, a rotation, and a scaling by $|a|$. Thus, f maps figures to *similar* figures. The set of all such maps forms a group called the group of *similitudes* or *similarity transformations* of the plane. If $b = 0$ $(f(z) = az, a \neq 0)$, we call f a *dilation* of the plane. Most similarity transformations cannot be isometries of the Poincaré model since most similarities (like translations and scalings) do not fix the boundary circle of the Poincaré disk.

Clearly, we must expand our set of possible transformations. One way to do this is to consider the set of all one-to-one and onto conformal maps of the *extended* complex plane to itself. In Theorem 3.18 we saw that these maps have the form

$$f(z) = \frac{az+b}{cz+d}, \quad ad - bc \neq 0 \tag{8.2}$$

8.1 Möbius Transformations

Definition 8.1. A *Möbius transformation* is a function on the extended complex plane defined by equation 8.2. The set of Möbius transformations forms a group called the *Möbius group.*

Every Möbius transformation is composed of simpler transformations.

Theorem 8.1. *Let T be a Möbius transformation. Then T is the composition of translations, dilations, and inversion ($g(z) = \frac{1}{z}$).*

Proof: If $c = 0$, then $f(z) = \frac{a}{d}z + \frac{b}{d}$, which is the composition of a translation with a dilation.

If $c \neq 0$, then $f(z) = \frac{a}{c} - \frac{ad-bc}{c^2}\frac{1}{z+\frac{d}{c}}$. Thus, f is the composition of a translation (by $\frac{d}{c}$), an inversion, a dilation (by $-\frac{ad-bc}{c^2}$), and a translation (by $\frac{a}{c}$). $\square$

Note that the Möbius group includes the group of Euclidean rigid motions ($a = 1, c = 0, d = 1$), and the group of similarities ($a \neq 0, c = 0, d = 1$) as subgroups. Also note that we could define Möbius transformations as those transformations of the form in equation 8.2 with $ad - bc = 1$, by dividing the numerator by an appropriate factor.

Within the group of Möbius transformations, can we find a subgroup that will serve as the group of orientation-preserving isometries for the Poincaré model? We shall see that there is indeed such a sub-group. Before we can prove this, we need to develop a toolkit of basic results concerning Möbius transformations.

8.1.1 Fixed Points and the Cross Ratio

How many fixed points can a Möbius transformation f have? Suppose $f(z) = z$. Then

$$z = \frac{az+b}{cz+d}$$

So, $cz^2 + (d-a)z - b = 0$. This equation has at most two roots. Thus, we have

Lemma 8.2. *If a Möbius transformation f has three or more fixed points, then $f = id$, where id is the identity Möbius transformation.*

We saw in Chapter 5 that an isometry is uniquely defined by its effect on three non-collinear points. For Möbius transformations we can relax the condition on collinearity.

Theorem 8.3. *Given any three distinct complex numbers z_1, z_2, z_3, there is a unique Möbius transformation f that maps these three values to a specified set of three distinct complex numbers w_1, w_2, w_3.*

Proof: Let $g_1(z) = \frac{z-z_2}{z-z_3}\frac{z_1-z_3}{z_1-z_2}$. Then g_1 is a Möbius transformation (the proof is an exercise) and g_1 maps z_1 to 1, z_2 to 0, and z_3 to the point at infinity.

Let $g_2(w) = \frac{w-w_2}{w-w_3}\frac{w_1-w_3}{w_1-w_2}$. We see that g_2 is a Möbius transformation mapping w_1 to 1, w_2 to 0, and w_3 to ∞.

Then $f = g_2^{-1} \circ g_1$ will map z_1 to w_1, z_2 to w_2, and z_3 to w_3.

Is f unique? Suppose f' also mapped z_1 to w_1, z_2 to w_2, and z_3 to w_3. Then $f^{-1} \circ f'$ has three fixed points, and so $f^{-1} \circ f' = id$ and $f' = f$. □

Corollary 8.4. *If two Möbius transformations f, g agree on three distinct points, then $f = g$.*

Proof: This is an immediate consequence of the preceding theorem. □

The functions g_1 and g_2 used in the proof of Theorem 8.3 are called *cross ratios.*

Definition 8.2. The cross ratio of four complex numbers z_0, z_1, z_2, and z_3 is denoted by (z_0, z_1, z_2, z_3) and is the value of

$$\frac{z_0 - z_2}{z_0 - z_3}\frac{z_1 - z_3}{z_1 - z_2}$$

The cross ratio is an important invariant of the Möbius group.

Theorem 8.5. *If z_1, z_2, and z_3 are distinct points and f is a Möbius transformation, then $(z, z_1, z_2, z_3) = (f(z), f(z_1), f(z_2), f(z_3))$ for any z.*

Proof: Let $g(z) = (z, z_1, z_2, z_3)$. Then $g \circ f^{-1}$ will map $f(z_1)$ to 1, $f(z_2)$ to 0, and $f(z_3)$ to ∞. But, $h(z) = (z, f(z_1), f(z_2), f(z_3))$ also maps $f(z_1)$ to 1, $f(z_2)$ to 0, and $f(z_3)$ to ∞. Since $g \circ f^{-1}$ and h are both Möbius transformations, and both agree on three points, then $g \circ f^{-1} = h$. Since $g \circ f^{-1}(f(z)) = (z, z_1, z_2, z_3)$ and $h(f(z)) = (f(z), f(z_1), f(z_2), f(z_3))$, the result follows. □

8.1.2 Geometric Properties of Möbius Transformations

Of particular interest to us will be the effect of a Möbius transformation on a circle or line.

Definition 8.3. A subset of the plane is a *cline* if it is either a circle or a line.

The cross ratio can be used to identify clines.

Theorem 8.6. *Let z_0, z_1, z_2, and z_3 be four distinct points. Then the cross ratio (z_0, z_1, z_2, z_3) is real if and only if the four points lie on a cline.*

Proof: Let $f(z) = (z, z_1, z_2, z_3)$. Then since f is a Möbius transformation, we can write

$$f(z) = \frac{az+b}{cz+d}$$

Now $f(z)$ is real if and only if

$$\frac{az+b}{cz+d} = \frac{\overline{az}+\overline{b}}{\overline{cz}+\overline{d}}$$

Multiplying this out, we get

$$(a\overline{c} - c\overline{a})|z|^2 + (a\overline{d} - c\overline{b})z - (d\overline{a} - b\overline{c})\overline{z} + (b\overline{d} - d\overline{b}) = 0 \tag{8.3}$$

If $(a\overline{c} - c\overline{a}) = 0$, let $\alpha = (a\overline{d} - c\overline{b})$ and $\beta = b\overline{d}$. Equation 8.3 simplifies to

$$Im(\alpha z + \beta) = 0$$

This is the equation of a line (proved as an exercise).

If $(a\overline{c} - c\overline{a}) \neq 0$, then dividing through by this term we can write equation 8.3 in the form

$$|z|^2 + \frac{a\overline{d} - c\overline{b}}{a\overline{c} - c\overline{a}}z - \frac{d\overline{a} - b\overline{c}}{a\overline{c} - c\overline{a}}\overline{z} + \frac{b\overline{d} - d\overline{b}}{a\overline{c} - c\overline{a}} = 0$$

Let $\gamma = \frac{a\overline{d}-c\overline{b}}{a\overline{c}-c\overline{a}}$ and $\delta = \frac{b\overline{d}-d\overline{b}}{a\overline{c}-c\overline{a}}$. Since $a\overline{c} - c\overline{a}$ is pure imaginary, we have that

$$\overline{\gamma} = (-)\frac{d\overline{a} - b\overline{c}}{a\overline{c} - c\overline{a}} = \frac{d\overline{a} - b\overline{c}}{c\overline{a} - a\overline{c}}$$

Equation 8.3 becomes

$$|z|^2 + \gamma z + \overline{\gamma z} + \delta = 0.$$

Or,

$$|z + \overline{\gamma}|^2 = -\delta + |\gamma|^2$$

After multiplying and regrouping on the right, we get

$$|z + \overline{\gamma}|^2 = \left|\frac{ad - bc}{a\overline{c} - c\overline{a}}\right|^2$$

Since $ad - bc \neq 0$, this gives the equation of a circle centered at $-\overline{\gamma}$. □

Theorem 8.7. *A Möbius transformation f will map clines to clines. Also, given any two clines c_1 and c_2, there is a Möbius transformation f mapping c_1 to c_2.*

Proof: Let c be a cline and let z_1, z_2, and z_3 be three distinct points on c. Let $w_1 = f(z_1)$, $w_2 = f(z_2)$, and $w_3 = f(z_3)$. These three points will lie on a line or determine a unique circle. Thus, w_1, w_2, and w_3 will lie on a cline c'. Let z be any point on c different than z_1, z_2, or z_3. By the previous theorem we have that (z, z_1, z_2, z_3) is real. Also, $(f(z), w_1, w_2, w_3) = (f(z), f(z_1), f(z_2), f(z_3)) = (z, z_1, z_2, z_3)$, and thus $f(z)$ is on the cline through w_1, w_2, and w_3.

For the second claim of the theorem, let z_1, z_2, and z_3 be three distinct points on c_1 and w_1, w_2, and w_3 be three distinct points on c_2. By Theorem 8.3 there is a Möbius transformation f taking z_1, z_2, z_3 to w_1, w_2, w_3. It follows from the first part of this proof that f maps all points on c_1 to points on c_2. □

So, Möbius transformations map circles to circles. They also preserve *inversion* through circles. Recall from Chapter 2 that the inverse of a point P with respect to a circle c centered at O is the point P' on the ray $\overrightarrow{OP}$ such that $(\overline{OP'})(\overline{OP}) = r^2$, where r is the radius of c (Fig. 8.2).

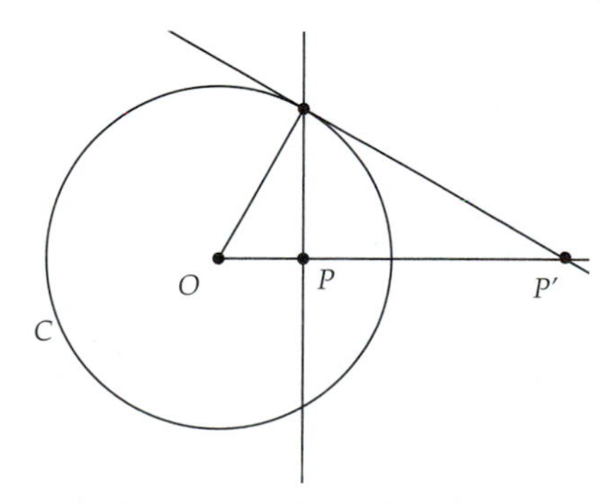

Fig. 8.2 Circle Inversion

If $z = P$ and $a = O$, then the defining equation for the inverse z^* of z with respect to a circle c with radius r is

$$|z^* - a||z - a| = r^2$$

Since z^* is on the ray through a and z, we get that $z^* - a = r_1 e^{i\theta}$ and $z - a = r_2 e^{i\theta}$, and $|z^* - a||z - a| = r_1 * r_2 = r_1 e^{i\theta} r_2 e^{-i\theta} = (z^* - a)(\overline{z} - \overline{a})$. Thus,

$$z^* - a = \frac{r^2}{\overline{z} - \overline{a}} \tag{8.4}$$

It turns out that inversion can also be defined using the cross ratio.

Lemma 8.8. *Let z_1, z_2, and z_3 be distinct points on a circle c. Then, z^* is the inverse of z with respect to c if and only if $(z^*, z_1, z_2, z_3) = \overline{(z, z_1, z_2, z_3)}$.*

Proof: Let c have center a and radius r. Then, since the cross ratio is invariant under translation by $-a$, we have

$$\begin{aligned}
\overline{(z, z_1, z_2, z_3)} &= \overline{(z - a, z_1 - a, z_2 - a, z_3 - a)} \\
&= (\overline{z - a}, \overline{z_1 - a}, \overline{z_2 - a}, \overline{z_3 - a}) \\
&= (\overline{z} - \overline{a}, \frac{r^2}{z_1 - a}, \frac{r^2}{z_2 - a}, \frac{r^2}{z_3 - a})
\end{aligned}$$

Since the cross ratio is invariant under the transformation $f(z) = \frac{r^2}{z}$, we have

$$\overline{(z, z_1, z_2, z_3)} = (\frac{r^2}{\overline{z} - \overline{a}}, z_1 - a, z_2 - a, z_3 - a)$$

Finally, translation by a yields

$$\overline{(z, z_1, z_2, z_3)} = (a + \frac{r^2}{\overline{z} - \overline{a}}, z_1, z_2, z_3)$$

So, if $(z^*, z_1, z_2, z_3) = \overline{(z, z_1, z_2, z_3)}$, we see immediately that $z^* = a + \frac{r^2}{\overline{z} - \overline{a}}$ and z^* is the inverse to z.

On the other hand, if $z^* = a + \frac{r^2}{\overline{z} - \overline{a}}$, then $(z^*, z_1, z_2, z_3) = \overline{(z, z_1, z_2, z_3)}$.

□

Definition 8.4. Two points z and z^* are *symmetric with respect to a circle* c if $(z^*, z_1, z_2, z_3) = \overline{(z, z_1, z_2, z_3)}$ for points z_1,z_2,z_3 on c.

By the Lemma, this definition is not dependent on the choice of points z_1,z_2,z_3.

Theorem 8.9 (The Symmetry Principle). *If a Möbius transformation f maps circle c to circle c', then it maps points symmetric with respect to c to points symmetric with respect to c'.*

Proof: Let z_1,z_2,z_3 be on c. Since

$$\begin{aligned}(f(z^*), f(z_1), f(z_2), f(z_3)) &= (z^*, z_1, z_2, z_3) \\ &= \overline{(z, z_1, z_2, z_3)} \\ &= \overline{(f(z), f(z_1), f(z_2), f(z_3))}\end{aligned}$$

the result follows. □

8.2 Isometries in the Poincaré Model

We now return to our quest of finding one-to-one and onto maps of the Poincaré disk to itself that are orientation-preserving and will preserve the Poincaré distance function. Our earlier idea was to search within the group of Möbius transformations for such functions. It is clear that any candidate Möbius transformation must map the Poincaré disk to itself and so must leave the boundary (unit) circle invariant.

Theorem 8.10. *A Möbius transformation f mapping $|z| < 1$ onto $|w| < 1$ and $|z| = 1$ onto $|w| = 1$ has the form*

$$f(z) = \beta\frac{z-\alpha}{\overline{\alpha}z - 1}$$

where $|\alpha| < 1$ and $|\beta| = 1$.

Proof: Let α be the point which gets sent to 0 by f. Then by equation 8.4, the inverse to $z = \alpha$, with respect to the unit circle, is the point $z^* = \frac{1}{\overline{\alpha}}$. By Theorem 8.9, this inverse point gets sent to the inverse of 0, which must be ∞.

If $\alpha \neq 0$, then $c \neq 0$ (∞ does not map to itself), and

$$f(z) = \left(\frac{a}{c}\right)\frac{z + \frac{b}{a}}{z + \frac{d}{c}}$$

Since α maps to 0, then $\frac{b}{a} = -\alpha$ and since $\frac{1}{\overline{\alpha}}$ gets mapped to ∞, then $\frac{d}{c} = -\frac{1}{\overline{\alpha}}$. Letting $\beta = \overline{\alpha}\left(\frac{a}{c}\right)$, we get

$$f(z) = \beta\frac{z-\alpha}{\overline{\alpha}z - 1}$$

Now $1 = |f(1)| = |\beta|\frac{|1-\alpha|}{|\overline{\alpha}-1|} = |\beta|$. Thus, $|\beta| = 1$.

If $\alpha = 0$, then $f(z) = \frac{a}{d}z + \frac{c}{d}$. Since $f(0) = 0$, $\frac{c}{d} = 0$, and since $|f(1)| = 1$, we get $|\beta| = |\frac{a}{d}| = 1$. □

Will transformations of the form given in this theorem preserve orientation and distance? Since such transformations are Möbius transformations, and thus conformal maps, they automatically preserve orientation. To determine if they preserve the Poincaré distance function, we need to evaluate the distance function for points represented as complex numbers.

Theorem 8.11. *The* hyperbolic distance *from z_0 to z_1 in the Poincaré model is given by*

$$\begin{aligned} d_P(z_0, z_1) &= |\ln((z_0, z_1, w_1, w_0))| \\ &= |\ln(\frac{z_0 - w_1}{z_0 - w_0}\frac{z_1 - w_0}{z_1 - w_1})| \end{aligned} \tag{8.5}$$

where w_0 and w_1 are the points where the hyperbolic line through z_0 and z_1 meets the boundary circle.

Proof: From our earlier development of the Poincaré model (equation 7.2), we have

$$d_P(z_0, z_1) = |\ln(\frac{|z_0 - w_1|}{|z_0 - w_0|}\frac{|z_1 - w_0|}{|z_1 - w_1|})|$$

Since $|zw| = |z||w|$ and $|\frac{z}{w}| = \frac{|z|}{|w|}$, we have

$$d_P(z_0, z_1) = |\ln(|\frac{z_0 - w_1}{z_0 - w_0}\frac{z_1 - w_0}{z_1 - w_1}|)|$$

By Theorem 8.6, we know that $\frac{z_0-w_1}{z_0-w_0}\frac{z_1-w_0}{z_1-w_1}$, which is the cross ratio of z_0, z_1, w_1, and w_0, is real since all four points lie on a circle. Also, this cross ratio is non-negative (proved as an exercise). Thus,

$$d_P(z_0, z_1) = |\ln(\frac{z_0 - w_1}{z_0 - w_0}\frac{z_1 - w_0}{z_1 - w_1})|$$

□

Corollary 8.12. *Transformations of the form*

$$f(z) = \beta\frac{z - \alpha}{\overline{\alpha}z - 1}$$

where $|\alpha| < 1$ and $|\beta| = 1$ preserve the Poincaré distance function.

Proof: Let f be a transformation of the form described in the corollary, and let z_0, z_1 be two points in the Poincaré disk, with w_0, w_1 the points where the hyperbolic line through z_0, z_1 meets the boundary circle. Then, since f is a Möbius transformation, it will map clines to clines and will preserve angles. Thus, $f(z_0)$, $f(z_1)$ will be points in the Poincaré disk and $f(z_0)$, $f(z_1)$, $f(w_0)$, $f(w_1)$ will all lie on a cline that meets the boundary circle at right angles. That is, these points will lie on a hyperbolic line. Also, since f maps the boundary to itself, we know that $f(w_0)$ and $f(w_1)$ will lie on the boundary.

Thus, by Theorem 8.5, we have

$$\begin{aligned} d_P(f(z_0), f(z_1)) &= |\ln((f(z_0), f(z_1), f(w_1), f(w_0)))| \\ &= |\ln(z_0, z_1, w_1, w_0))| \\ &= d_P(z_0, z_1) \end{aligned}$$

□

What types of transformations are included in the set defined by

$$f(z) = \beta \frac{z - \alpha}{\overline{\alpha} z - 1} \tag{8.6}$$

where $|\alpha| < 1$ and $|\beta| = 1$?

If $|\beta| = 1$, then $\beta = e^{i\theta}$. So, multiplication by β has the geometric effect of rotation about the origin by an angle of θ. Thus, if $\alpha = 0$ in equation 8.6, then $T(z) = -\beta z$ is a simple rotation about the origin by an angle of $\pi + \theta$.

On the other hand, if $\beta = 1$ in equation 8.6, consider the line $t\alpha$ passing through the origin and α. We have that $f(t\alpha) = \alpha \frac{t-1}{|\alpha|^2 - 1}$, which is again a point on the line through α. The map f can be considered a *translation* along this line.

Thus, we see that orientation- and distance-preserving maps contain rotations and translations, similar to what we saw in the Euclidean case. However, translations are not going to exhibit the nice parallel properties that they did in the Euclidean plane.

What about orientation-*reversing* isometries? Since the cross ratio appears in the distance function and is always real on Poincaré lines, then simple complex conjugation ($\overline{f}(z) = \overline{z}$) of Poincaré points will be a distance-preserving transformation in the Poincaré model and will reverse orientation.

In fact, $\overline{f}$, which is a Euclidean reflection, is also a hyperbolic reflection, as it fixes a hyperbolic line. Similarly, Euclidean reflection in any diameter will be a hyperbolic reflection. This is most easily seen by the fact that Euclidean reflection about a diameter can be expressed as the conjugation of $\overline{f}$ by a rotation R about the origin by $-\theta$, where θ is the angle the diameter makes with the x-axis. Since rotation about the origin is an isometry in the Poincaré model, then $R^{-1} \circ \overline{f} \circ R$ is also an isometry.

All other hyperbolic reflections about lines that are not diameters can be expressed as inversion through the circle which defines the line (proved as an exercise).

We can now determine the structure of the complete group of isometries of the Poincaré model. Let g be any orientation-reversing isometry of the Poincaré model. Then, $h = \overline{f} \circ g$ will be an orientation-preserving isometry, and so h must be a transformation of the type in equation 8.6. Since $\overline{f}^{-1} = \overline{f}$, we have that g can be expressed as the product of $\overline{f}$ with an orientation-preserving isometry. This same conjugation property would be true for *any* hyperbolic reflection. Thus, we have

Theorem 8.13. *The orientation-preserving isometries of the Poincaré model can be expressed in the form $g = \beta\frac{z-\alpha}{\overline{\alpha}z-1}$. Also, if r is hyperbolic reflection about some hyperbolic line, then all orientation-reversing isometries can be expressed as $r \circ g$ for some orientation preserving g.*

Hyperbolic isometries can be used to prove many interesting results in hyperbolic geometry. Using Klein's Erlanger Program approach, in order to prove any result about general hyperbolic figures, it suffices to transform the figure to a "nice" position and prove the result there. For example, we can prove the following theorems on distance quite easily using this transformational approach.

Theorem 8.14. *Let z be a point in the Poincaré disk. Then*

$$d_H(0, z) = \ln\left(\frac{1+|z|}{1-|z|}\right)$$

Proof: Let $z = re^{i\theta}$ and T be rotation about the origin by $-\theta$. Then $d_H(0, z) = d_H(T(0), T(z)) = d_H(0, r)$. Now,

$$\begin{aligned} d_H(0,r) &= \left|\ln\left(\frac{0-1}{0-(-1)}\,\frac{r-1}{r-(-1)}\right)\right| \\ &= \left|\ln\left(\frac{1-r}{1+r}\right)\right| \\ &= \ln\left(\frac{1+r}{1-r}\right) \end{aligned}$$

Since $r = |z|$, the result follows. □

This result lets us convert from hyperbolic to Euclidean distance.

Corollary 8.15. *Let z be a point in the Poincaré disk. If $|z| = r$ and if δ is the hyperbolic distance from 0 to z, then*

$$\delta = \ln\left(\frac{1+r}{1-r}\right)$$

and

$$r = \frac{e^{\delta}-1}{e^{\delta}+1}$$

Proof: The first equality is a re-statement of the preceding theorem. The second equality is proved by solving for r in the first equality. □

Exercise 8.2.1. Show that the set of Euclidean rigid motions $f(z) = e^{i\phi}z + b$, with b complex and ϕ real, forms a group.

Exercise 8.2.2. Show that the equation $Im(\alpha z + \beta) = 0$, with α and β complex constants, defines a line in the plane.

Exercise 8.2.3. Find a Möbius transformation mapping the circle $|z = 1|$ to the x-axis.

The next four exercises prove that the set of Möbius transformations forms a group.

Exercise 8.2.4. Let $f(z) = \frac{az+b}{cz+d}$, where $(ad - bc) \neq 0$, and let $g(z) = \frac{ez+f}{gz+h}$, where $(eh - fg) \neq 0$, be two Möbius transformations. Show that the composition $f \circ g$ is again a Möbius transformation.

Exercise 8.2.5. Show that the set of Möbius transformations has an identity element.

Exercise 8.2.6. Let $f(z) = \frac{az+b}{cz+d}$, where $(ad - bc) \neq 0$, be a Möbius transformation. Show that $f^{-1}(w)$ has the form $f^{-1}(w) = \frac{dw-b}{-cw+a}$ and show that f^{-1} is a Möbius transformation.

Exercise 8.2.7. Why does the set of Möbius transformations automatically satisfy the associativity requirement for a group?

Exercise 8.2.8. Show that the set of Möbius transformations that fix the unit circle is a group.

Exercise 8.2.9. Show that in the Poincaré model there is a hyperbolic isometry taking any point P to any other point Q. [Hint: Can you find an isometry taking any point to the origin?]

Exercise 8.2.10. Let $T(z) = \frac{z-\alpha}{1-\overline{\alpha}z}$ be a hyperbolic transformation, with $\alpha \neq 0$. Show that T has two fixed points, both of which are on the boundary circle, and thus T has no fixed points inside the Poincaré disk. Why would it make sense to call T a translation?

Exercise 8.2.11. Show that the cross ratio term used in the definition of hyperbolic distance is always real and non-negative. [Hint: Use transformations to reduce the calculation to one that is along the x-axis.]

Exercise 8.2.12. Use the idea of conjugation of transformations to derive the formula for reflection across the diameter $y = x$ in the Poincaré model. [Hint: Refer to Exercise 5.7.3.]

Exercise 8.2.13. In the definition of hyperbolic distance given by equation 8.5, we need to determine boundary points w_0 and w_1. Show that we can avoid this boundary calculation by proving that

$$d_H(z_0, z_1) = \ln\left(\frac{|1 - z_0\overline{z_1}| + |z_0 - z_1|}{|1 - z_0\overline{z_1}| - |z_0 - z_1|}\right)$$

[Hint: Use the hyperbolic transformation $g(z) = \frac{z - z_1}{1 - \overline{z_1}z}$ and the fact that d_H is invariant under g.]

Exercise 8.2.14. Prove that if z_0, z_1, and z_2 are collinear in the Poincaré disk with z_1 between z_0 and z_2, then $d_P(z_0, z_2) = d_P(z_0, z_1) + d_P(z_1, z_2)$. This says that the Poincaré distance function is additive along Poincaré lines.

Exercise 8.2.15. Let l be a Poincaré line. Define a map f on the Poincaré disk by $f(P) = P'$, where P' is the inverse point to P with respect to the circle on which l is defined. We know by the results at the end of Chapter 2 that f maps the Poincaré disk to itself. Prove that f is an isometry of the Poincaré disk. Then, show that f must be a reflection. That is, inversion in a Poincaré line is a reflection in the Poincaré model. [Hint: Review the proof of Lemma 8.8.]

8.3 Isometries in the Klein Model

At the end of the last section, we saw that reflections play an important role in describing the structure of isometries in the Poincaré model. This should not be too surprising. In Chapter 5 we saw that all Euclidean isometries could be built from one, two, or three reflections. The proof of this fact used only *neutral* geometry arguments. That is, the proof used no assumption about Euclidean parallels. Thus, any isometry in hyperbolic geometry must be similarly built from one, two, or three hyperbolic reflections.

In the Poincaré model, the nature of a reflection depended on whether the Poincaré line of reflection was a diameter or not. If the line of reflection was a diameter, then the hyperbolic reflection across that line was simple Euclidean reflection across the line. If the line was not a diameter, then hyperbolic reflection across the line was given by inversion of points through the circle that defined the line.

Let's consider the first class of lines in the Klein model. Since the Klein distance function is defined in terms of the cross ratio and since the cross ratio is invariant under complex conjugation and rotation about the origin, then these two transformations will be isometries of the Klein model. By the same argument that we used in the last section, we see that any Euclidean reflection about a diameter must be a reflection in the Klein model.

What about reflection across a Klein line that is not a diameter? Let's recall the defining properties of a reflection. By Theorem 5.6, we know that if P, P' are two Klein points, then there is a unique reflection taking P to P'. The line of reflection will be the perpendicular bisector of $\overline{PP'}$.

Thus, given a Klein line l in the Klein disk and given a point P, we know that the reflection of P across l can be constructed as follows. Drop a perpendicular line from P to l intersecting l at Q. Then, the reflected point P' will be the unique point on the ray opposite $\overrightarrow{QP}$ such that $\overline{PQ} \cong \overline{QP'}$. This point can be found by the following construction.

Theorem 8.16. *Let l be a Klein line that is not a diameter. Let P be a Klein point not on l. Let t be the Klein line through P that meets l at Q at right angles. Let $\overleftrightarrow{P\Omega}$ be the Klein line through P perpendicular to t, with omega point Ω. Let $\Omega\Omega'$ be the Klein line through Q and Ω. Finally, let P' be the point on t where the (Euclidean) line through the pole of t and Ω' passes through t. Then P' is the reflection of P across l (Fig. 8.3).*

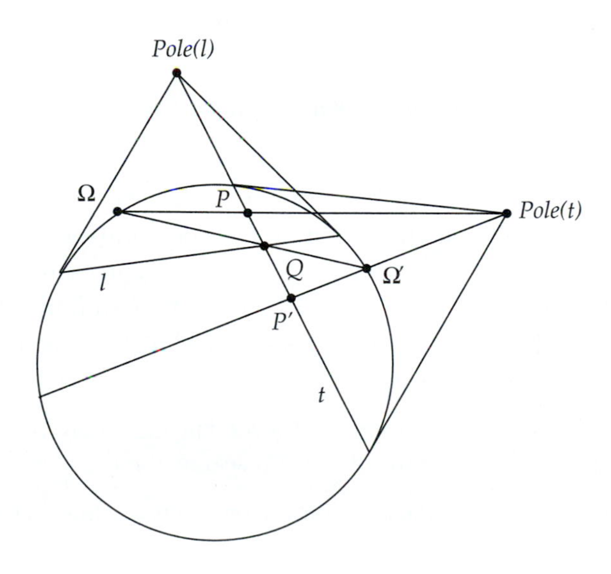

Fig. 8.3

Proof: To construct t we draw a line from the pole of l through P. To construct $\overleftrightarrow{P\Omega}$ we draw a line from the pole of t through P. Both of these perpendiculars are guaranteed to exist by results from neutral geometry.

Now the line through Q and Ω must intersect the boundary at a point Ω' that is on the other side of t from Ω (in the Euclidean sense). Thus, the line through the pole of t and Ω' must intersect t at a point P'. If we can show that $\overline{PQ} \cong \overline{QP'}$, we are done.

Since $\overleftrightarrow{P'\Omega'}$ passes through the pole of t, then it is perpendicular to t (in the Klein sense). By Angle-Angle congruence of omega triangles (see Exercise 7.3.10), we know that $\overline{PQ} \cong \overline{QP'}$. □

The construction described in the theorem is quite important for many other constructions in the Klein model. In the exercises at the end of this section, we will investigate other constructions based on this one.

We can use this theorem to show that any two Klein lines are congruent. That is, there is an isometry of the Klein model taking one to the other. To prove this result, we will need the following fact.

Lemma 8.17. *Given a Klein line l that is not a diameter of the Klein disk, we can find a reflection r that maps l to a diameter of the Klein disk.*

Proof: Let P be a point on l and let O be the center of the Klein disk. Let Q be the midpoint of $\overline{PO}$ (in the sense of Klein distance). Let n be the perpendicular to $\overline{PO}$ at Q. Then Klein reflection of l about n will map P to O and thus must map l to a diameter, since Klein reflections map Klein lines to Klein lines. □

Corollary 8.18. *Let l and m be Klein lines. Then there is an isometry in the Klein model taking l to m.*

Proof: By the lemma we know there is a reflection r_l taking l to a diameter d_l. If d_l is not on the x-axis, let R_l be rotation by $-\theta$, where θ is the angle made by d_l and the x-axis. Then the Klein isometry $h_l = R_l \circ r_1$ maps l to the diameter on the x-axis. Likewise, we can find a Klein isometry h_m taking m to a diameter on the x-axis. Then $h_m^{-1} \circ h_l$ is a Klein isometry mapping l to m. □

Exercise 8.3.1. Use the construction ideas of Theorem 8.16 to devise a construction for the perpendicular bisector of a Klein segment PP'.

Exercise 8.3.2. Devise a construction for producing a line l that is orthogonal to two parallel (but not limiting parallel) lines m and n.

Exercise 8.3.3. Show that in the Klein model there exists a pentagon with five right angles. [Hint: Start with two lines having a common perpendicular.]

Exercise 8.3.4. Let Ω and Ω' be two omega points in the Klein disk and let P be a Klein point. Let the Euclidean ray from the pole of $\Omega\Omega'$ through P intersect the unit circle at omega point Ω''. Show that $\overrightarrow{P\Omega''}$ is the Klein model angle bisector of $\angle\Omega P\Omega'$. [Hint: Use omega triangles.]

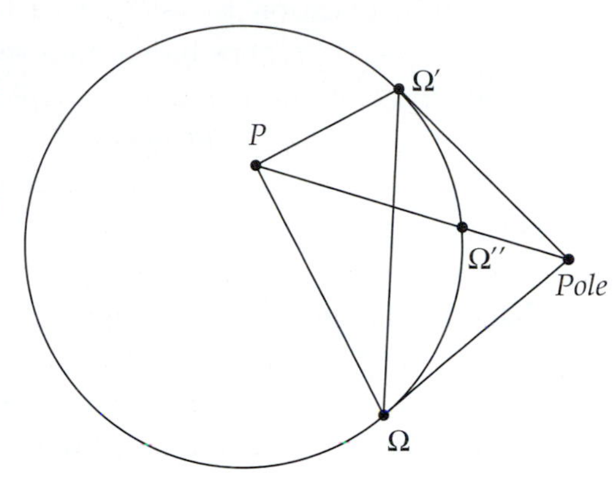

Exercise 8.3.5. Define a hyperbolic translation T in the Klein model as the product of two reflections r_l and r_m where l and m have a common perpendicular t. Show that t is invariant under T.

Exercise 8.3.6. Define a hyperbolic *parallel displacement* D in the Klein model as the product of two reflections r_l and r_m where l and m are limiting parallel to each other. Show that no Klein point P is invariant under D. [Hint: Assume that P is invariant (that is, $r_l(P) = r_m(P)$) and consider the line joining P to $r_l(P)$.]

8.4 Mini-Project - The Upper Half-Plane Model

In this chapter we have looked in detail at the isometries of the Poincaré and Klein models of hyperbolic geometry. Both of these models are based on the unit disk.

There is really nothing special about using the unit disk in these models. In the Poincaré model, for example, we could just as well have used any circle in the plane and defined lines as either diameters or arcs meeting the boundary at right angles. In fact, if c was any circle, we know there is a Möbius transformation f that will map the unit disk to f. Since Möbius transformations preserve angles, we could define new lines in c to be the image under f of lines in the Poincaré model. Likewise, circles in c would be images of Poincaré model circles, and the distance could be defined in terms of f as well.

A natural question to ask is whether there are models for hyperbolic geometry other than ones based on a set of points inside a circle.

If we think of the extended complex plane as being equivalent to the sphere via stereographic projection, then lines in the plane are essentially circles that *close up* at the point at infinity.

Is there a model for hyperbolic geometry that uses as its boundary curve a Euclidean line? To build a model using a line as a boundary curve, we use the Klein Erlanger Program approach and determine the transformations that leave the line invariant. This is analogous to the work we did earlier to find the transformations that fix the unit circle in the Poincaré model. For simplicity's sake, let's assume our line boundary is the x-axis. Then we want to find Möbius transformations

$$f(z) = \frac{az+b}{cz+d}$$

that fix the x-axis.

Exercise 8.4.1. Show that if f maps the x-axis to itself, then a, b, c, and d must all be real. [Hint: Use the fact that 0 and ∞ must be mapped to real numbers and that ∞ must also be the image of a real number.]

What about the half-planes $y < 0$ and $y > 0$? Let's restrict our attention to those transformations that move points within one half-plane, say the upper half-plane $y < 0$.

Exercise 8.4.2. Show that if f fixes the x-*axis* and maps the upper half-plane to itself, then $ad - bc > 0$. [Hint: Consider the effect of f on $z = i$.]

We have now proved the following result.

Theorem 8.19. *If f fixes the x-axis and maps the upper half-plane to itself, then*

$$f(z) = \frac{az+b}{cz+d}$$

where a,b,c,d are all real and $ad - bc > 0$.

We will call the group of transformations in the last theorem the *Upper Half-Plane* group, denoted by U.

Exercise 8.4.3. What curves should play the role of lines in the geometry defined by U? [Hint: Refer to Fig. 8.4.]

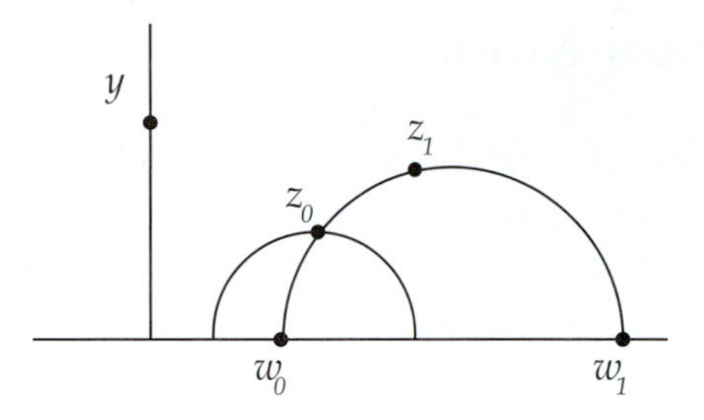

Fig. 8.4 Upper Half-Plane Model

We can define distance just as we did for the Poincaré model in terms of the cross ratio. For example, in Fig. 8.4 the distance from z_0 to z_1 will be defined as

$$d_U(z_0, z_1) = \ln((z_0, z_1, w_1, w_0))$$

Exercise 8.4.4. What are the values of w_0 and w_1 in the distance formula if z_0 and z_1 are on a piece of a Euclidean line?

Exercise 8.4.5. Discuss why this geometry will satisfy the hyperbolic parallel postulate. [Hint: Argue that it is enough to show the postulate is satisfied for the y-axis and z_0 as shown in Fig. 8.4.]

The geometry defined by the group U will be another model for hyperbolic geometry. This model (the upper half-plane model) is very similar to the Poincaré model. In fact, there is a conformal map taking one to the other. This map is defined by

$$g(z) = -i\frac{z+i}{z-i} \tag{8.7}$$

Exercise 8.4.6. Show that g maps the unit circle onto the x-axis. [Hint: Consider $z = i, -i, 1$.]

Since g is a Möbius transformation, then g will preserve angles. So, it must map Poincaré lines to upper half-plane lines. Also, since the distance function is defined in terms of the cross ratio and the cross ratio is invariant under g, then Poincaré circles will transform to upper half-plane circles.

We conclude that the upper half-plane model is *isomorphic* to the Poincaré model. Any property of one can be moved to the other by g or g^{-1}.

For future reference, we note that the inverse transformation to g is given by

$$g^{-1} = i\frac{w-i}{w+i} \tag{8.8}$$

8.5 Weierstrass Model

There are other models of hyperbolic geometry. One of the most interesting is the *Weierstrass model.* Here points are defined to be Euclidean points on one sheet of the *hyperboloid* $x^2 + y^2 - z^2 = -1$ (see Fig. 8.5). Lines in this model are the curves on the top sheet of the hyperboloid that are created by intersecting the surface with planes passing through the origin. Each such line will be one branch of a hyperbola. For a complete development of this model, see [15].

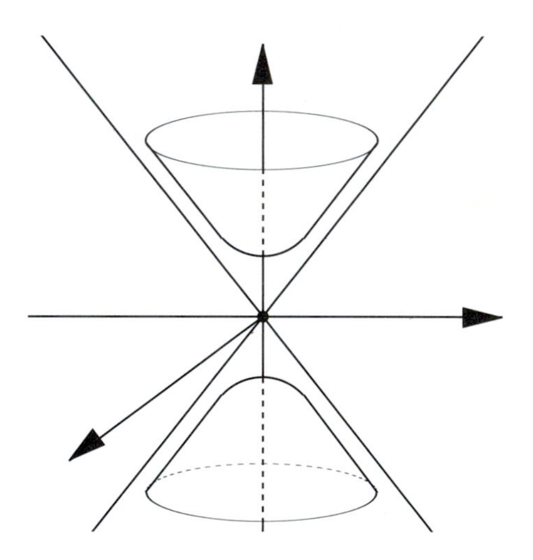

Fig. 8.5 Weierstrass Model

8.6 Hyperbolic Calculation

In the last chapter we explored many of the basic results of hyperbolic geometry, illustrating these results within the Poincaré and Klein models.

Except for the formula relating the area of a triangle to its defect, we have not yet developed ways of *calculating* within hyperbolic geometry. For example, in Exercises 7.3.11 and 7.3.12 we analyzed the *angle of parallelism* function, which relates lengths of segments to angles of limiting parallels. We saw that it was order-reversing and one-to-one, but as yet, we have no explicit formula for this function.

Likewise, we do not have any way of calculating lengths and areas for figures that are not constructed from segments.

In this section we will use transformations in the Poincaré model to develop explicit formulas for these and other quantities of hyperbolic geometry.

We begin our development with perhaps the most fundamental of all calculations—that of arclength.

8.6.1 Arclength of Parameterized Curves

Recall the definition for distance between points z_0 and z_1 in the Poincaré disk (refer to Theorem 8.11):

$$\begin{aligned} d_H(z_0, z_1) &= |\ln((z_0, z_1, w_1, w_0))| \qquad (8.9) \\ &= |\ln(\frac{z_0 - w_1}{z_0 - w_0}\frac{z_1 - w_0}{z_1 - w_1})| \end{aligned}$$

where w_0 and w_1 are the points where the hyperbolic line through z_0 and z_1 meets the boundary circle of the Poincaré disk.

If $z_0 = 0$ and $z_1 = z$, we saw in Theorem 8.14 that this could be simplified to

$$d_H(0, z) = \ln\left(\frac{1 + |z|}{1 - |z|}\right) \qquad (8.10)$$

Let $\gamma = z(t)$, $a \leq t \leq b$, be a smooth curve in the Poincaré disk, as shown in Fig. 8.6. Consider the arclength of γ between points $z(t)$ and $z(t + \triangle t)$.

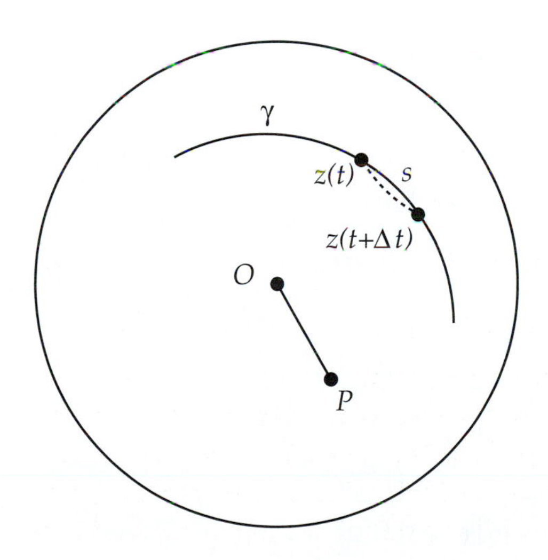

Fig. 8.6

Let

$$g(z) = \frac{z - z(t)}{1 - \overline{z(t)}z}$$

This transformation is a hyperbolic transformation that will map $z(t)$ to the origin O and $z(t+\triangle t)$ to a point P. If $\triangle z = z(t+\triangle t) - z(t)$, then

$$\begin{aligned} P &= \frac{z(t+\triangle t) - z(t)}{1 - \overline{z(t)}z(t+\triangle t)} \\ &= \frac{\triangle z}{1 - \overline{z(t)}z(t+\triangle t)} \end{aligned} \tag{8.11}$$

If we approximate the arclength s along γ between $z(t)$ and $z(t+\triangle t)$ by a hyperbolic segment (the dashed line in Fig. 8.6), then that segment will be mapped by g to a segment from O to P and the length of the original segment will be preserved by g. By equation 8.10, the hyperbolic length of the segment is thus

$$\begin{aligned} d_H(O,P) &= \ln\left(\frac{1+|P|}{1-|P|}\right) \\ &= \ln\left(\frac{1+\frac{|\triangle z(t)|}{|1-\overline{z(t)}z(t+\triangle t)|}}{1-\frac{|\triangle z(t)|}{|1-\overline{z(t)}z(t+\triangle t)|}}\right) \\ &= \ln\left(1+\frac{|\triangle z(t)|}{|1-\overline{z(t)}z(t+\triangle t)|}\right) - \ln\left(1-\frac{|\triangle z(t)|}{|1-\overline{z(t)}z(t+\triangle t)|}\right) \end{aligned}$$

For x small, $\ln(1+x) \approx x$, and this approximation turns to an equality as x goes to zero. Thus, we have that

$$\begin{aligned} d_H(O,P) &\approx \frac{|\triangle z(t)|}{|1-\overline{z(t)}z(t+\triangle t)|} - (-)\frac{|\triangle z(t)|}{|1-\overline{z(t)}z(t+\triangle t)|} \\ &= \frac{2|\triangle z(t)|}{|1-\overline{z(t)}z(t+\triangle t)|} \\ &= \frac{2|\frac{\triangle z(t)}{\triangle t}|}{|1-\overline{z(t)}z(t+\triangle t)|}\triangle t \end{aligned} \tag{8.12}$$

As $\triangle t$ goes to zero, $|\triangle z(t)|$ goes to zero, and the approximation in equation 8.12 becomes an equality. Also, $z(t+\triangle t)$ becomes $z(t)$. Thus, if we approximate γ by a series of (hyperbolic) linear approximations of the form given by equation 8.12, and take the limit as $\triangle t$ goes to zero, we have the following:

Theorem 8.20. *Let $\gamma = z(t)$ be a smooth curve in the Poincaré disk for $a \le t \le b$. Then the arclength of γ, defined in terms of the distance function d_H, is given by*

$$\int_a^b \frac{2|z'(t)|}{1-|z|^2} dt \tag{8.13}$$

Corollary 8.21. *Let T be a hyperbolic transformation and γ a smooth curve. Then the arclength of $T(\gamma)$ is the same as the arclength of γ.*

The proof is left as an exercise.

8.6.2 Geodesics

In Euclidean geometry a segment joining point A to point B is the shortest length path among all paths joining the two points. We call such a shortest length path a *geodesic*. What are the geodesics in hyperbolic geometry?

Theorem 8.22. *Given two points z_0 and z_1 in the Poincaré disk, the geodesic joining the points is the portion of the hyperbolic line between z_0 and z_1.*

Proof: By using a suitable hyperbolic transformation, we can assume that z_0 is at the origin O and z_1 is along the x-axis.

Then, it is sufficient to prove that the geodesic (shortest *hyperbolic* length path) from O to X is the Euclidean segment from O to X. Let $\gamma = z(t) = x(t) + iy(t)$ be a smooth curve from O to X, $a \le t \le b$. Then, $z(a) = 0$ and $z(b) = X$, and thus $x(a) = 0$ and $x(b) = X$. The length of γ is

$$\begin{aligned}
\int_a^b \frac{2|z'|}{1-|z|^2} dt &= \int_a^b \frac{2\sqrt{x'^2+y'^2}}{1-x^2-y^2} dt \\
&\ge \int_a^b \frac{2\sqrt{x'^2}}{1-x^2} dt \\
&= \int_0^X \frac{2}{1-x^2} dx \\
&= \ln\left(\frac{1+X}{1-X}\right)
\end{aligned}$$

Since $\ln(\frac{1+X}{1-X})$ is also the hyperbolic distance along the line from 0 to X, we see that the line segment must be the shortest length path. □

8.6.3 The Angle of Parallelism

Recall the definition of the *angle of parallelism function.*

Let $\overline{PQ}$ be a segment of length h. Let l be a perpendicular to $\overline{PQ}$ at Q and $\overleftrightarrow{PR}$ the limiting parallel to l at P. The angle of parallelism, $a(h)$, is defined as $a(h) = \angle QPR$.

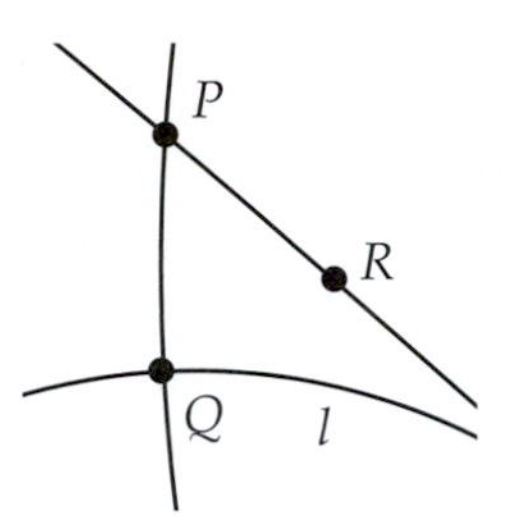

By use of appropriate hyperbolic transformations we can assume that P is at the origin and l is symmetrically located across the x-axis, as shown in Fig. 8.7.

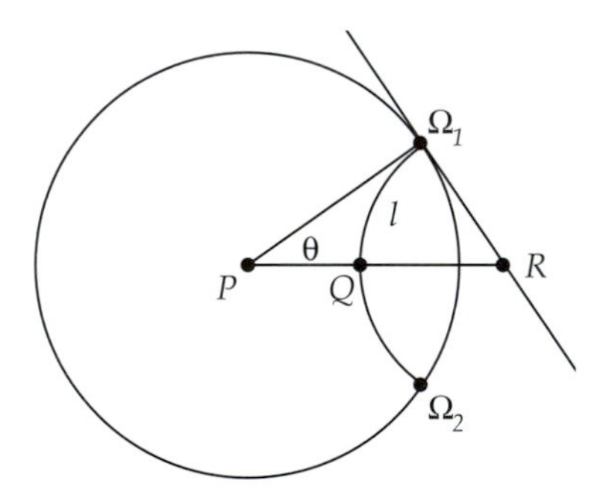

Fig. 8.7

Let Ω_1 and Ω_2 be the ideal points of l and let $\overleftrightarrow{\Omega_1 R}$ be the tangent to the Poincaré disk at Ω_1, with R the intersection of the tangent with $\overleftrightarrow{PQ}$. Then, R is the center of the (Euclidean) arc through Ω_1, Q, and Ω_2.

Since $\angle P\Omega_1 R$ is a right angle and the Poincaré disk has unit radius, then

$$\begin{aligned} QR &= \Omega_1 R = \tan(\theta) \\ PR &= \sec(\theta) \\ PQ &= \sec(\theta) - \tan(\theta) = \frac{1 - \sin(\theta)}{\cos(\theta)} \end{aligned}$$

Also, we can convert between Euclidean length and hyperbolic length by using the formula of Corollary 8.15:

$$h = ln\left(\frac{1+PQ}{1-PQ}\right)$$

This is equivalent to

$$\begin{aligned} e^{-h} &= \frac{1-PQ}{1+PQ} \\ &= \frac{\cos(\theta)+\sin(\theta)-1}{\cos(\theta)-\sin(\theta)+1} \end{aligned} \tag{8.14}$$

After multiplying the numerator and denominator of this last equation by $\cos(\theta)+\sin(\theta)+1$, we get

$$\begin{aligned} e^{-h} &= \frac{\cos^2(\theta)+2\cos(\theta)\sin(\theta)+\sin^2(\theta)-1}{\cos^2(\theta)+2\cos(\theta)+1-\sin^2(\theta)} \\ &= \frac{2\cos(\theta)\sin(\theta)}{2\cos^2(\theta)+2\cos(\theta)} \\ &= \frac{\sin(\theta)}{\cos(\theta)+1} \\ &= \frac{2\sin(\frac{\theta}{2})\cos(\frac{\theta}{2})}{2\cos^2(\frac{\theta}{2})-1+1} \\ &= \tan(\frac{\theta}{2}) \end{aligned} \tag{8.15}$$

Solving this equation for θ gives the following formula for the angle of parallelism:

Theorem 8.23. *Let $\overline{PQ}$ be a hyperbolic segment of length h. Then the angle of parallelism function $a(h)$ is given by*

$$a(h) = 2\tan^{-1}(e^{-h})$$

8.6.4 Right Triangles

In Fig. 8.8 we have a triangle with a right angle at the origin and vertices along the x- and y- axes at points x and iy.

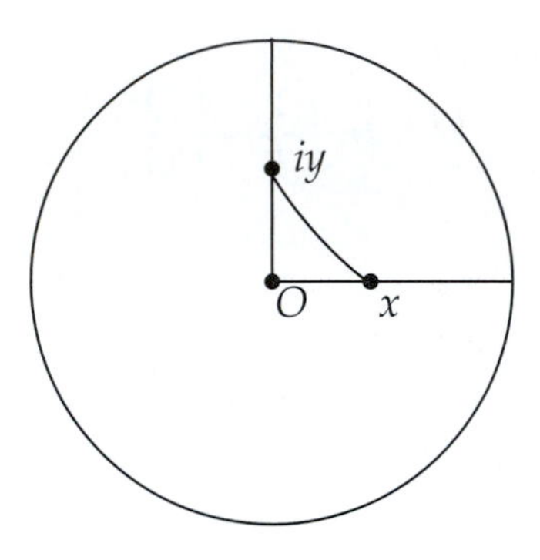

Fig. 8.8

In order to develop a hyperbolic version of the Pythagorean Theorem, we will use the hyperbolic trigonometric functions *cosh* and *sinh.*

The hyperbolic functions are defined as:

$$\sinh(x) = \frac{e^x - e^{-x}}{2}, \quad \cosh(x) = \frac{e^x + e^{-x}}{2}$$

The hyperbolic functions share many of the same properties as the *circular* functions sine and cosine. For example, the identity $\cos^2(x)+\sin^2(x) = 1$ has as its counterpart the hyperbolic identity $\cosh^2(x) - \sinh^2(x) = 1$. The proof of this fact is left as an exercise.

To compare the lengths of the sides of a hyperbolic right triangle, as shown in Fig. 8.8, we will make use of the following relationship:

Lemma 8.24. *Let z and w be two points in the Poincaré disk. Then*

$$\cosh(d_H(z, w)) = \frac{|1 - z\overline{w}|^2 + |z - w|^2}{|1 - z\overline{w}|^2 - |z - w|^2}$$

Proof: From Exercise 8.2.13, we know that

$$d_H(z, w) = \ln\left(\frac{|1 - z\overline{w}| + |z - w|}{|1 - z\overline{w}| - |z - w|}\right)$$

The result then follows from the definition of cosh and simple algebra and is left as an exercise. □

In the case of our right triangle, we have

$$\begin{aligned}
\cosh(d_H(x, iy)) &= \frac{|1+ixy|^2 + |x-iy|^2}{|1+ixy|^2 - |x-iy|^2} \\
&= \frac{1+x^2y^2+x^2+y^2}{1+x^2y^2-x^2-y^2} \\
&= \left(\frac{1+x^2}{1-x^2}\right)\left(\frac{1+y^2}{1-y^2}\right) \\
&= \cosh(d_H(0,x))\cosh(d_H(0,iy))
\end{aligned}$$

Since any right triangle can be moved into the position in Fig. 8.8 by a suitable hyperbolic transformation, we have the following hyperbolic version of the Pythagorean Theorem.

Theorem 8.25. (Hyperbolic Pythagorean Theorem) *Let ΔABC be a hyperbolic right triangle with hypotenuse of length c and base lengths of a and b. Then*

$$\cosh(c) = \cosh(a)\cosh(b)$$

8.6.5 Area

Let R be an area in the Poincaré disk. Since the differential of arclength is

$$ds = \frac{2|dz|}{1-|z|^2}$$

and since hyperbolic geometry is approximately Euclidean in very small regions, then it makes sense to *define* the area integral as

$$Area(R) = \int\int_R \frac{4}{(1-|z|^2)^2} dy\, dx \tag{8.16}$$

The polar form of this area formula is as follows:

Theorem 8.26. *Let R be a region in the plane, r the hyperbolic length of z, and θ the angle z makes with the x-axis. Then*

$$Area(R) = \int\int_R \sinh(r) dr\, d\theta$$

Proof: Let x and y have the polar form

$$\begin{aligned}
x &= \rho\cos(\theta) \\
y &= \rho\sin(\theta)
\end{aligned}$$

where ρ is the Euclidean length of the point z. The polar form of the integral expression in area definition 8.16 becomes

$$Area(R) = \int\int_R \frac{4}{(1-\rho^2)^2} \rho \, d\rho \, d\theta$$

Now hyperbolic length and Euclidean length are connected by equation 8.10, so if r is the hyperbolic length of z we have

$$r = \ln\left(\frac{1+\rho}{1-\rho}\right)$$

Solving for ρ and differentiating, we get

$$\begin{aligned} \rho &= \tanh(\frac{r}{2}) \\ d\rho &= \frac{1}{2}\text{sech}^2(\frac{r}{2})dr \end{aligned}$$

Substituting these values into the area integral and using the fact that $1 - \tanh^2(a) = \text{sech}^2(a)$, we get

$$Area(R) = \int\int_R 2\sinh(\frac{r}{2})\cosh(\frac{r}{2}) \, dr \, d\theta$$

Since $2\ \sinh(\frac{r}{2})\cosh(\frac{r}{2}) = \sinh(r)$ (proved as an exercise), we get

$$Area(R) = \int\int_R \sinh(r) dr \, d\theta$$

and the proof is complete. □

Exercise 8.6.1. Show that $\cosh^2(x) - \sinh^2(x) = 1$.

Exercise 8.6.2. Finish the proof of Lemma 8.24.

Exercise 8.6.3. Show that $2\sinh(\frac{r}{2})\cosh(\frac{r}{2}) = \sinh(r)$.

Exercise 8.6.4. Let c be a circle in the Poincaré Model of hyperbolic radius hr. Show that the circumference of the circle is $s = 2\pi\ \sinh(hr)$ by using the formula for the arclength of a parameterized curve $\gamma = z(t)$. [Hint: Use a transformation to simplify the calculation.]

Exercise 8.6.5. Let $S(w)$ be the model isomorphism from the upper half-plane model of hyperbolic geometry to the Poincaré Model given by

$$z = S(w) = i\frac{w-i}{w+i} \tag{8.17}$$

[Refer to equation 8.8.]

If $w(t)$, $a \leq t \leq b$ is a smooth parameterized curve in the upper half-plane, discuss why $z(t) = S(w(t))$ will be a smooth curve in the Poincaré disk having the same length. Show that $|z'| = \frac{2}{|w+i|^2}|w'|$, and use this and the integral for arclength in the Poincaré Model, to show that the length of $w(t) = u(t) + iv(t)$ in the upper half-plane is given by

$$\int_a^b \frac{|w'(t)|}{v(t)} dt \tag{8.18}$$

Exercise 8.6.6. Prove Corollary 8.21. [Hint: Use the chain rule.]

8.7 Project 12 - Infinite Real Estate?

Now that we have experience with transformations and calculations in hyperbolic geometry, let's see if we have developed any intuition about how it would feel to live in such a world.

Suppose we had a triangular plot of land defined by three stakes placed at A, B, and C.

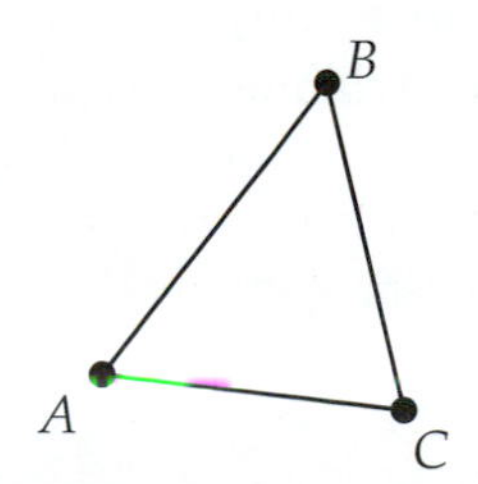

Now suppose we pull up the stakes and move out to new positions A', B', and C'.

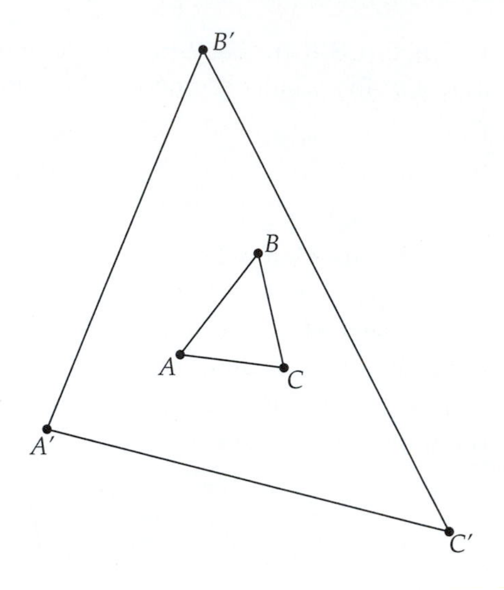

Clearly, the area of our plot of land will increase as we move the three stakes farther and farther apart, and the area will increase *without bound.* Theoretically, we could have a plot of land with area approaching infinity!

Will the same be true for a triangular plot of land in hyperbolic space?

Start *Geometry Explorer* and create a hyperbolic canvas by choosing **New** (**File** menu). Then, create a hyperbolic triangle with vertices A, B, and C.

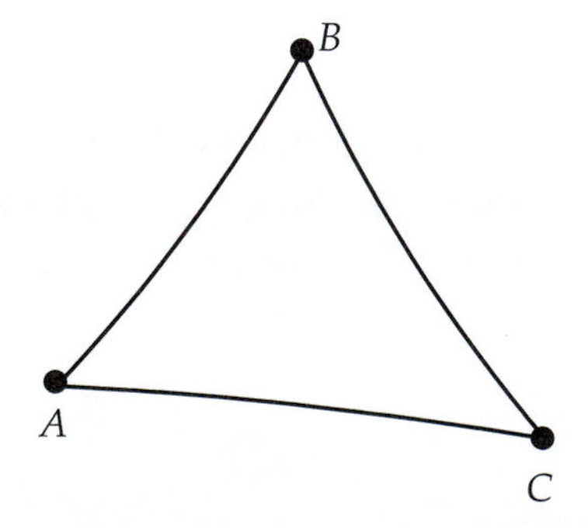

Multi-select points A, B, and C, and click on the Filled-Polygon button in the Construct panel (third button in third row). A filled area will be constructed. Select the area by clicking on the colored region, and then choose **Area** (**Measure** menu) to calculate the area of the triangle.

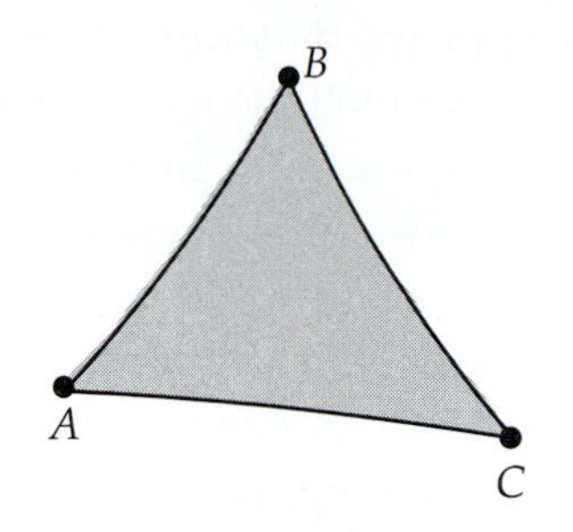

Area(polygon(A,C,B)) = 0.20

Now, let's pull up the stakes and move the points toward infinity, that is, toward the boundary of the Poincaré disk.

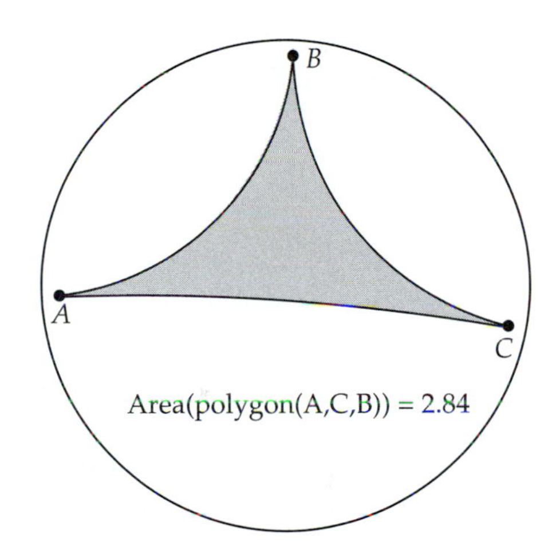

The area has grown a bit, but not that much, actually. Let's move as close to the boundary as we can.

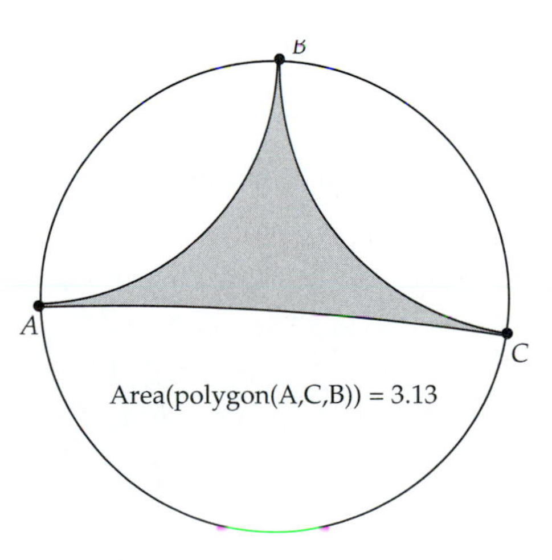

It appears that our plot of land is *not* becoming infinitely large, even though it would take us a very long time to walk from one corner to the other. In fact, as you were moving the points closer and closer to the boundary, you may have become curious about the value of the area. The number 3.13 is very close to another significant mathematical constant.

To settle once and for all what happens to this hyperbolic triangle, it will be convenient to convert the area integral from the last section to an area integral in the upper half-plane model. Let $S(w)$ be the model isomorphism from the upper half-plane model of hyperbolic geometry to the Poincaré Model as given in equation 8.8. That is, if $w = u + iv$ is a point in the upper half-plane, then $z = x + iy$ in the Poincaré disk is given by

$$z = S(w) = i\frac{w - i}{w + i} \tag{8.19}$$

Exercise 8.7.1. Show that

$$\begin{aligned} x &= \frac{2u}{u^2 + (v+1)^2} \\ y &= \frac{u^2 + (v^2 - 1)}{u^2 + (v+1)^2} \end{aligned} \tag{8.20}$$

Consider S as a map from (x, y) to (u, v). By the change of basis formula from calculus, we know that the area integral in x and y will change by the *Jacobian* of the map S, which is the determinant of the matrix of partial derivatives.

Due to the Cauchy-Riemann equations of complex variables, the Jacobian J for the change of variables defined by $x(u, v)$ and $y(u, v)$ will be equal to $x_u^2 + y_u^2$.

Exercise 8.7.2. Show that

$$J = x_u^2 + y_u^2 = \frac{4}{(u^2 + (v+1)^2)^2} = \frac{4}{|w + i|^4} \tag{8.21}$$

The area integral for the upper half-plane model then becomes

$$\begin{aligned} Area(R) &= \int\int_R \frac{4}{(1 - \frac{|w-i|^2}{|w+i|^2})^2} J \, dv \, du \\ &= \int\int_R \frac{4|w+i|^4}{(|w+i|^2 - |w-i|^2)^2} \frac{4}{|w+i|^4} \, dv \, du \\ &= \int\int_R \frac{16}{16v^2} \, dv \, du \\ &= \int\int_R \frac{1}{v^2} \, dv \, du \end{aligned}$$

Theorem 8.27. *The area of a region R in the upper half-plane is given by*

$$Area(R) = \int\int_R \frac{1}{v^2}\, dv\, du$$

Before we tackle the question of a triangle where all vertices are at infinity, we will consider a *doubly limiting* triangle, one where only two vertices are at infinity.

Given a doubly limiting triangle, we can find an isometry that will move it to the triangle with vertices at -1, $P = (\cos(\theta), \sin(\theta))$, and at infinity in the upper half-plane, as illustrated by the shaded region at right. The points -1 and ∞ will be Ω points; that is, points at infinity.

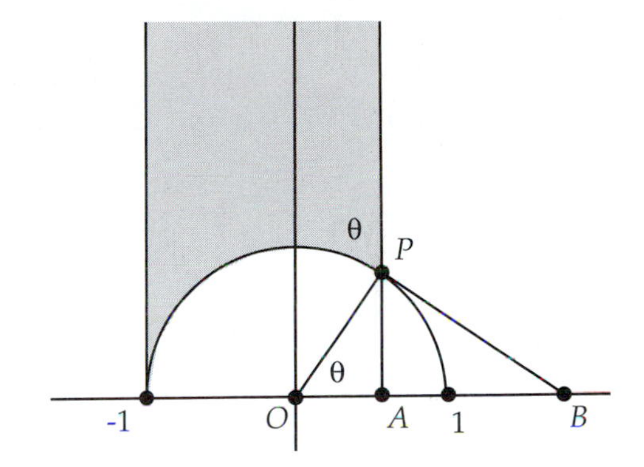

Exercise 8.7.3. Let $\overleftrightarrow{PB}$ be the tangent to the unit circle at P. (In the figure above, the tangent intersects the x-axis, but this is not necessarily always the case.) Show that the interior angle of the triangle at P must be θ.

Exercise 8.7.4. Use the area integral and the fact that the triangle is bounded by $v = \sqrt{1-u^2}$, $v = \infty$, and $-1 \leq u \leq \cos(\theta)$ to show that the area of this triangle is actually $\pi - \theta$.

We have now proved that

Theorem 8.28. *The area of a doubly limiting triangle of angle θ is $\pi - \theta$. The area of a triply limiting triangle is π.*

What conclusions can you make about triangular real estate in hyperbolic geometry? What about four-sided real estate? Five-sided real estate? And n-sided real estate?

For your project report give a careful and complete summary of your work done on this project.

Chapter 9

Fractal Geometry

> Why is geometry often described as "cold" and "dry"? One reason lies in its inability to describe the shape of a cloud, a mountain, a coastline, or a tree. Clouds are not spheres, mountains are not cones, coastlines are not circles, and bark is not smooth, nor does lightning travel in a straight line.
>
> —Benoit Mandelbrot in *The Fractal Geometry of Nature* [30]

9.1 The Search for a "Natural" Geometry

Classical Euclidean geometry had its roots in ancient Babylonian and Egyptian calculations of land areas and architectural designs. The word *geometry* means "earth measurement." It is said that Aristotle came to believe that the earth was a sphere by watching ships disappear over the ocean's horizon.

Intuitively, we like to think our notions of Euclidean geometry are the results of our interactions with nature, but is that really the case?

As mentioned in the quote above, most objects in nature are not really regular in form. A cloud may look like a lumpy ball from far away, but as we get closer, we notice little wisps of vapor jutting out in every direction. As we move even closer, we notice that the cloud has no real boundary. The solidness of the form dissolves into countless filaments of vapor. We may be tempted to replace our former notion of the cloud being three-dimensional with a new notion of the cloud being a collection of one-dimensional curves. But, if we look closer, these white curves dissolve into tiny water droplets. The cloud now appears to be a collection of tiny three-dimensional balls. As we move even closer, to the molecular level, these droplets dissolve into tiny whirling masses of hydrogen and oxygen, shapes akin to the original cloud itself.

So, a cloud in the sky cannot *really* be described as a classical geometric figure. This seems to contradict the commonly held intuition about Euclidean geometry being a product of our natural environment. How can we make geometric sense of objects like clouds, mountains, trees, atoms, planets, and so on?

We saw in Chapters 5 and 6 that there is one geometric idea that does seem to resonate with our experience of the natural world—the idea of *symmetry*. Symmetry is the idea that an object is invariant under some transformation of that object. In Chapter 5 we looked at the notion of symmetry as invariance under Euclidean isometries. Such symmetries include bilateral symmetry, rotational symmetry, translation symmetry, and glide symmetry.

In the discussion of the cloud as a geometric object, we noticed that when we viewed a section of the cloud at the molecular level, we saw a shape similar to the original cloud itself. That is, the molecular "cloud" appeared to be the same shape as the original cloud, when *scaled* up by an appropriate scale factor.

Similarly, consider the fern in Fig. 9.1. Each leaf is made up of sub-leaves that look similar to the original leaf, and each sub-leaf has sub-sub-leaves similar to the sub-leaves, and so on.

Fig. 9.1 Fern Leaf

Thus, many natural objects are similar to parts of themselves, once you scale up the part to the size of the whole. That is, they are symmetric under a change of scale. We call such objects *self-similar* objects. Such objects will be our first example of *fractals*, a geometric class of objects that we will leave undefined for the time being.

9.2 Self-Similarity

An object will be called *self-similar* if a part of the object, when scaled by a factor $c > 0$, is equivalent to the object itself.

We can be more precise in this definition by making use of *similarity transformations.*

Definition 9.1. A *similarity transformation* S, with *ratio* $c > 0$, is a transformation (i.e., one-to-one and onto map) from Euclidean n-dimensional space ($\mathbb{R}^n$) to itself such that

$$|S(x) - S(y)| = c\,|x - y| \tag{9.1}$$

A self-similar set will be a set that is invariant under one or more similarity transformations.

Definition 9.2. A *self-similar* set F in $\mathbb{R}^n$ is a set that is invariant under a finite number of non-identity similarity transformations.

Which classical Euclidean objects are self-similar? Consider a circle C. We know that the closer we "look" at the circle, the flatter the curve of the circle becomes. Thus, a circle cannot be self-similar. Similarly, any differentiable curve in the plane will not be self-similar, with one exception—a line. Lines are perhaps the simplest self-similar figure.

Self-similarity is a concept foreign to most of the 2-dimensional geometry covered in calculus and Euclidean geometry. To develop some intuition for self-similarity, we need some examples.

9.2.1 Sierpinski's Triangle

Our first example is generated from a simple filled-in triangle, $\triangle ABC$.

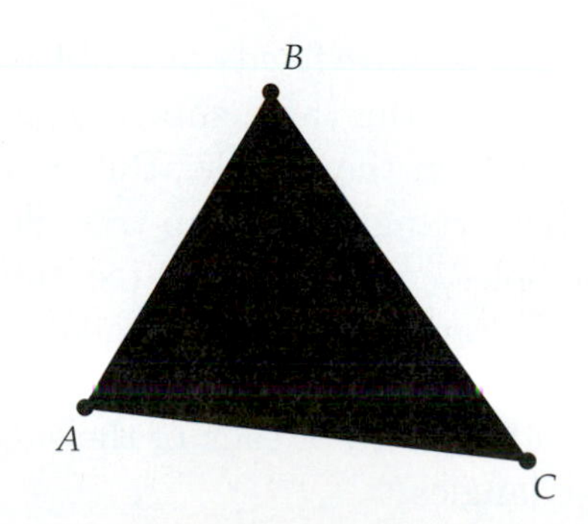

Let L, M, and N be the midpoints of the sides and remove the middle third triangle, $\triangle LMN$.

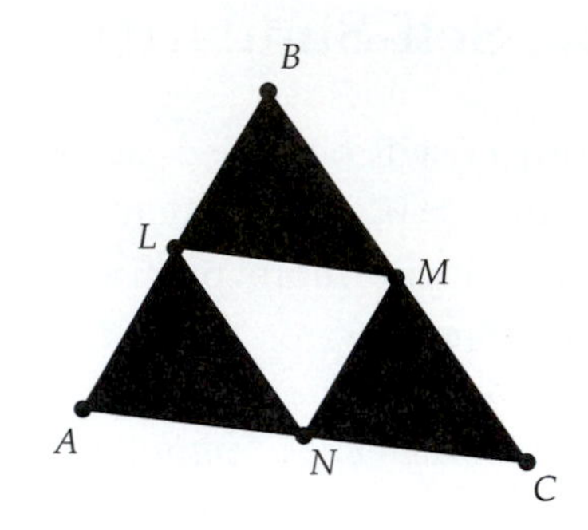

Now, each of the three smaller triangles is *almost* the same figure as the original triangle, when scaled up by a factor of three, except for the "hole" in the middle of the big triangle, which is missing in the smaller triangles. To make the smaller triangles similar to the original, let's remove the middle third of each triangle.

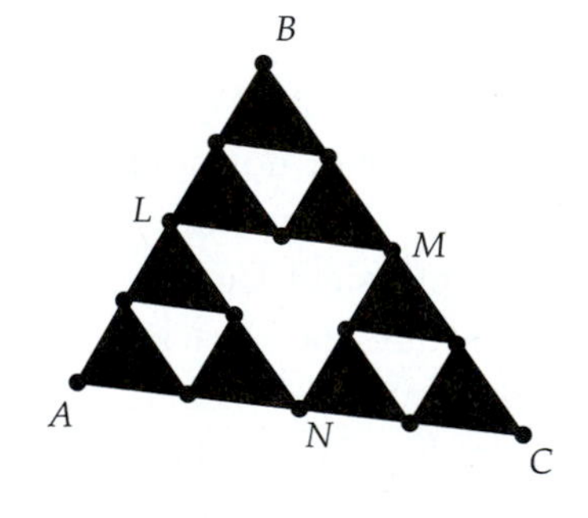

Now, we have fixed our problem, and each of the three sub-triangles has a hole in the middle. But, now the big triangle has sub-triangles with holes, and so the three sub-triangles are, again, not similar to the big triangle. So, we will remove the middle third of each of the sub-sub-triangles.

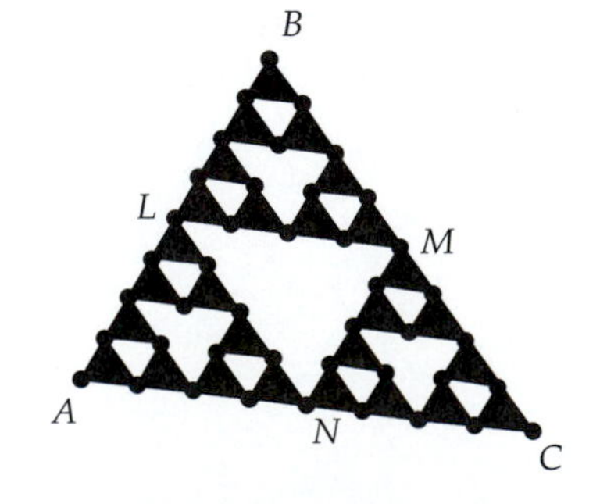

Again, we have *almost* self-similarity. To create a truly self-similar figure, we need to continue this middle third removal process to infinity! Then each sub-triangle will be *exactly* similar to the original triangle, each sub-sub-triangle will be similar to each sub-triangle, and so on, with the scale factor being 3 at each stage.

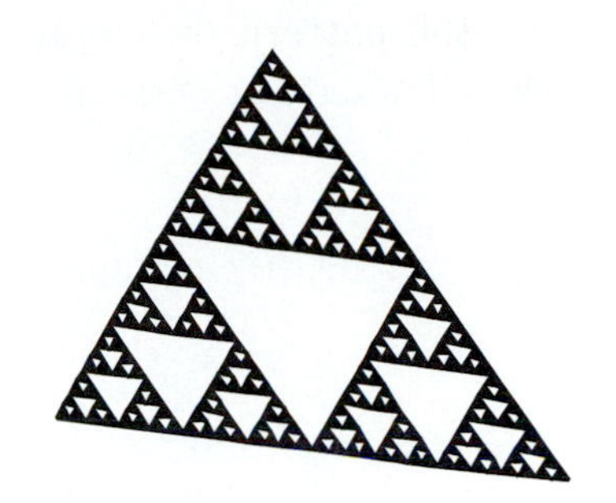

The figure that is left after carrying on the middle third removal process infinitely often is called the *Sierpinski Triangle*, or *Sierpinski Gasket*, in honor of Waclaw Sierpinski (1882–1969), a Polish mathematician who is known for his work in set theory, topology, and analysis.

Note that Sierpinski's Triangle is invariant under three basic similarities, scaling transformations by a factor of $\frac{1}{2}$ toward points A, B, and C.

Since the process of creating Sierpinski's Triangle starts with a simple 2-dimensional filled-in triangle, it is natural to try to calculate the area of the final figure.

Let's denote by *stages* the successive process of removing middle thirds from sub-triangles. At stage 0 we have the original filled-in triangle, $\triangle ABC$. At stage 1 we have removed the middle third. At stage 2 we have removed the middle third of each of the remaining triangles, and so on.

We can assume the area of the first triangle to be anything we like, so we will assume it equal to 1. At stage 1 the area remaining in the figure will be

$$Area(stage\ 1) = 1 - \frac{1}{4}$$

since all of the four sub-triangles are congruent (proved as an exercise), and thus have an area $\frac{1}{4}$ the area of the original triangle.

At stage 2 we remove three small triangles from each of the remaining sub-triangles, each of area $\frac{1}{16}$. The area left is

$$Area(stage\ 2) = 1 - \frac{1}{4} - \frac{3}{16}$$

At stage 3 we remove nine areas, each of area $\frac{1}{64}$. Thus,

$$Area(stage\ 2) = 1 - \frac{1}{4} - \frac{3}{16} - \frac{9}{64}$$

Seeing the pattern developing here, we conclude that the area left for Sierpinski's Triangle at stage n is

$$\begin{aligned} Area(stage\ n) &= 1 - \frac{1}{4}\sum_{k=0}^{n}\left(\frac{3}{4}\right)^k \\ \lim_{n\to\infty} Area &= 1 - \frac{1}{4}\frac{1}{(1-\frac{3}{4})} \\ &= 0 \end{aligned}$$

This is truly an amazing result! The Sierpinski Triangle has had all of its area removed, but still exists as an infinite number of points. Also, it has all of the boundary segments of the original triangle remaining, plus all the segments of the sub-triangles. So, it must be at "least" a 1-dimensional object. We will make this idea of "in-between" dimension more concrete in the next section.

9.2.2 Cantor Set

What kind of shape would we get if we applied the middle-third removal process to a simple line segment?

Here we have a segment $\overline{AB}$.

A B

Remove the middle third to get two segments $\overline{AC}$ and $\overline{DB}$.

A C D B

Perform this process again and again as we did for the Sierpinski Triangle. Here is a collection of the stages of the process, where we have hidden some of the points for clarity.

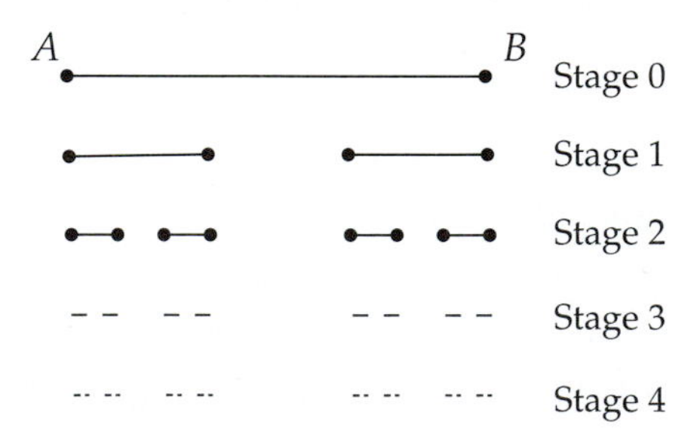

The limiting figure, which has an infinite number of middle thirds removed, is called the *Cantor Set*, in honor of Georg Cantor (1845–1918). Cantor is known for his work in set theory and in particular for his investigation of orders of infinity and denumerable sets. It is left as an exercise to show that the Cantor Set has length 0, although it is made up of an infinite number of points.

Both of our examples so far, Sierpinski's Triangle and the Cantor Set, do not seem to fit into our classical notion of 1- and 2- dimensional objects.

Sierpinski's Triangle is more than a 1-dimensional curve, yet certainly less than an area, which is 2-dimensional, while the Cantor Set lies somewhere between dimensions 0 and 1. How can this be? How can an object have *fractional* dimension? This idea of fractional dimension was the critical organizing principle for Mandelbrot in his study of natural phenomena and is why he coined the term *fractal* for objects with non-integer dimension.

9.3 Similarity Dimension

The dimension of a fractal object turns out to be quite difficult to define precisely. This is because there are several different definitions of dimension used by mathematicians, all having their positive as well as negative aspects. In this section we will look at a fairly simple definition of dimension for self-similar sets, the *similarity dimension.*

To motivate this definition, let's look at some easy examples.

Consider our simplest self-similar object, a line segment $\overline{AB}$.

It takes two segments of size $\frac{AB}{2}$ to cover this segment.

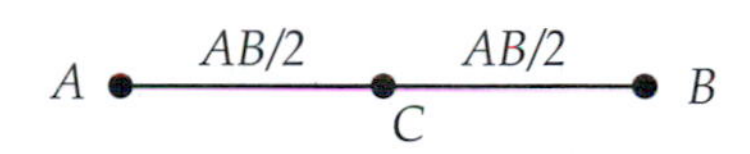

In general, it takes N sub-segments of the original segment of size $\frac{AB}{N}$ to cover the original segment. Or, if we think of the sub-segments as being similar to the original, it takes N sub-segments of *similarity ratio* $\frac{1}{N}$ to cover $\overline{AB}$.

We will define the function $r(N)$ to be the similarity ratio between a figure and its parts.

Suppose we have a rectangle $ABCD$ of side lengths $AB = a$ and $BC = b$. If we subdivide this region into N parts that are similar to the original region, then the area of each subpart will be $\frac{ab}{N}$, and the length of each side will be scaled by the similarity ratio $r(N) = \frac{1}{\sqrt{N}} = \frac{1}{N^{\frac{1}{2}}}$. In the figure at right, we have $N = 9$.

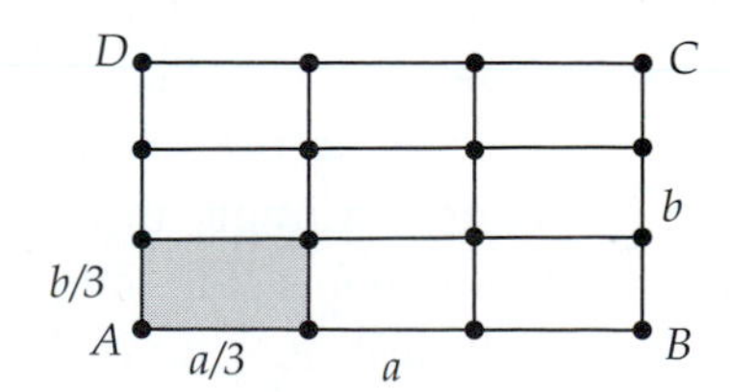

For a rectangular solid in three dimensions, the similarity ratio would be $r(N) = \frac{1}{N^{\frac{1}{3}}}$.

Note that these three similarity ratios hold no matter the size of the segment, rectangle, or solid under consideration. In higher dimensions, the similarity ratio for a d-dimensional rectangular object would be $r(N) = \frac{1}{N^{\frac{1}{d}}}$.

Equivalently, if we let $r = r(N)$, we have

$$Nr^d = 1$$
$$log(N) + d\, log(r) = 0$$
$$d = \frac{log(N)}{-log(r)} = \frac{log(N)}{log(\frac{1}{r})} \tag{9.2}$$

Since the value of d matches our Euclidean notion of dimension for simple self-similar objects, like segments and rectangles, we will define the similarity dimension to be the value of d given by equation 9.2.

Definition 9.3. The *similarity dimension* d of a self-similar object is given by

$$d = \frac{log(N)}{log(\frac{1}{r})} \tag{9.3}$$

where the value of $r = r(N)$ is the ratio by which a ruler measuring a side of the object will change under the assumption that each of N sub-objects can be scaled to form the original object. The ratio r is called the *similarity ratio* of the object.

For example, consider the Cantor Set. Each sub-object is $\frac{1}{3}$ the scale of the original segment, and it takes two sub-objects to make up the whole (after stage 0). Thus, the similarity dimension of the Cantor Set is

$$\frac{log(2)}{log\left(\frac{1}{\frac{1}{3}}\right)} = \frac{log(2)}{log(3)} \approx .6309$$

This result agrees with our earlier intuition about the dimension of the Cantor Set being somewhere between 0 and 1.

As another example, it takes three sub-triangle shapes to scale up and cover the bigger triangle in Sierpinski's Triangle (again, ignoring stage 0). Each sub-triangle has side-length scaled by $\frac{1}{2}$ of the original side-length, thus the similarity ratio is $\frac{1}{2}$, and the similarity dimension is

$$\frac{log(3)}{log\left(\frac{1}{\frac{1}{2}}\right)} = \frac{log(3)}{log(2)} \approx 1.58496$$

Exercise 9.3.1. Show that all sub-triangles at stage 1 of the Sierpinski Triangle process are congruent.

Exercise 9.3.2. Show that the length left after all of the stages of construction of Cantor's Set is 0.

Exercise 9.3.3. *Sierpinski's Carpet* is defined by starting with a square and removing the middle third sub-square, each side of which is $\frac{1}{3}$ the size of the original square.

Here is stage 1 for Sierpinski's Carpet.

The Carpet is defined as the limiting process of successively removing middle-third squares. Does Sierpinski's Carpet have non-zero area? Sketch the next iteration of this figure.

Exercise 9.3.4. Show that the similarity dimension of Sierpinski's Carpet is $\frac{log(8)}{log(3)}$. Thus, Sierpinski's Carpet is much more "area-like" than Sierpinski's Triangle.

Exercise 9.3.5. In the construction of the Sierpinski Carpet, instead of removing just the middle-third square, remove this square and all four squares with which it shares an edge. What is the similarity dimension of the limiting figure?

Exercise 9.3.6. The *Menger Sponge* is defined by starting with a cube and sub-dividing it into 27 sub-cubes. Then, we remove the center cube and all six cubes with which the central cube shares a face. Continue this process repeatedly, each time sub-dividing the remaining cubes and removing pieces. The Menger Sponge is the limiting figure of this process. Show that the dimension of the Menger Sponge is $\frac{log(20)}{log(3)}$. Is the sponge more of a solid or more of a surface? Devise a cube removal process that would produce a limiting figure with fractal dimension closer to a surface than a cube. The Menger Sponge is named for Karl Menger (1902–1985). He is known for his work in geometry and on the definition of dimension.

Exercise 9.3.7. Starting with a cube, is there a removal process that leads to a fractal dimension of 2, yet with a limiting figure that is fractal in nature?

9.4 Project 13 - An Endlessly Beautiful Snowflake

In this project we will use *Geometry Explorer* to create a self-similar fractal. Self-similar fractals are ideal for study using computational techniques, as their construction is basically a *recursive* process—one that loops back upon itself. To construct a recursive process using *Geometry Explorer*, we will make use of the program's ability to record a sequence of geometric constructions and then play the constructions back recursively.

The self-similar fractal we will construct is "Koch's snowflake curve," named in honor of Helge von Koch (1870–1924), a Swedish mathematician who is most famous for the curve that bears his name. It is an example of a continuous curve that is not differentiable at any of its points. In the next few paragraphs we will discuss how to use a "template" curve to recursively build the Koch self-similar snowflake curve. Please read this discussion carefully (you do not need to construct anything yet).

To construct the template curve for the Koch snowflake, we start with a segment $\overline{AB}$. Just as we removed the middle third for the Cantor Set construction, we remove the middle third of $\overline{AB}$, but replace it with the two upper segments of an equilateral triangle of side length equal to the middle third we removed (Fig. 9.2).

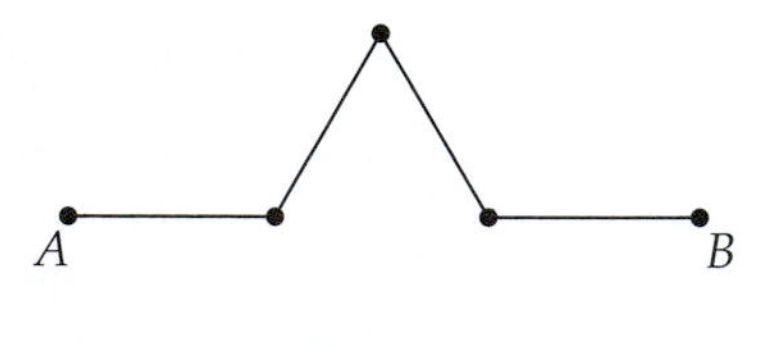

Fig. 9.2

Now think of the process just described as a *replacement* process, where we take a segment and replace it with a new curve, the template curve. Each segment of the new template curve can then be replaced with a copy of the template that is scaled by a factor of $\frac{1}{3}$. If we carry out this replacement process for each of the four small segments in the template, we get the curve shown in Fig. 9.2.

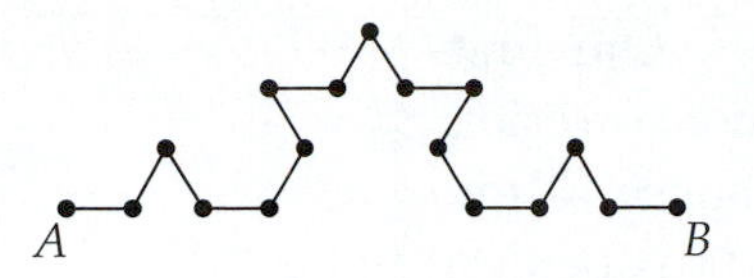

Fig. 9.3

The process just outlined can be made recursive. We could take each of the new segments in the curve just described and replace them with a scaled-down copy of the template. We could then repeatedly take each new set of segments at stage n and replace them with copies of the template to get a curve at stage $n + 1$. Thus, the replacing of segments by copies of the template loops back on itself indefinitely.

At the point where we stopped the replacement process, the curve had 16 small segments, each of length $\frac{1}{9}$ of the original segment $\overline{AB}$. Replacing each of these segments with a $\frac{1}{27}$ scale copy of our template, we get a new curve with 64 segments, each a length $\frac{1}{27}$ of the original. The new curve is shown in Fig. 9.4 where we have hidden all points except A and B for clarity.

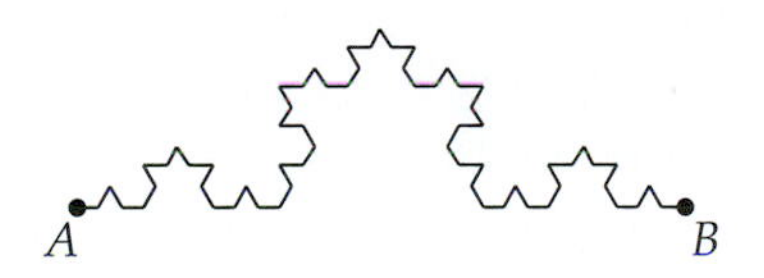

Fig. 9.4

The Koch curve is the curve that results from applying this template replacement process an *infinite* number of times. The curve is self-similar in the sense that if you took a piece of the curve and magnified it by a factor of 3, you would see the same curve again.

Let's see how to use the recording capability of *Geometry Explorer* to construct the Koch curve.

Start *Geometry Explorer*. Our first task is to record the construction of the template. Choose **New Recording** (**File** menu) to open up the Recording window. Click the button labeled "Record" to begin recording. Then, create a segment $\overline{AB}$ on the Canvas. (In this series of figures, we show only that portion of the screen necessary to illustrate recording.)

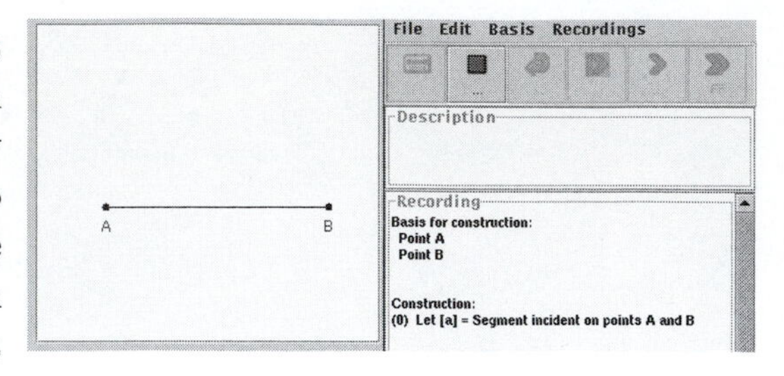

Notice how the Recording window records what we have done. Next, we will divide segment $\overline{AB}$ into three equal parts. Select point A and choose **Center** from the **Mark** pop-up menu in the Transform panel. Then choose **Dilation** from the **Custom** pop-up menu and type in 1 and 3 for the numerator and denominator in the dialog box that will appear. Select point B and click the Dilate button to scale B by $\frac{1}{3}$ toward A.

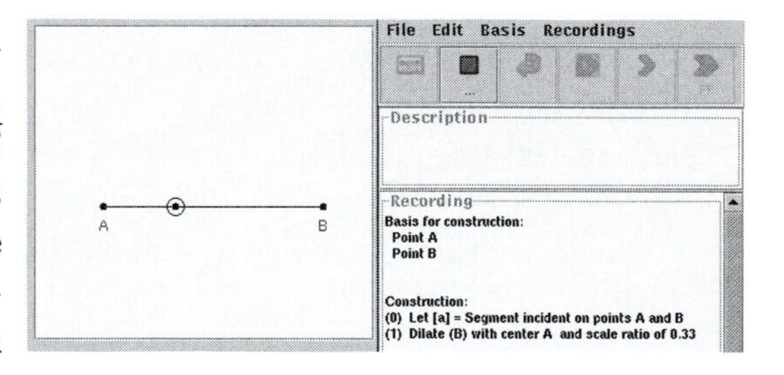

Similarly, carry out the necessary steps to dilate point B by a ratio of $\frac{2}{3}$ toward point A. Then hide segment $\overline{AB}$. We have now split $\overline{AB}$ into equal thirds.

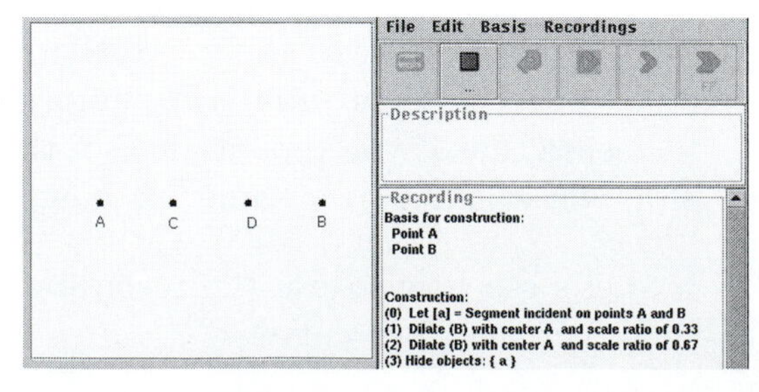

Next we create the "bump" in the middle of the template. As we did before, set C as a new center of rotation/dilation and define a custom rotation of 60 degrees. Then select point D and click on the Rotate button in the Transform panel.

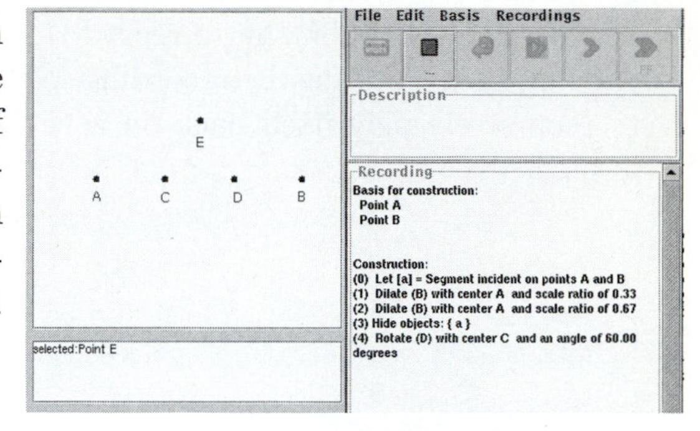

Finally, select points A, C, E, D, and B (in that order) and click on the Open Polygon button in the Construct panel (first button in third row). At this point our template curve is complete. Note how the Recorder has kept track of our constructions. However, do not stop the Recorder yet.

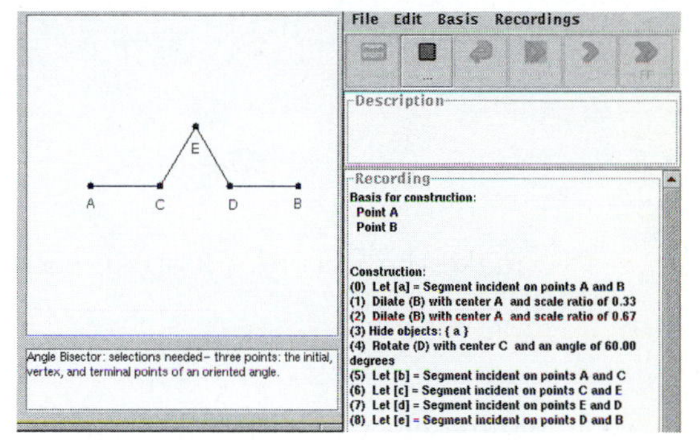

At this point in the Koch curve construction, the template should be used to replace each of the four segments that are in the template itself. That is, we need to make our construction a *recursive* process. We need to apply the same recording that we just completed to each of the four segments of the template curve that is currently in the Canvas. We can do this using the *Loop* button in the Recorder window.

Multi-select points A and C. The Loop button (the one with the arrow looping back on itself) will now be active in the Recording window. In general the Loop button will become active whenever a selection is made that matches the set of basis elements used in the recording. Since this recording has two points (A and B) as basis elements, then any selection of two points will activate the Loop button.

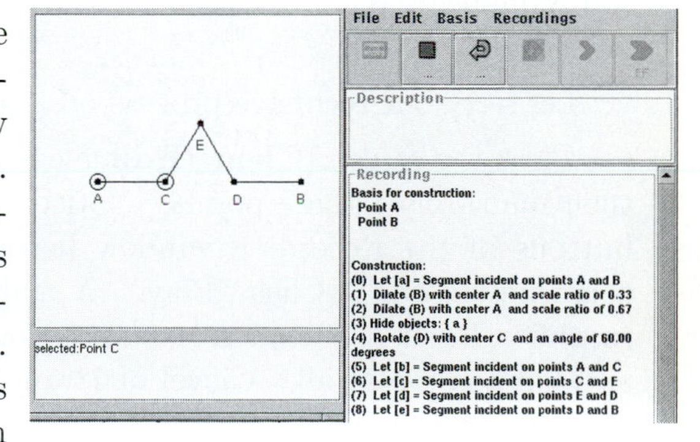

Click the Loop button to record the fact that we want the recording to recursively play itself back on A and C.

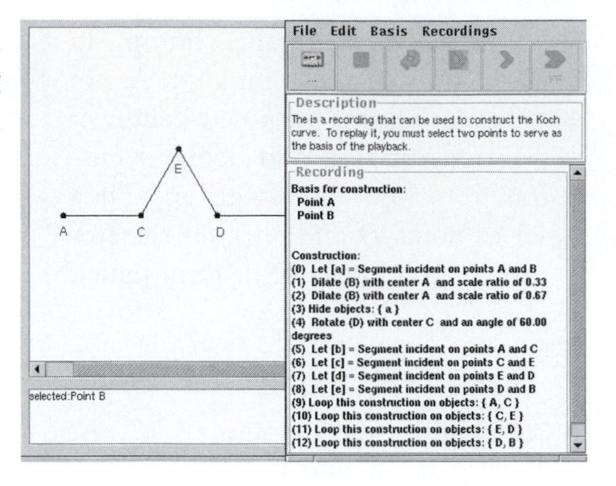

To make the Recorder loop on each of the other three segments, do the following:

1. Multi-select C and E and click Loop.

2. Multi-select E and D and click Loop.

3. Multi-select D and B and click Loop.

We have now completed recording the looping process for the Koch curve, so we should stop the Recorder by clicking on the Stop button in the Recording window.

Let's try our recursive process on a line segment.

Clear the screen (**Clear** (**Edit** menu)). Create two points A and B and then multi-select these points. Notice that the "Step," "Play," and "FF" buttons in the Recording window become active. We can now play our construction back. Click "Play." A dialog box will pop up requesting the recursion level. A recursion level of 0 would mean to just play the recording and not to loop at all. A level of 1 would mean to play the recording back and loop the recording on all sub-segments. Level 2 would mean that sub-sub-segments would be replaced with templates, and so on for level 3, 4, etcetera. Type in 2 and hit Return.

The recording will now play back, recursively descending down segment levels as the template is used to replace smaller and smaller segments. Watch closely how the recording gets played back to get a feel for this recursive process.

The curve is densely packed together, so we will stretch it out by grabbing one of the endpoints and dragging it.

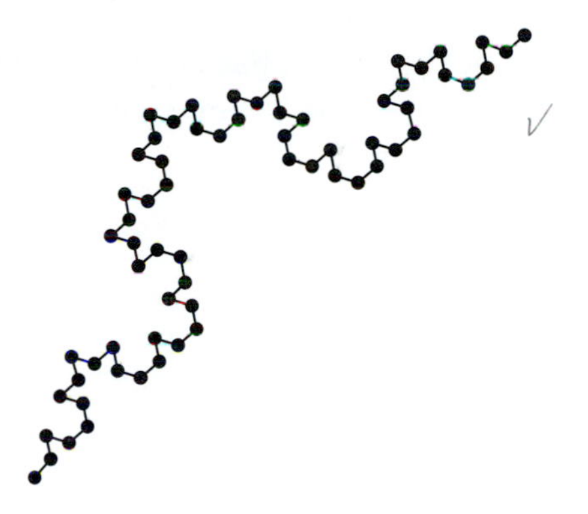

Exercise 9.4.1. Find a formula for the length of the Koch curve at stage n of its construction, with stage 0 being the initial segment $\overline{AB}$, which you can assume has length 1. Is the Koch curve of finite or infinite length?

The Koch Snowflake curve is the result of applying this recursive process to each of three segments of a triangle.

Create a triangle, $\triangle ABC$, in the *Geometry Explorer* Canvas. Then run the Recording we just created on each edge, setting the recursion level at depth 1. We can see why this curve is called the "snowflake" curve.

The Koch snowflake curve is the result of applying the template replacement process an infinite number of times to each of the three edges of $\triangle ABC$.

Exercise 9.4.2. Suppose the area of $\triangle ABC$ is equal to 1. Show that the area enclosed inside the Koch snowflake curve is given by the infinite series

$$Area = 1 + \frac{3}{9} + 4 \cdot \frac{3}{9^2} + 4^2 \cdot \frac{3}{9^3} + \cdots$$

Find the sum of this series.

Exercise 9.4.3. Find the similarity dimension of the Koch snowflake curve.

Let the "Koch hat curve" be the curve obtained by running the recursive process described, but replacing segments at each stage with a different template.

Instead of the triangle template, use the square template at right, where each side of the square is $\frac{1}{3}$ the length of the original segment.

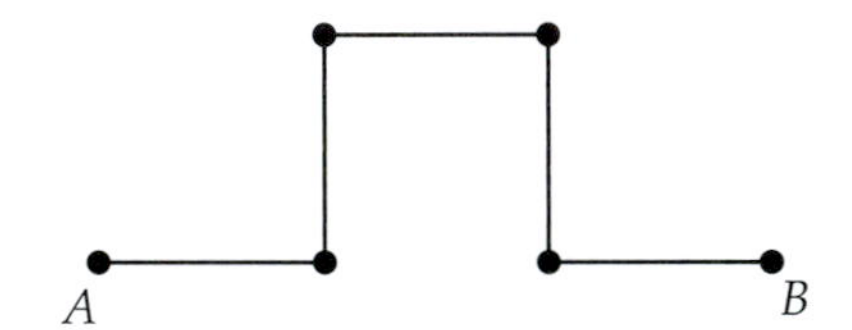

Exercise 9.4.4. Find the next stage of this fractal curve and show that the curve intersects itself. Find the dimension of the Koch hat curve.

Exercise 9.4.5. Design a template different from the two used above, and use it to create a fractal curve. Find the dimension of this new curve.

9.5 Contraction Mappings and the Space of Fractals

In the last few sections we focused our attention on the simplest kinds of self-similar sets, those that are invariant under similarities S_n, all of which share the same scaling ratio c.

For example, the Cantor Set, defined using the interval $[0, 1]$ as the initial figure, will be invariant under the similarities S_1 and S_2, defined by

$$\begin{aligned} S_1(x) &= \frac{x}{3} \\ S_2(x) &= \frac{x}{3} + \frac{2}{3} \end{aligned}$$

Both S_1 and S_2 are special cases of more general functions called *contraction mappings.*

Definition 9.4. A *contraction mapping* S is a function defined on a set D of $(\mathbb{R}^n)$ such that there is a number $0 \leq c < 1$ with

$$|S(x) - S(y)| \leq c\,|x - y|$$

for all x, y in D.

One of the great insights in the subject of fractal geometry is the fact that not only are many fractals invariant under contraction mappings, but the fractal itself can be *generated* from *iterating* the contraction mappings on an initial shape.

Let's consider again the Cantor Set on the interval $[0, 1]$, with the contraction mappings S_1 and S_2 as just defined. Let B_n be the result of applying these contractions repeatedly n times. That is,

$$\begin{aligned}
B_0 &= [0,1] \\
B_1 &= S_1(B_0) \cup S_2(B_0) = [0,\frac{1}{3}] \cup [\frac{2}{3},1] \\
B_2 &= S_1(B_1) \cup S_2(B_1) = S_1([0,\frac{1}{3}]) \cup S_1([\frac{2}{3},1]) \cup S_2([0,\frac{1}{3}]) \cup S_2([\frac{2}{3},1]) \\
&= [0,\frac{1}{9}] \cup [\frac{2}{9},\frac{1}{3}] \cup [\frac{2}{3},\frac{7}{9}] \cup [\frac{8}{9},1] \\
&\;\;\vdots \\
B_n &= S_1(B_{n-1}) \cup S_2(B_{n-1})
\end{aligned}$$

Note that B_0 is the figure from the Cantor Set construction at stage 0; B_1 is the figure at stage 1 (the middle third is gone); B_2 is the figure at stage 3; and so on. Why is this the case? Think of what the two contraction mappings are doing geometrically. The effect of S_1 is to contract everything in the interval $[0, 1]$ into the first third of that interval, while S_2 also contracts by $\frac{1}{3}$, but then shifts by a distance of $\frac{2}{3}$. Thus, the effect of iterating these two maps on stage $k - 1$ of the Cantor Set construction is to shrink the previous set of constructed segments by $\frac{1}{3}$ and then copying them to the intervals $[0, \frac{1}{3}]$ and $[\frac{2}{3}, 1]$. This has an effect equivalent to the removal of all middle-thirds from the previous segments at stage $k - 1$.

Suppose we started our construction with the interval $[2,4]$ instead of $[0,1]$. Then

$$\begin{aligned}
B_0 &= [2,4] \\
B_1 &= S_1(B_0) \cup S_2(B_0) = [\frac{2}{3}, \frac{4}{3}] \cup [\frac{4}{3}, 2] = [\frac{2}{3}, 2] \\
B_2 &= S_1(B_1) \cup S_2(B_1) = [\frac{2}{9}, \frac{2}{3}] \cup [\frac{8}{9}, \frac{4}{3}] \\
B_3 &= [\frac{2}{27}, \frac{2}{9}] \cup [\frac{8}{27}, \frac{4}{9}] \cup [\frac{20}{27}, \frac{8}{9}] \cup [\frac{26}{27}, \frac{10}{9}] \\
B_4 &= [\frac{2}{81}, \frac{2}{27}] \cup \cdots [\frac{80}{81}, \frac{28}{27}]
\end{aligned}$$

and so on.

Note what is happening to the initial and final intervals at each stage of iteration. For example, at stage 3 we have points ranging from $\frac{2}{27}$ to $\frac{10}{9}$, and at stage 4 we have points ranging from $\frac{2}{81}$ to $\frac{28}{27}$. It looks like repeatedly applying S_1 and S_2 to the interval $[2,4]$ is closing in on the interval $[0,1]$. This should not be surprising, as 0 and 1 are *fixed points* of S_1 and S_2. In fact, for any x we have

$$|S_2(x) - S_2(1)| = |S_2(x) - 1| = \frac{1}{3}|x-1|$$

Thus,

$$\begin{aligned}
|S_2^n(x) - 1| &= |S_2^n(x) - S_2^n(1)| = \frac{1}{3}|S_2^{n-1}(x) - S_2^{n-1}(1)| \\
&= \frac{1}{3}^2 |S_2^{n-2}(x) - S_2^{n-2}(1)| \\
&\quad \cdot \\
&\quad \cdot \\
&\quad \cdot \\
&= \frac{1}{3}^n |x-1|
\end{aligned}$$

Clearly, as n grows without bound, $S_2^n(x)$ must approach 1. By a similar argument, we can show $S_1^n(x)$ approaches 0. On the other hand, points inside $[0,1]$ can "survive" forever by the *combined* action of S_1 and S_2. For example, the points $\frac{1}{3}$ and $\frac{2}{3}$ are pulled back and forth by S_1 and S_2, but always by the same amount toward each of the fixed points, and thus they survive all stages of the construction process. No point outside $[0, 1]$ will have this prospect of surviving.

Thus, it appears that the points in the Cantor Set are not uniquely tied to the starting interval $[0, 1]$. We would get the same set of limiting points if we started with *any* interval. The Cantor Set is thus *attracting* the iterates of the two contraction mappings S_1 and S_2.

From this brief example one might conjecture that something like this attracting process is ubiquitous to fractals, and such a result is, in fact, the case.

To prove this, we need to develop some tools for handling the iteration of sets of functions on subsets of Euclidean space. We will first need a few definitions.

Definition 9.5. The distance function in $\mathbb{R}^n$ will be denoted by d. Thus, $d(x, y)$ measures the Euclidean distance from x to y.

Definition 9.6. A set D in $\mathbb{R}^n$ is *bounded* if it is contained in some sufficiently large ball. That is, there is a point a and radius R such that for all $x \in D$, we have $d(x, a) < R$.

Definition 9.7. A sequence $\{x_n\}$ of points in $\mathbb{R}^n$ *converges* to a point x as n goes to infinity, if $d(x_n, x)$ goes to zero. The point x is called the *limit point* of the sequence.

Definition 9.8. A set D in $\mathbb{R}^n$ is *closed* if it contains all of its limit points, that is, if every convergent sequence of points from D converges to a point inside D.

Definition 9.9. An *open ball* of radius ϵ in $\mathbb{R}^n$ is the set of points $\vec{x}$ in $\mathbb{R}^n$ such that $||\vec{x}|| < \epsilon$.

Definition 9.10. A set D in $\mathbb{R}^n$ is *open* if for every x in D, there is an open ball of non-zero radius that is entirely contained in D.

The unit disk $x^2 + y^2 \leq 1$ is closed. The interval $(0, 1)$ is not. The set $x^2 + y^2 < 1$ is open, as is the interval $(0, 1)$. The interval $[0, 1)$ is neither open nor closed.

Definition 9.11. A set D in $\mathbb{R}^n$ is *compact* if any collection of open sets that covers D (i.e., with the union of the open sets containing D) has a finite sub-collection of open sets that still covers D.

It can be shown that if D is compact, then it is also closed and bounded [4, pages 20–25].

We will be working primarily with compact sets and so will define a structure to contain all such sets.

Definition 9.12. The space $\mathcal{H}$ is defined as the set of all compact subsets of $\mathbb{R}^n$.

The "points" of $\mathcal{H}$ will be compact subsets. In the preceding example of the Cantor Set, we can consider the stages B_0, B_1, and so forth, of the construction as a sequence B_n of compact sets, that is, a sequence of "points" in $\mathcal{H}$. It appeared that this sequence *converged* to the Cantor Set. However, to speak of a sequence converging, we need a way of measuring the distance between points in the sequence. That is, we need a way of measuring the distance between compact sets.

Definition 9.13. Let $B \in \mathcal{H}$ and $x \in \mathbb{R}^n$. Then

$$d(x, B) = min\{d(x, y)|y \in B\}$$

Lemma 9.1. *The function $d(x, B)$ is well defined. That is, there always exists a minimum value for $d(x, y)$ where y is any point in B.*

Proof: Let $f(y) = d(x, y)$. Since the distance function is continuous (by definition!), we have that f is a continuous function on a compact set and thus must achieve a minimum and maximum value. (For the proof of this extremal property of continuous function on a compact set, see [4, page 31].) □

Definition 9.14. Let $A, B \in \mathcal{H}$. Then

$$d(A, B) = max\{d(x, B)|x \in A\}$$

The value of $d(A, B)$ will be well defined by a similar continuity argument to the one given in the proof of the last lemma.

As an example, let A be a square and B a triangle as shown at the right. Given $x \in A$, it is clear that $d(x, B) = d(x, y_1)$, where y_1 is on the left edge of triangle B. Then $d(A, B) = d(x_1, y_1)$, where x_1 is on the left edge of the square. On the other hand, $d(B, A) = d(y_2, x_2) \neq d(A, B)$.

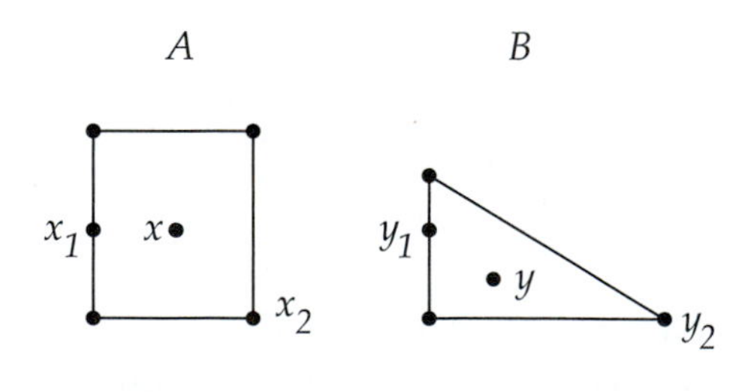

Thus, in general $d(A, B)$ need not equal $d(B, A)$. Since one of the defining conditions for a distance function (or *metric*) is that it be symmetric, we need to modify the definition a bit.

Definition 9.15. The *Hausdorff distance* between A and B in $\mathcal{H}$ is given by

$$d_{\mathcal{H}}(A, B) = max\{d(A, B), d(B, A)\}$$

The Hausdorff distance function satisfies all of the requirements for a metric. That is,

1. $d_{\mathcal{H}}(A, B) = d_{\mathcal{H}}(B, A)$
2. $d_{\mathcal{H}}(A, A) = 0$
3. $d_{\mathcal{H}}(A, B) > 0$ if $A \neq B$
4. $d_{\mathcal{H}}(A, B) \leq d_{\mathcal{H}}(A, C) + d_{\mathcal{H}}(C, B)$ for A, B, C in $\mathcal{H}$. This is called the *Triangle Inequality.*

Condition 1 is true by the way we defined the Hausdorff distance. Conditions 2 and 3 are left as exercises.

Lemma 9.2. *The Hausdorff distance function satisfies the Triangle Inequality.*

Proof: Let a, b, and c be points in A, B, and C, respectively. Using the Triangle Inequality for Euclidean distance, we have that $d(a, b) \leq d(a, c) + d(c, b)$. Thus,

$$\begin{aligned}
d(a, B) &= min\{d(a, b)|b \in B\} \\
&\leq min\{d(a, c) + d(c, b)|b \in B\} \; for \; all \; c \in C \\
&\leq d(a, c) + min\{d(c, b)|b \in B\} \; for \; all \; c \in C \\
&\leq min\{d(a, c)|c \in C\} + min\{min\{d(c, b)|b \in B\}|c \in C\} \\
&\leq d(a, C) + max\{min\{d(c, b)|b \in B\}|c \in C\} \\
&\leq d(a, C) + d(C, B)
\end{aligned}$$

And so,

$$\begin{aligned}
d(A, B) &= max\{d(a, B)|a \in A\} \\
&\leq max\{d(a, C) + d(C, B)|a \in A\} \\
&\leq d(A, C) + d(C, B)
\end{aligned}$$

Likewise, $d(B,A) \leq d(B,C) + d(C,A)$. So,

$$\begin{aligned} d_{\mathcal{H}}(A,B) &= max\{d(A,B), d(B,A)\} \\ &\leq max\{d(A,C)+d(C,B), d(B,C)+d(C,A)\} \\ &\leq max\{d(A,C), d(C,A)\} + max\{d(B,C), d(C,B)\} \\ &\leq d_{\mathcal{H}}(A,C) + d_{\mathcal{H}}(C,B) \end{aligned}$$

□

We are now in a position to prove that contraction mappings on $\mathbb{R}^n$ generate contraction mappings on $\mathcal{H}$.

Theorem 9.3. *Let S be a contraction mapping on $\mathbb{R}^n$, with ratio $c < 1$. Then S is a contraction mapping on compact sets in $\mathcal{H}$.*

Proof: First, since a contraction map is continuous (proved as an exercise), then if D is a non-empty compact set, $S(D)$ must also be a non-empty compact set. So, S is a well-defined mapping from $\mathcal{H}$ to itself.

Now, for A and B in $\mathcal{H}$:

$$\begin{aligned} d(S(A), S(B)) &= max\{min\{d(S(x), S(y)) | y \in B\} | x \in A\} \\ &\leq max\{min\{c\, d(x,y) | y \in B\} | x \in A\} \\ &\leq c\, max\{min\{d(x,y) | y \in B\} | x \in A\} \\ &\leq c\, d(A,B) \end{aligned}$$

Likewise, $d(S(B), S(A)) \leq c\, d(B,A)$, and so

$$\begin{aligned} d_{\mathcal{H}}(S(A), S(B)) &= max\{d(S(A), S(B)), d(S(B), S(A))\} \\ &\leq max\{c\, d(A,B), c\, d(B,A)\} \\ &\leq c\, d_{\mathcal{H}}(A,B) \end{aligned}$$

□

Theorem 9.4. *Let $S_1, S_2, \ldots, S_n$ be contraction mappings in $\mathbb{R}^n$ with ratios $c_1, c_2, \ldots, c_n$. Define a transformation S on $\mathcal{H}$ by*

$$S(D) = \bigcup_{i=1}^{n} S_i(D)$$

Then, S is a contraction mapping on $\mathcal{H}$, with contraction ratio $c = max\{c_i | i = 1, \ldots, n\}$, and S has a unique fixed set $F \in \mathcal{H}$ given by

$$F = \lim_{k \to \infty} S^k(E)$$

for any non-empty $E \in \mathcal{H}$.

To prove this theorem, we will use the following lemma.

Lemma 9.5. *Let A, B, C, and D be elements of $\mathcal{H}$. Then*

$$d_{\mathcal{H}}(A \cup B, C \cup D) \leq max\{d_{\mathcal{H}}(A, C), d_{\mathcal{H}}(B, D)\}$$

Proof: First, $d(A \cup B, C) = max\{d(A, C), d(B, C)$ (proved as an exercise). Second, $d(A, C \cup D) \leq d(A, C)$ and $d(A, C \cup D) \leq d(A, D)$ (proved as an exercise). Thus,

$$\begin{aligned}
d_{\mathcal{H}}(A \cup B, C \cup D) &= max\{d(A \cup B, C \cup D),\ d(C \cup D, A \cup B)\} \\
&= max\{max\{d(A, C \cup D),\ d(B, C \cup D)\}, \\
&\qquad max\{d(C, A \cup B),\ d(D, A \cup B)\}\} \\
&\leq max\{max\{d(A, C), d(B, D)\},\ max\{d(C, A), d(D, B)\}\} \\
&\leq max\{max\{d(A, C), d(C, A)\},\ max\{d(B, D), d(D, B)\}\} \\
&\leq max\{d_{\mathcal{H}}(A, C),\ d_{\mathcal{H}}(B, D)\}
\end{aligned}$$

□

Now for the proof of Theorem 9.4. We will prove the result in the case where $n = 2$. Let A and B be in $\mathcal{H}$. Then, by the previous lemma, we have

$$\begin{aligned}
d_{\mathcal{H}}(S(A), S(B)) &= d_{\mathcal{H}}(S_1(A) \cup S_2(A),\ S_1(B) \cup S_2(B)) \\
&\leq max\{d_{\mathcal{H}}(S_1(A), S_1(B)),\ d_{\mathcal{H}}(S_2(A), S_2(B))\} \\
&\leq max\{c_1\, d_{\mathcal{H}}(A, B),\ c_2\, d_{\mathcal{H}}(A, B)\} \\
&\leq c\, d_{\mathcal{H}}(A, B)
\end{aligned}$$

Now let E be a non-empty set in $\mathcal{H}$. Then assuming $m \leq n$, we have

$$\begin{aligned}
d_{\mathcal{H}}(S^m(E), S^n(E)) &\leq c d_{\mathcal{H}}(S^{m-1}(E), S^{n-1}(E)) \\
&\quad \vdots \\
&\leq c^{n-m} d_{\mathcal{H}}(E, S^{n-m}(E))
\end{aligned}$$

Also, by the Triangle Inequality, we have

$$\begin{aligned}
d_{\mathcal{H}}(E, S^k(E)) &\leq d_{\mathcal{H}}(E, S(E)) + d_{\mathcal{H}}(S(E), S^2(E)) + \cdots \\
&\quad + d_{\mathcal{H}}(S^{k-1}(E), S^k(E)) \\
&\leq (1 + c + c^2 + \cdots + c^{k-1})\, d_{\mathcal{H}}(E, S(E)) \\
&\leq \left(\frac{c^k}{1 - c}\right) d_{\mathcal{H}}(E, S(E)) \\
&\leq \left(\frac{1}{1 - c}\right) d_{\mathcal{H}}(E, S(E))
\end{aligned}$$

Thus,

$$d_{\mathcal{H}}(S^m(E), S^n(E)) \leq \left(\frac{c^{n-m}}{1-c}\right) d_{\mathcal{H}}(E, S(E))$$

Since $d_{\mathcal{H}}(E, S(E))$ is fixed and $0 \leq c < 1$, we can make the term $d_{\mathcal{H}}(S^m(E), S^n(E))$ as small as we want. A sequence having this property is called a *Cauchy sequence.* It is a fact from real analysis that Cauchy sequences converge in $\mathbb{R}^n$. Thus, this sequence must converge to some set F in $\mathcal{H}$.

If F' were another compact fixed set of S, then the sequence $\{S^k(F')\}$ must converge to F, which means the distance from F to F' goes to zero, and so $F = F'$. □

Definition 9.16. The set F that is the limit set of a system of contraction mappings S_k on $\mathbb{R}^n$ is called the *attractor* of the system. A system of contraction mappings that is iterated on a compact set is called an *iterated function system* or *IFS.*

Note how Theorem 9.4 confirms our conjecture arising from the Cantor Set construction. This theorem guarantees that the Cantor Set is actually the attractor for the IFS consisting of the two contractions S_1 and S_2 that were used in the set's construction.

For another example, let's return to the Sierpinski Triangle, defined on an initial triangle ΔABC. This figure is fixed under three contractions: scaling by $\frac{1}{2}$ toward A, scaling by $\frac{1}{2}$ toward B, and scaling by $\frac{1}{2}$ toward C.

Suppose that points A, B, and C are at positions $(0,0)$, $(0,1)$, and $(1,1)$. Then the three contractions are defined as $S_1(x,y) = \frac{1}{2}(x,y)$, $S_2(x,y) = \frac{1}{2}(x,y) + (0, \frac{1}{2})$, and $S_3(x,y) = \frac{1}{2}(x,y) + (\frac{1}{2}, \frac{1}{2})$.

Suppose we iterate these three contractions on the unit square shown at the right.

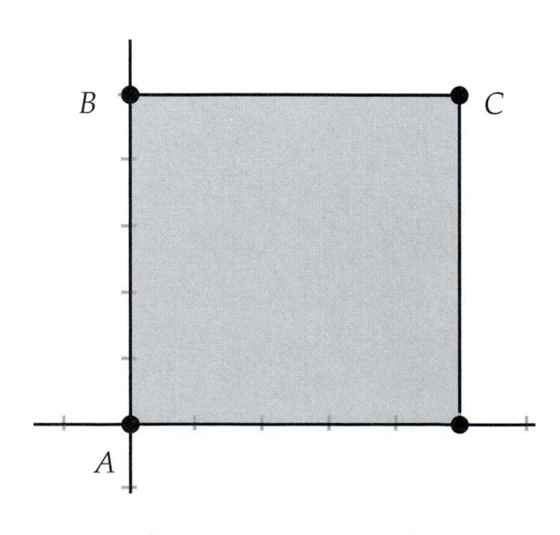

After one iteration of the three contractions, we would have three sub-squares as shown.

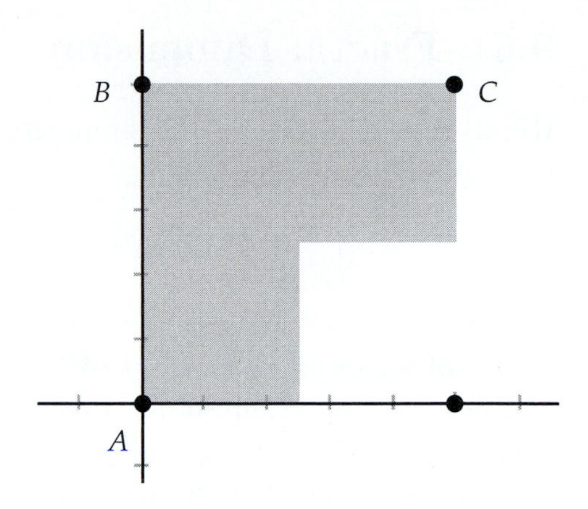

After two iterations, we would have nine sub-squares.

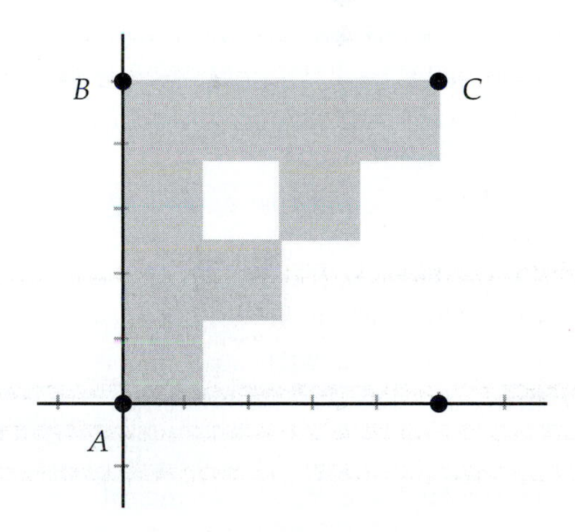

After one more iteration, we begin to see the Sierpinski Triangle take shape, as this is the attractor of the set of three contractions.

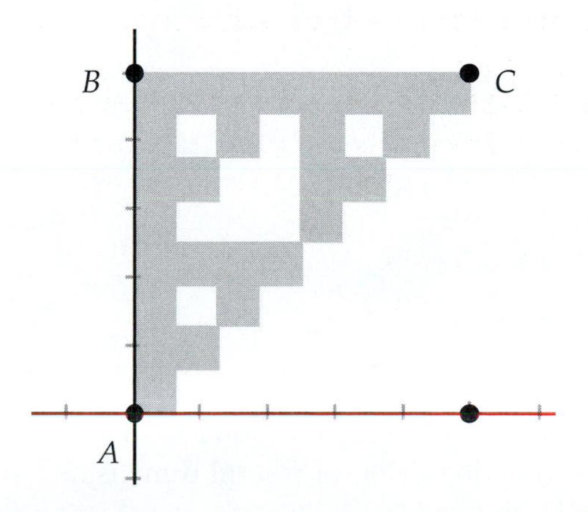

9.6 Fractal Dimension

Recall our definition of the similarity dimension of a self-similar fractal as the valueof

$$d = \frac{log(N)}{log(\frac{1}{r})}$$

where $r = r(N)$ was the similarity ratio, the ratio by which a ruler measuring a side of an object at stage i of the construction would change under the assumption that N sub-objects can be scaled to exactly cover the object at stage i.

This definition assumes that the scaling of sub-parts of the fractal *exactly* matches the form of the fractal itself, the scaling factor is uniform throughout the fractal, and there is only one such scaling factor.

It would be nice to have a definition of fractal dimension that does not suffer from all of these constraints. In this section we will expand the similarity dimension concept to a more general covering-scaling concept of dimension.

First, we need to explicitly define the notion of a *covering set.*

Definition 9.17. Let A be a compact set. Let $B(x, \epsilon) = \{y \in A | d(x, y) \leq \epsilon\}$. That is, $B(x, \epsilon)$ is a closed ball centered at x of radius ϵ. Then $\bigcup_{n=1}^{M} B(x_n, \epsilon)$, with $x_n \in A$, is an ϵ*-covering* of A if for every $x \in A$, we have $x \in B(x_n, \epsilon)$ for some n.

Here we have a $\frac{1}{2}$ covering of the unit square. Note that this covering uses eight disks (2-dimensional balls) to cover the square.

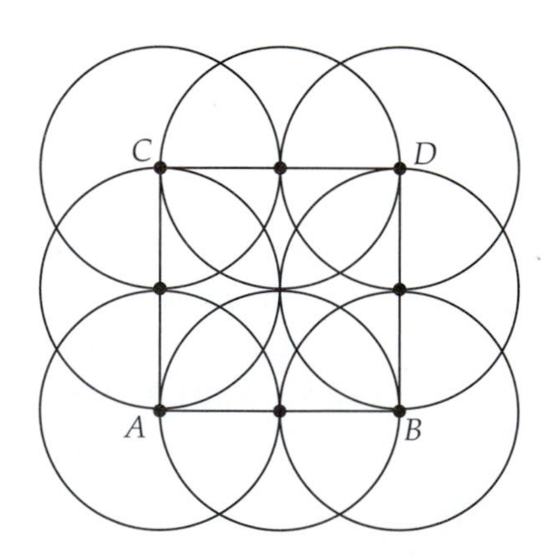

Clearly, there must be some minimal number of circles of radius $\frac{1}{2}$ that will cover the unit square. This minimal covering number will be used in our definition of fractal dimension.

Definition 9.18. Let A be a compact set. Then the *minimal ϵ covering number* for A is

$$\mathcal{N}(A,\epsilon) = min\{M | A \subset \bigcup_{n=1}^{M} B(x_n,\epsilon)\}$$

for some set of points $\{x_n\}$.

We note that the value of $\mathcal{N}(A,\epsilon)$ is well defined, as every set can be covered by some collection of *open* ϵ-balls. Since A is compact, then there is a finite sub-collection that still covers A, and the closure of this collection of open balls will also cover A. Thus, A has at least one finite ϵ-covering and, thus, must have one with the fewest number of elements.

Intuitively, as ϵ changes, the minimum number of balls needed to cover A should also change. For a line segment, the two should be directly related: if we decrease ϵ by a factor of $\frac{1}{2}$, the number of segments (1-dimensional balls) should go up by $2 = 2^1$. If A is a rectangle and we decrease ϵ by a factor of $\frac{1}{2}$, the number of disks should go up by $4 = 2^2$. For a cube, the increase in the number of balls should be $8 = 2^3$.

This was precisely the relationship we noticed when defining the similarity dimension, so it makes sense to define the fractal dimension to mirror the definition of similarity dimension.

Definition 9.19. Let A be a compact set. Then if

$$D_F = \lim_{\epsilon \to 0} \frac{log(\mathcal{N}(A,\epsilon))}{log(\frac{1}{\epsilon})}$$

exists, we call D_F the *fractal dimension* of A.

Using this definition, it is not hard to prove the following result.

Theorem 9.6. *(Box-Counting Theorem) Let A be compact. Cover $\mathbb{R}^n$ by just-touching square boxes of side-length $\frac{1}{2^n}$. Let $\mathcal{N}_n(A)$ be the number of boxes that intersect A. Then*

$$D_F(A) = lim_{n\to\infty} \frac{log(\mathcal{N}_n(A))}{log(2^n)}$$

The proof amounts to showing that the boxes in a covering of A can be trapped between two sequences of ϵ balls that both converge to the fractal dimension. The proof can be found in [4].

Finally, we have the following simplification of the calculation of fractal dimension in the case of an *iterated function system* (IFS).

Theorem 9.7. *(IFS Fractal Dimension Theorem) Let $\{S_n\}_{n=1}^M$ be an IFS and let A be its attractor. Assume that each S_n is a transformation with scale ratio $0 < c_n < 1$, and assume that, at each stage of its construction, portions of the fractal meet only at boundary points. Then the fractal dimension is the unique number $D = D_F(A)$ such that*

$$\sum_{n=1}^{M} c_n^D = 1$$

Again, the proof can be found in [4]. For example, for the Cantor Set, at each stage, the fractal consists of completely distinct segments, which do not intersect at all. Thus, we have that $\left(\frac{1}{3}\right)^D + \left(\frac{1}{3}\right)^D = 1$, and so $D = \frac{log(2)}{log(3)}$.

This theorem would also apply to Sierpinski's Triangle, as portions of the fractal at each stage in the construction are triangles that meet each other only at boundary segments.

Exercise 9.6.1. Show that a contraction mapping must be a continuous function on its domain.

Exercise 9.6.2. Give an example of two compact subsets, A and B of $\mathbb{R}^2$, with $d(A, B) = d(B, A)$.

Exercise 9.6.3. Prove that conditions 2 and 3 in the list of properties for a metric hold for the Hausdorff distance function.

Exercise 9.6.4. Prove the first statement in the proof of Lemma 9.5. That is, show that $d(A \cup B, C) = max\{d(A, C), d(B, C)\}$.

Exercise 9.6.5. Prove the second statement in the proof of Lemma 9.5. That is, show that $d(A, C \cup D) \leq d(A, C)$ and $d(A, C \cup D) \leq d(A, D)$.

Exercise 9.6.6. Use the construction for Sierpinski's Triangle based on the points $A = (0, 0)$, $B = (0, 1)$, and $C = (1, 1)$ to show that $\mathcal{N}_1(S) = 3$, $\mathcal{N}_2(S) = 3^2$, $\mathcal{N}_3(S) = 3^3$, and so on, where S is Sierpinski's Triangle. Use this to find the fractal dimension. How does the fractal dimension compare to the similarity dimension?

Exercise 9.6.7. Use the IFS Fractal Dimension Theorem to compute the fractal dimension of Sierpinski's Triangle.

Exercise 9.6.8. Show that the IFS consisting of $S_1(x) = \frac{1}{2}x$ and $S_2(x) = \frac{2}{3}x + \frac{1}{3}$ does not meet the overlapping criterion of Theorem 9.7. Nevertheless, by identifying the attractor for this IFS, show that its fractal dimension can be calculated and is equal to 1.

9.7 Project 14 - IFS Ferns

We saw in the last few sections that a compact set A can be realized as the *attractor* of an iterated function system (IFS), a system of contraction mappings whose fixed set is A.

It should not be too surprising, then, that natural shapes can arise as the attractors of IFS systems. In this project we will look at an especially pretty attractor—the 2-dimensional outline of a fern.

Consider the fern shown in Fig. 9.5. Each leaf is made up of sub-leaves similar to the original leaf, and each sub-leaf has sub-sub-leaves similar to the sub-leaves, and so on.

Fig. 9.5 Fern Leaf

How can we find the contraction mappings that lead to the fern? Michael Barnsley in [4] proved that to find an IFS whose attractor approximates a given shape F, one only needs to find *affine mappings* that when applied to F yield a union, or *collage* of shapes that approximate F. Barnsley called this the *Collage Theorem.*

Definition 9.20. An *affine mapping* S on the plane is a function of the form

$$S(x, y) = (ax + by + e, cx + dy + f) \tag{9.4}$$

where a, b, c, d, e, and f are real constants.

Affine mappings are generalizations of the planar isometries we studied in Chapter 5. For example, if $a = \cos(\theta)$, $b = -\sin(\theta)$, $c = \sin(\theta)$, $d = \cos(\theta)$, and $e = f = 0$, then S will be a rotation of angle θ about the origin. Note that affine mappings are not necessarily invertible. If they are invertible, we will call them *affine transformations.*

In this project we will discover a set of affine mappings that will split the fern into a collage of sub-ferns that can be reassembled into the original shape of the fern.

Start *Geometry Explorer* and choose **Image** (**View** menu). A file dialog box will pop up (Fig. 9.6). On the right side of the dialog box, there is an Image Preview area and a button labeled "Examples." Click on this button to go to the directory where the fern image is stored. Scroll down (if necessary) and click on the file labeled "fern.jpg." In a few seconds, the image of the file will appear in the preview area of the dialog box.

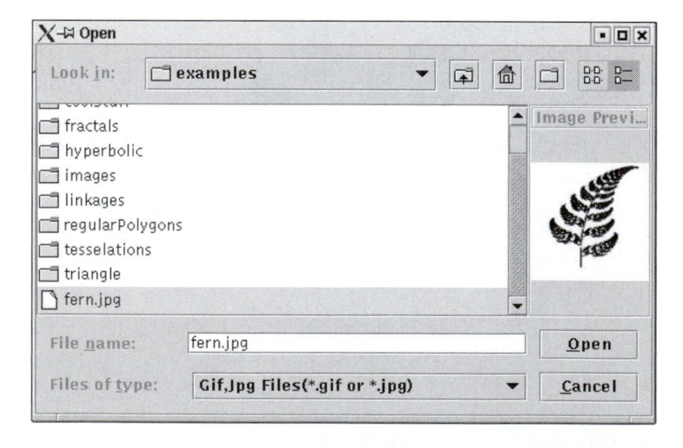

Fig. 9.6

Click "Open" in the dialog box to load the fern image into the main *Geometry Explorer* window. Create point A at the very tip of the fern, C at the second branch point, and B at the base, as shown at right.

To split the fern into a collage of similar pieces, we will start with the set of branches totally enclosed in the polygon $ADECF$. Clearly, this sub-fern of branches is similar to the entire fern, if we scale the sub-fern by an appropriate scaling factor.

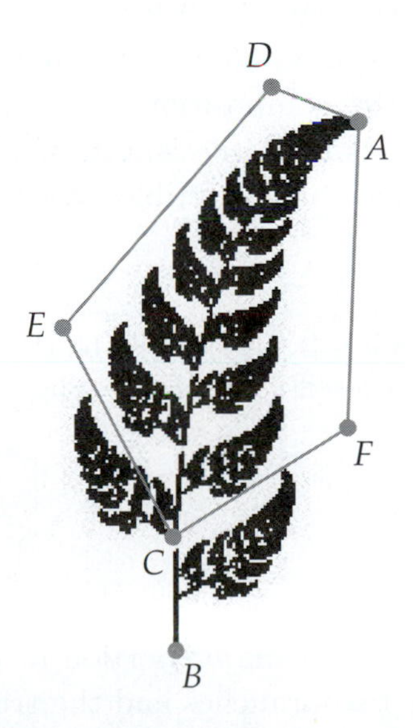

To calculate the scaling factor, we will measure the ratio of AC to AB. This should match the scaling factor of the sub-fern to the original fern. Multi-select A and C and choose **Distance** (**Measure** menu). Similarly, calculate AB and use the Calculator (**Calculator** (**View** menu)) to compute the ratio of these two distances. (Note: Your distances may vary from the ones shown. This is okay—we are interested only in the *ratio* of these distances.)

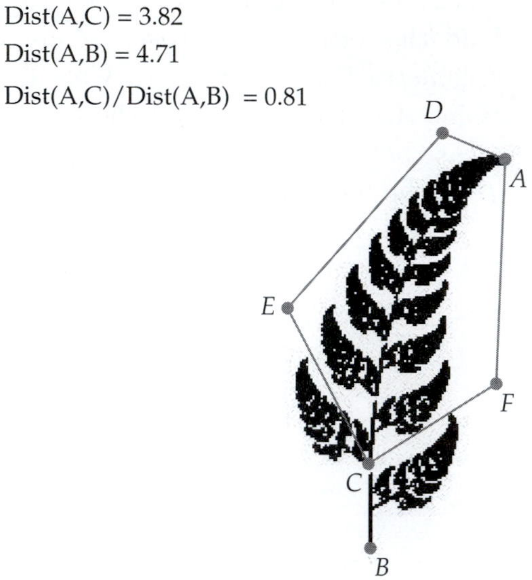

Dist(A,C) = 3.82
Dist(A,B) = 4.71
Dist(A,C)/Dist(A,B) = 0.81

It appears that the ratio of the sub-fern to the entire fern is about 0.8. This is the scaling factor we sought. However, the sub-fern is not just a scaled-down version of the bigger fern. After scaling the big fern by a scale factor of 0.8, we need to turn it slightly, about 5 degrees clockwise, and translate it upward by the length of $\overline{BC}$ to exactly match the outline of the big fern. Measure the length of $\overline{BC}$. (Again, your measurement may vary from the value shown.)

The affine mapping that will take the bigger fern to the smaller fern is then the composition of a rotation by 5 degrees, a scaling by 0.8, and a translation by the length of $\overline{BC}$, which we will denote by h. All of these functions are invertible, so the composition will be a transformation of the plane.

Exercise 9.7.1. Let T_1 be the transformation taking the entire fern to the sub-fern enclosed by $ADECF$. Show that

$$T_1(x,y) \approx \begin{bmatrix} 0.8 & 0.07 \\ -0.07 & 0.8 \end{bmatrix} \begin{pmatrix} x \\ y \end{pmatrix} + \begin{pmatrix} 0.0 \\ h \end{pmatrix}$$

We now have an affine transformation, T_1, that when applied to the fern will cover a major portion of the fern itself. But, we are still missing the lower two branches and the trunk.

Consider the lower branches and in particular the one enclosed by the polygon $CHGJ$. Measure appropriate distances to verify that the scaling factor for this sub-fern is about 0.3.

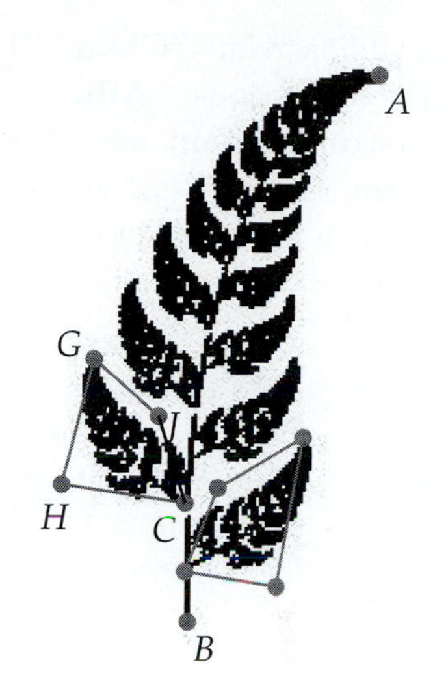

Exercise 9.7.2. Let T_2 be the transformation taking the entire fern to the sub-fern enclosed by $CHGJ$. The angle of rotation for T_2 appears to be about 50 degrees. Show that

$$T_2(x, y) \approx \begin{bmatrix} 0.19 & -0.23 \\ 0.23 & 0.19 \end{bmatrix} \begin{pmatrix} x \\ y \end{pmatrix} + \begin{pmatrix} 0.0 \\ h \end{pmatrix}$$

Exercise 9.7.3. Let T_3 be the transformation taking the entire fern to the bottom right sub-fern. Show that the scaling factor for T_3 is about 0.3, the angle of rotation for T_3 is about -60 degrees, and T_3 includes a reflection about the y-axis. Use this information to show that

$$T_3(x, y) \approx \begin{bmatrix} -0.15 & 0.26 \\ 0.26 & 0.15 \end{bmatrix} \begin{pmatrix} x \\ y \end{pmatrix} + \begin{pmatrix} 0.0 \\ \frac{h}{2} \end{pmatrix}$$

Using the affine transformations T_1, T_2, and T_3, we can *almost* cover the original fern with copies of itself. The only piece of the fern missing from this collage is the small piece of trunk between B and C. Since this is about $\frac{1}{5}$ of the height of the fern, we can just squash the fern down into a vertical line of length 0.2. The affine mapping that will accomplish this is

$$T_4(x, y) \approx \begin{bmatrix} 0.0 & 0.0 \\ 0.0 & 0.2 \end{bmatrix} \begin{pmatrix} x \\ y \end{pmatrix} + \begin{pmatrix} 0.0 \\ 0.0 \end{pmatrix}$$

We note here that T_4, while a valid affine mapping, is not a transformation. It is not one-to-one and, thus, not invertible.

We have now completed our splitting of the fern into a collage of four sub-pieces. According to the Collage Theorem, an IFS consisting of T_1, T_2, T_3, and T_4 should have as its attractor a shape that approximates the original fern. Let's see if that is true.

Clear the screen (**Clear** (**Edit** menu)), and choose **Affine** from the **Custom** pop-up menu in the Transform panel. A dialog box will appear as shown at right.

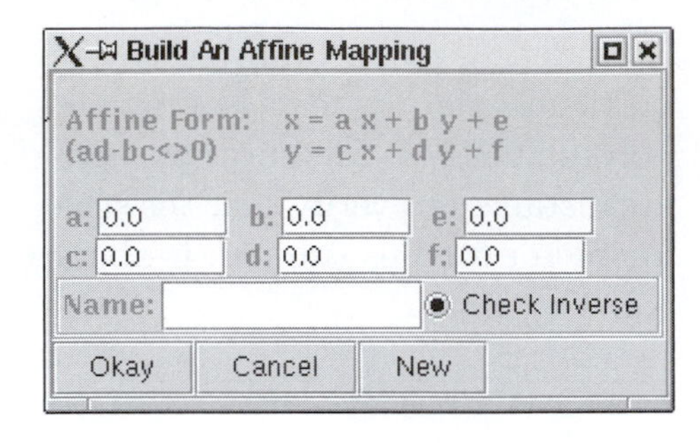

By using this dialog box, we can define an affine mapping for use in *Geometry Explorer*. The affine mapping will have the general form

$$T(x, y) = \begin{bmatrix} a & b \\ c & d \end{bmatrix} \begin{pmatrix} x \\ y \end{pmatrix} + \begin{pmatrix} e \\ f \end{pmatrix}$$

For example, the transformation T_1 defined earlier has $a = 0.8$, $b = 0.07$, $c = -0.07$, $d = 0.8$, $e = 0.0$, and $f = 1.0$. (We are using $f = h = 1.0$ from the earlier measurements.)

Type these values into the dialog box text fields and name the map $T1$ as shown. Then, hit the New button to have *Geometry Explorer* store this function.

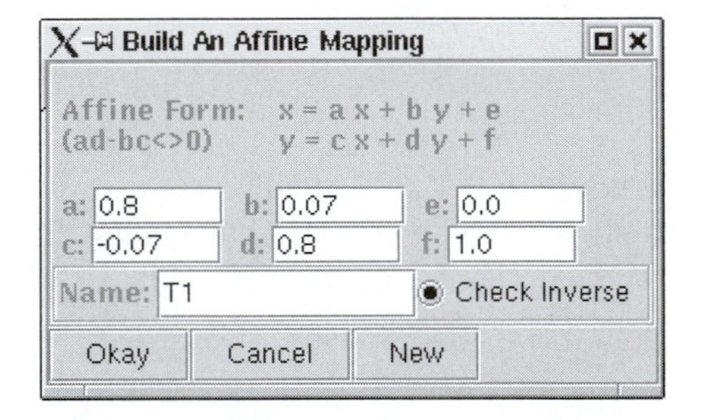

Similarly, type in the values for T_2 and T_3, each time hitting the New button to complete the definition of each transformation. Finally, type in the value of $d = 0.2$ for T_4 as shown, but don't hit the New button.

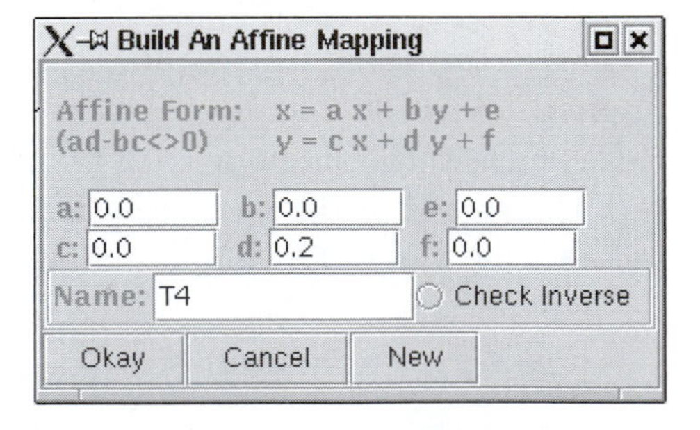

Since T_4 is a *non-invertible* function, we need to turn off the inverse checking that *Geometry Explorer* does for affine transformations. (This is the only place in *Geometry Explorer* where non-invertible maps of the plane are allowed.) Click the toggle button labeled "Check Inverse" to disable inverse checking. Then hit the Okay button as we are finished defining our affine maps.

At this point we should have four new affine mappings that we can apply to arbitrary geometric objects. Click on the **Custom** pop-up menu in the Transform panel. The four affine mappings should appear.

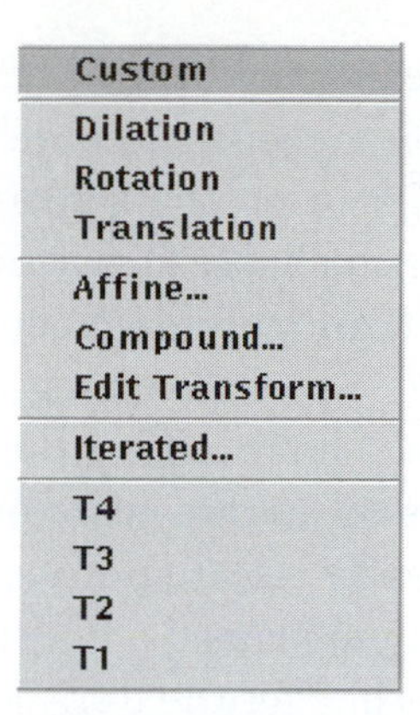

We will now use these four mappings to define an iterated function system (IFS). Choose **Compound** from the **Custom** pop-up menu.

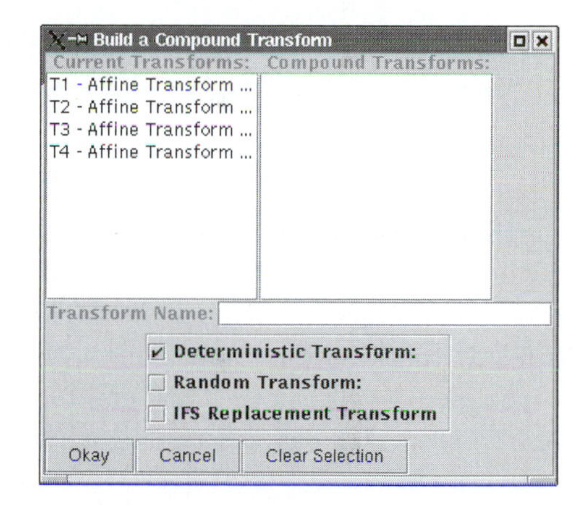

Currently defined transformations (and other mappings) appear in the left column. We want to define a *compound* function using all four of our new affine mappings. To do this, click on each mapping's name, placing it into the right column. Then name the compound function. In this example we have named it "ifs-fern." Finally, click on the checkbox labeled "IFS Replacement Transform" to define the new compound function as an IFS system.

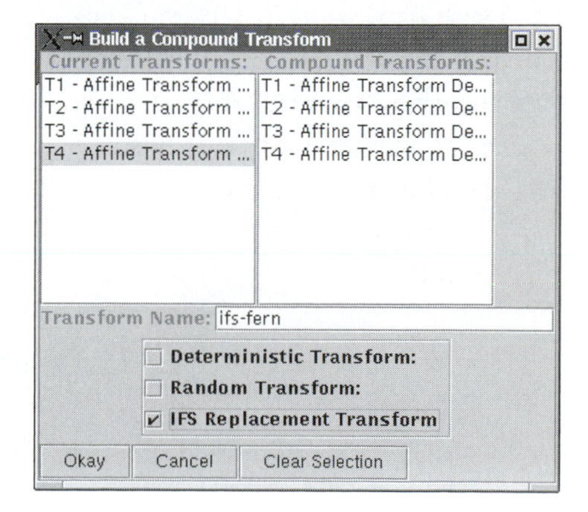

We can now iterate this IFS system on a geometric object. Create a point A on the screen. Then select A and choose **Iterated** under the **Custom** pop-up menu in the Transform panel. A dialog box will pop up as shown. This dialog box allows us to iterate a mapping on the currently selected object. Click the list item named "ifs-fern" and then type in 4 for the iteration level.

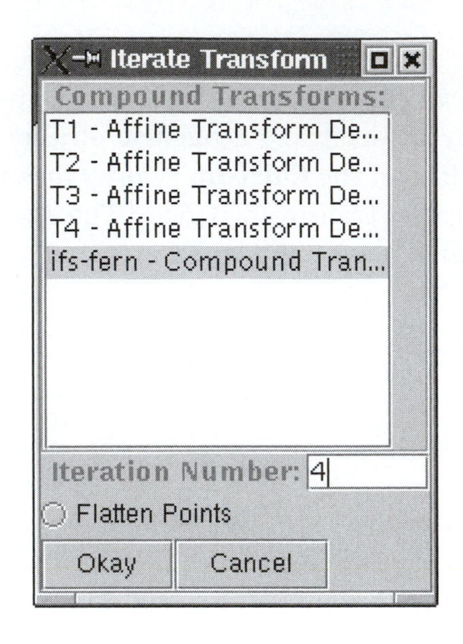

The iteration level is the number of times the IFS system is applied *recursively* to the point A. The first time it is applied, four new points are created, the second time it is applied, each of the four transformations is applied to these four new points, yielding $16 = 4^2$ points. The third time through, $64 = 4^3$ points are created, and the fourth time $256 = 4^4$ points are created. We can see that this recursive process grows in size *very* fast.

Click "Okay" in the dialog box to run the IFS on point A.

The shape that appears is *something* like the fern. One problem is that our points are too large. Choose **Preference** (**Edit** menu) and reduce the point size to 3 pixels.

Select one of the points on the screen and run the Iterated Transform dialog (choose **Iterated** under the **Custom** pop-up menu) again. This time, click the toggle button labeled "Flatten Points." This will speed up *Geometry Explorer*'s drawing routines by making points non-interactive. Finally, choose a higher level of recursion, say 7.

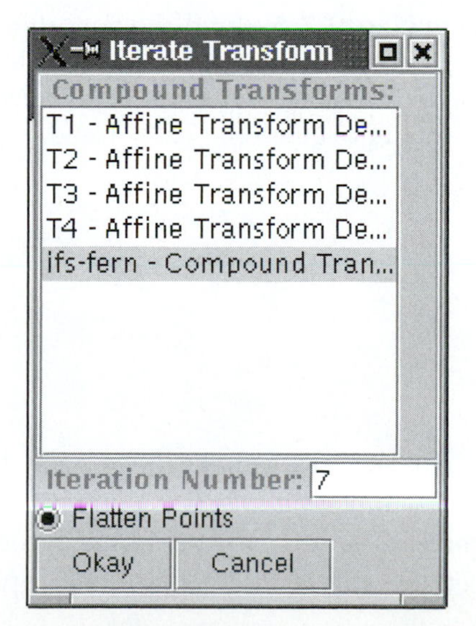

Click "Okay" in the dialog box to run the IFS. The completion of the process may take awhile, so wait until the "Stop" button returns to its "Okay" state. After the process finishes we see that, indeed, the fern reappears as the attractor of the IFS system.

Exercise 9.7.4. The affine maps T_1, T_2, and T_3 are valid transformations (i.e., are invertible), and thus the IFS Fractal Dimension Theorem (Theorem 9.7) applies to the shape of the fern minus its trunk. Use the contraction scale factors for T_1, T_2, and T_3 to estimate the fractal dimension of the portion of the fern that does not include the main trunk.

Exercise 9.7.5. Find the four affine maps that will create the shape at right. You may assume that shape is bounded by the unit square ($0 \leq x \leq 1$ and $0 \leq y \leq 1$).

In this project we have seen how a small number of affine maps can represent or *encode* the shape of a complex natural object. In theory, any complex object can be similarly encoded once the affine maps for that object are discovered. This has important implications for the transmission of complex data, such as images over a computer network. If we can encode each image with just a small number of affine maps, we can greatly reduce the transmission time for images and video. The Collage Theorem guarantees that we can get arbitrarily close to the original image by a good choice of affine maps. Once we have these affine maps, the reproduction of the original image is completely straightforward and fast, using modern computers. The major draw-back to an encoding scheme using IFS systems is the discovery of which affine maps will produce a good approximation to an image.

9.8 Algorithmic Geometry

In our discussion of self-similar fractals, we saw that their construction required a looping process, whereby each level of the construction was built from specific rules using the results of previous levels. This created a *recursive* sequence of constructions and transformations to produce the fractal.

In the last section we saw how we could use a system of affine maps, an IFS system, to generate a fractal. Transformations are applied recursively to an initial object, with the fractal appearing as the *attracting set* of the IFS system.

In both cases, fractals are created using a set of *instructions* which generate a recursive procedure. Such a set of instructions is called an *algorithm*. Since the construction of many fractals requires hundreds or even thousands of calculations, fractal algorithms are most often carried out by computers.

In the construction of self-similar fractals, we used the computer as a powerful bookkeeping device to record the steps whereby edges were replaced by template curves. For IFS systems, the computer carried out the numerous iterations of a set of affine maps, leading to a fractal attractor appearing as out of a mist. In both of these fractal construction algorithms, we use the computer to make the abstract ideas of self-similarity or attracting sets a concrete reality.

9.8.1 Turtle Geometry

The notion of utilizing computing technology to make abstract mathematical ideas more "real" was the guiding principle of Seymour Papert's work on turtle geometry and the programming language *LOGO*. In his book *Mindstorms*, Papert describes turtle geometry as the "tracings made on a display screen by a computer-controlled turtle whose movements can be described by suitable computer programs" [34, page xiv].

Papert's work has inspired thousands of teachers and children to use turtle geometry in the classroom as a means of exploring geometry (and computer programming) in a way that is very accessible to young (and old) students.

Turtle geometry has also proved to be an ideal way in which to explore fractal shapes. To see how fractals can be constructed using turtle geometry, we first have to create a scheme for controlling the turtle on screen.

We will direct the behavior of the turtle with the following set of commands.

1. f Forward: The turtle moves forward a specified distance without drawing.
2. | Back: The turtle moves backward a specified distance without drawing.
3. F Draw Forward: The turtle moves forward and draws as it moves.
4. $+$ Turn Left: The turtle rotates counterclockwise through a specified angle.
5. $-$ Turn Right: The turtle rotates clockwise through a specified angle.
6. [Push: The current state of the turtle is pushed onto a state stack.
7.] Pop: The state of the turtle is set to the state on top of a state stack.

As an example, suppose we have specified that the turtle move 1 unit and turn at an angle of 90 degrees. Also, suppose the initial heading of the turtle is vertical. Then the set of symbols

$$F + F + F + F$$

can be considered a *program* that will generate a square of side-length 1 when interpreted, or carried out, by the turtle.

Here we have labeled the edges drawn by the turtle as a, b, c, and d, in the order that the turtle created them.

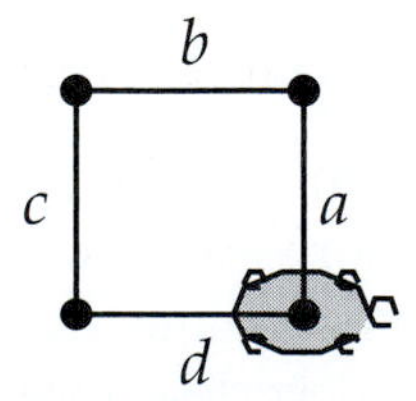

Here is an example illustrating the use of the turtle-state stack. The turtle was given the program $F[+FF+FF+FF]+$. After the first edge is drawn (edge a), the position and heading of the turtle (direction it is pointing) are stored as a *turtle-state*. This turtle-state is placed on the top of a virtual *stack* of possible turtle-states. Then the turtle interprets the symbols $+FF+FF+FF$, ending with edge g. At the symbol] the turtle "pops" the stored turtle-state off the stack and the turtle resets itself to that position and heading.

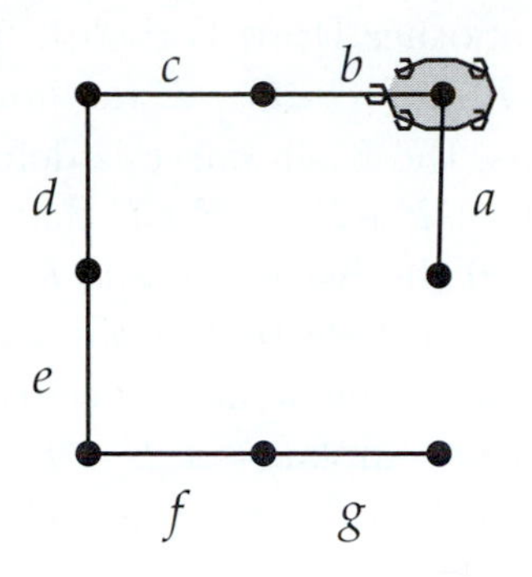

Let's see how we can use this set of symbols to represent the stages of construction of the Koch snowflake curve.

Recall that the Koch snowflake curve is a fractal that is constructed by beginning with an initial segment. This segment is then replaced by a template curve made up of four segments as shown at the right. The angles inside the peak are 60 degrees, making the triangle formed by the peak an equilateral triangle.

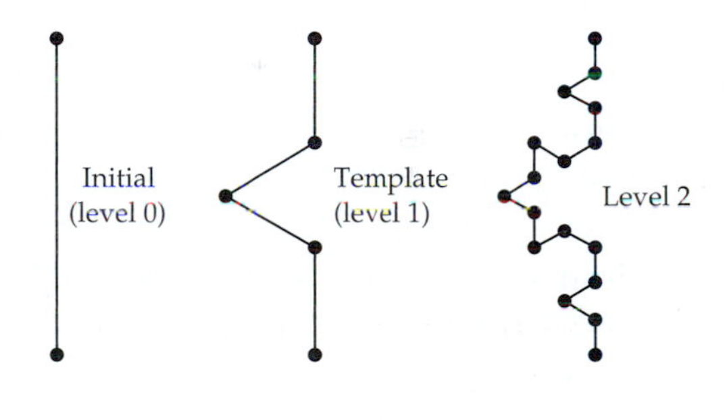

The initial segment is the Koch curve at level 0. The template is the Koch curve at level 1. If we replace each of the segments in the template with a copy of the template at a reduced scale, we get the Koch curve at level 2, and so on.

To model the Koch snowflake curve using our set of symbols, we could say that the initial segment is a Draw Forward, or the symbol F. The template is (forgetting for now the problem of scaling) a Draw Forward followed by a Turn Left of 60 degrees, then a Draw Forward followed by two Turn Rights of 60 degrees, then a Draw Forward followed by a Turn Left of 60 degrees, and finally another Draw Forward. The symbol set that describes the template is thus $F + F - -F + F$ (assuming our turns are always 60 degrees).

Thus, the Koch curve is defined by an initial symbol F and a template symbol set $F+F--F+F$ that governs how the initial segment is replaced. At level 0 the Koch curve is F. At level 1 we replace F by $F + F - -F + F$. At level 2 we replace all segments of level 1 by the template. This is equivalent to replacing all occurrences of the symbol F in the level 1 symbol set by the template set $F + F - -F + F$. The level 2 symbol set is then $F+F--F+F+F+F--F+F--F+F--F+F+F+F--F+F$.

9.9 Grammars and Productions

The set of turtle symbols described in the last section was used in the classic work by Prusinkiewicz and Lindenmayer, titled *The Algorithmic Beauty of Plants* [29]. This beautiful book describes how one can use turtle geometry to model plants and plant growth. Their method of describing natural fractal shapes has been called *Lindenmayer Systems*, or *L-systems*, in honor of Astrid Lindenmayer, who first pioneered the notion of using symbols to model plant growth.

Lindemayer's novel idea was to model plant structures through a *grammar rewriting* system. We can think of a set of symbols as a *word* in a grammar built from those symbols. The symbol set $F+F+F+F$ is a word in the grammar built on turtle command symbols.

In the Koch curve example, the curve is *grammatically* defined by an initial word F and a template word $F + F - -F + F$. At each level we *rewrite* the previous level's word by substituting in for each occurrence of the symbol F.

The Koch curve can be completely described by two words: the starter word F and template word $F + F - -F + F$. The Koch curve is then the limiting curve one gets by successively rewriting an *infinite* number of times, having the turtle interpret the final word, which is theoretically possible but physically impossible.

A major difference between the grammar rewriting system of describing the Koch curve and the recursive description given earlier in this chapter is that in the grammar-based system, we do not scale the template, whereas in the recursive description we scale the template by a factor of $\frac{1}{3}$ before looping. When the turtle draws succeeding levels of the rewritten start symbol, each successive curve will get longer than the one at the previous level.

The formal definition of an *L-system* is as follows.

Definition 9.21. A *Lindenmayer system* or *L-system* consists of

- a finite set Σ of symbols
- a set Ω of words over Σ
- a finite set P of *production rules* or *rewrite rules*, of the form $\sigma- > \omega$, where $\sigma \in \Sigma$ and $\omega \in \Omega$
- a symbol S in Σ that is called the *start symbol* or the *axiom* of the system

Lindenmayer Systems are special types of *formal grammars*. A formal grammar is used in computer science to describe a formal language, a set of strings made up of symbols from an alphabet. Formal grammars are used to express the syntax of programming languages such as Pascal or Java. These languages can be completely expressed as a set of production rules over a set of symbols and words.

For example, suppose $\Sigma = \{a, b, S\}$, and $P = \{S- > aSb, S- > ba\}$. That is, there are two production rules:

1. $S- > aSb$
2. $S- > ba$

Beginning with the start symbol S, we can rewrite using production rule 1 to get the new word aSb. Then using rule 2 on this word, we can rewrite to get $abab$. At this point, we can no longer rewrite, as there is no longer a start symbol in our word. We have reached a *terminal* word in the grammar. In fact, it is not hard to see that the set of all *producible* words (i.e., words that are the result of repeatedly rewriting the start symbol) consists of a subset of all strings containing an equal number of a's and b's. We only get a subset of such strings because the string $bbaa$, for example, is not producible. We will call the set of producible strings the *language* of the grammar.

9.9.1 Space-filling Curves

As another example, let's consider a grammar that classifies the Draw Forward segments of a turtle into two groups: left edges, which we will denote by F_l, and right edges, which we will denote by F_r.

By artificially creating such a classification, we can control which instances of the symbol F in a word will be replaced under a production.

Those edges labeled F_l will be replaced by a production rule for F_l, and those labeled F_r will be replaced by a production rule for F_r. When the turtle interprets a word that uses these two symbols, it will interpret both as a simple Draw Forward.

To see how this works, let's consider the problem of filling up a unit square by a path passing through the points in the square.

Here we have subdivided a square into 25 sub-squares. Look carefully at the path weaving through the square. It touches each of the corners of the internal sub-squares, will hit one or the other of the corners of the squares on the left or right sides, and will also do this on the top and bottom sides. Note that this path starts at the lower left corner of the square (P) and ends at the lower right corner (Q).

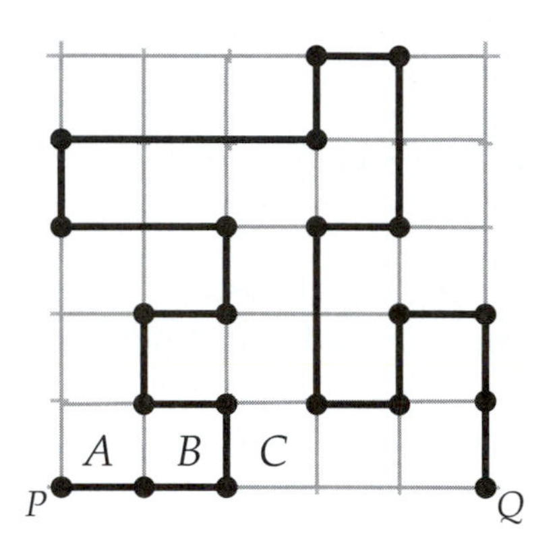

Consider the edge of this path in the sub-square labeled A. If we scaled down the path in the previous figure by a factor of $\frac{1}{5}$, we could replace this edge with the scaled-down copy of the original path. The original path can be considered a *template*, similar to the strategy we used for the Koch snowflake. Suppose we try to replace each edge of the original path with the template. In the figure at right, we have replaced the edges in sub-squares A and B.

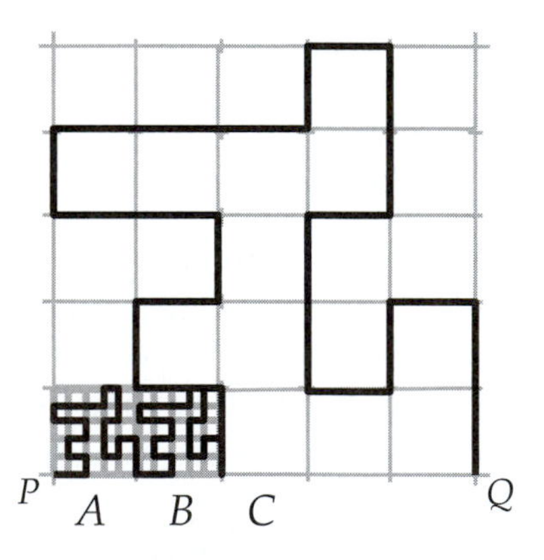

When we try to replace the edge in sub-square C, we have a problem. If we substitute our template on the left side of the edge in C, we will collide with the path we already have in B. Thus, we need to substitute to the *right* of this edge. However, our template is not oriented correctly to do this. We need to substitute a new path, one that first passes directly into square C, so that it does not "double-up" on any of the points already covered on the edge between B and C. Also, it must not "double-up" on any of the other edges it might possibly share with the previous template curve, if it is in another orientation.

Here is a picture of what the "right" template curve should be. Note that it is shown in relation to the original template curve. To see how it would appear in sub-square C, rotate the curve 90 degrees counter-clockwise. This template will not "double-up" with our old template, which we will call the "left" template. Also, if this template adjoins a left template, then all corner points on the edge where the two squares meet will be covered by one path or the other.

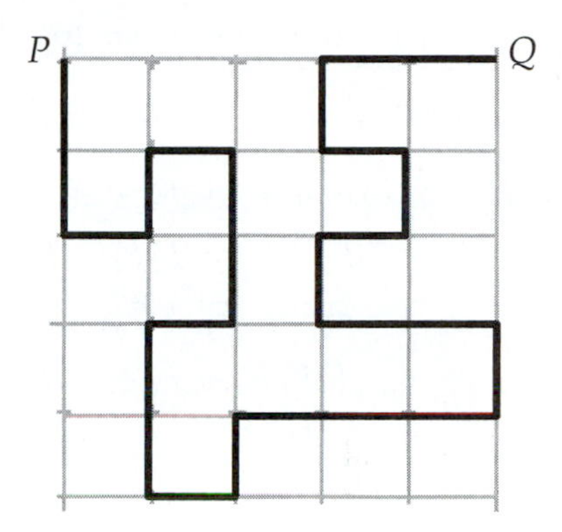

Now all we have to do to create the template replacement process is to label all edges in each of the two templates with "l" or "r" to designate which of the scaled-down templates will replace that edge. We start with the left template and label the first edge "l." Then the next edge is also "l," but the edge at C must be "r," as must be the next edge.

Continuing in this way, we get the following set of edge labels for the left template.

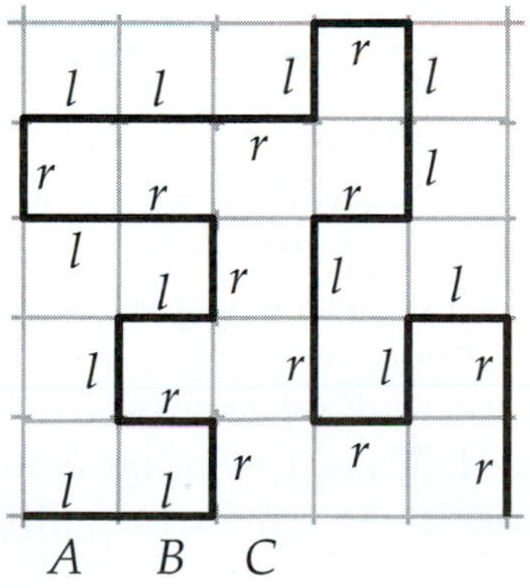

Note that every sub-square has a label in it. Thus, every sub-square will have a scaled-down template passing through it. Also, by the way we constructed the two templates, we are guaranteed that along an edge where two sub-squares meet, the new paths will not cross each other or intersect.

A simple way to keep track of the orientations of the sub-squares is that an edge labeled "l" will have the left template constructed to the left of the edge, and vice versa for an edge labeled "r."

The two production rules for the left and right templates are, thus,

$$\begin{aligned}
F_l -> & F_lF_l + F_r + F_r - F_l - F_l + F_r + F_rF_l - F_r - F_lF_lF_r + \\
& F_l - F_r - F_lF_l - F_r + F_lF_r + F_r + F_l - F_l - F_rF_r + \\
F_r -> & -F_lF_l + F_r + F_r - F_l - F_lF_r - F_l + F_rF_r + F_l + F_r - \\
& F_lF_rF_r + F_l + F_rF_l - F_l - F_r + F_r + F_l - F_l - F_rF_r
\end{aligned}$$

Note that for the left template, we need to add a final turn (+) to ensure that any new path starts in the same direction as the original.

Now all we need is an initial starting path. Since it is customary to have the turtle heading vertically at the start, we will turn the turtle so that it heads to the right initially. Thus, the path $-F_l$ will be our starting path.

If we don't want to use subscripts for the two types of F symbols, we can instead use the following production rules with the starter word $-Fl$.

$$\begin{aligned}
l -> & lFl + rF + rF - Fl - Fl + rF + rFFl - rF - FlFlrF + \\
& Fl - rF - FlFl - rF + FlrF + rF + Fl - Fl - rFrF + \\
r -> & -FlFl + rF + rF - Fl - FlrF - Fl + rFrF + Fl + rF - \\
& FlrFrF + Fl + rFFl - Fl - rF + rF + Fl - Fl - rFr
\end{aligned}$$

Note that we give productions only for how the symbols l and r are replaced. Thus, the initial F for the left production and the final F for the right production must be omitted from the replacement.

This set of productions will generate an equivalent rewriting system, if we interpret Fl as F_l and rF as F_r. The turtle, when interpreting a level 2 rewrite of this system, will generate the image in Fig. 9.7. We can start to see how this curve will fill up the space in the square, as the rewriting level increases without bound.

Fig. 9.7 Space-filling Curve

Exercise 9.9.1. Let an L-system be defined by $\Sigma = \{a, b, 1, S\}$, and $P = \{S- > aSb, S- > 1\}$, with start symbol S. Prove that this grammar generates the language $\{a^n b^n | n \geq 0\}$.

Exercise 9.9.2. Let an L-system be defined by $\Sigma = \{F, +, -, S\}$, and $P = \{S- > F - F - F - F, F- > F - F + F + FF - F - F + F\}$, with start symbol S. This system, when interpreted by a turtle, will generate a self-similar fractal called the *Quadratic Koch Island*, as first described in [30, page 50]. Sketch levels 1 (rewrite S once) and 2 (rewrite twice) of this curve (assume the turn angle is 90 degrees), and sketch the template for the fractal. Find the similarity dimension using the template.

Exercise 9.9.3. Let an L-system be defined by $\Sigma = \{F, +, -, L, R\}$, and $P = \{L- > +RF - LFL - FR+, R- > -LF + RFR + FL-\}$, with start symbol L. This curve is called the *Hilbert curve*. Find the level 1 and 2 rewrite words for this system and sketch them on a piece of paper using turtle geometry. Do you think this system will generate a space-filling curve?

Exercise 9.9.4. In our space-filling curve example, show that the right template is a simple rotation of the left template. On a 7x7 grid find a left template curve with edges labeled r and l, as we did earlier, such that all squares are visited by some edge of the curve. Also, create the curve so that a 180 degree rotation about the center of the grid will produce a right template curve with the same "doubling-up" properties as we had for the pair of curves in our example. That is, if the right and left templates would meet at an edge, then no edge of the left would intersect an edge of the right, except at the start and end of the paths.

9.10 Project 15 - Words Into Plants: The Geometry of Life

What makes many plants fractal-like is their branching structure. A branch of a tree often looks somewhat like the tree itself, and a branch's sub-branch system looks like the branch, and so forth. To model the development of branching structures, we will use the grammar rewriting ideas of the last section, plus the push and pop features of turtle geometry. This will be necessary to efficiently carry out the instructions for building a branch and then returning to the point where the branch is attached.

The grammar we will use consists of the turtle symbols described in the last section plus one new symbol "X." We can think of X as being a *virtual* node of the plant that we are creating. Initially, the start symbol for our grammar will be just the symbol X, signifying the potential growth of the plant.

For example, here is a very simplified branching system for a plant.

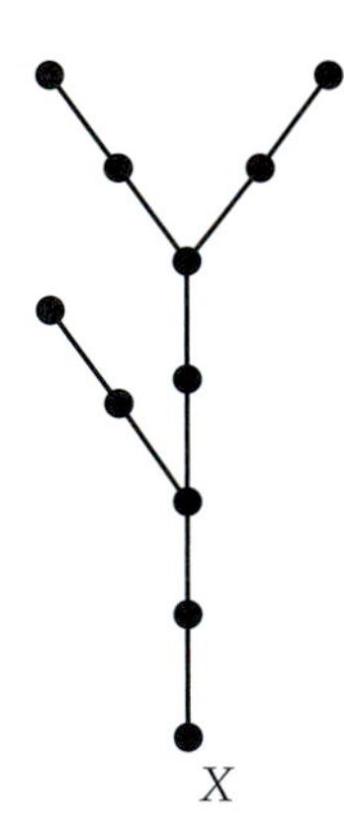

How can we represent this branching structure using our grammar rewriting system? It is clear that the plant grew in such a way that three new branch nodes were created from the original potential node X, which we represent here as a point. Thus, the start symbol X must be replaced with three new X's. Also, the branches were created at an angle to the main branch, so there needs to be some turning by the turtle. Finally, each new branch has length of two Draw Forward's, if we consider one Draw Forward to be the distance between the points on a branch. Which production rules will represent this plant? Consider the following set of productions:

$$X \; -> \; F[+X]F[-X]+X$$
$$F \; -> \; FF$$

The first production replaces a node X by the word $F[+X]F[-X]+X$. Thus, the replacement will produce a "stalk" of length 1, a new branch node turned to the left that is *independent* of the previous symbol F, due to the push and pop symbols, a second length of main stalk, a second independent branch node coming off the stalk at an angle to the right, and finally a third branch coming off the top at an angle to the left.

The second production says that a length of stalk will grow twice as long in the next generation. Putting this all together, we have in these two productions a blueprint for the growth of the plant.

Exercise 9.10.1. How many times was the start symbol X rewritten, using the production rules, to generate the preceding plant image?

Let's see how we can use *Geometry Explorer*'s turtle geometry capability to generate this branching structure.

To begin, we need to define a turn angle and a heading vector for our turtle. Start *Geometry Explorer* and create segments $\overline{AB}$, $\overline{BC}$, and $\overline{DE}$ as shown. Multi-select A, B, and C (in that order) and choose **Turtle Turn Angle** (**Turtle** menu) to define the turn angle for our turtle. Then, multi-select D and E and choose **Turtle Heading Vector** (**Turtle** menu) to define the turtle's initial direction of motion.

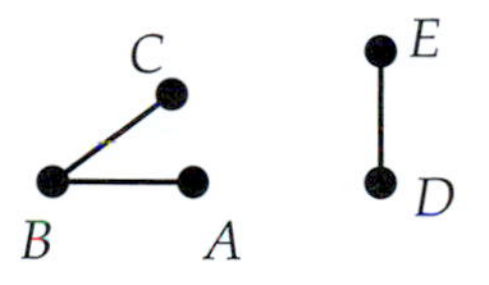

Now, create a point X on the screen. Select X and choose **Create Turtle at Point** (**Turtle** menu) to create a turtle based at X.

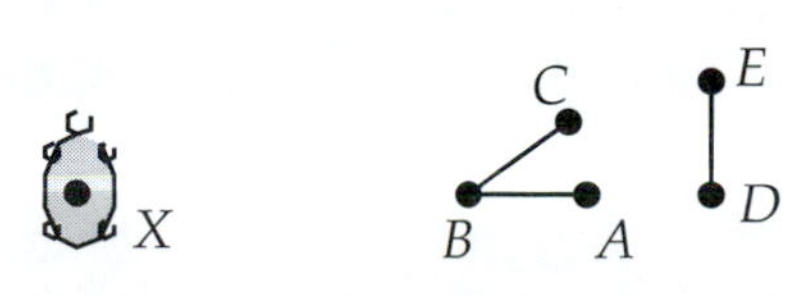

A dialog box will pop up to control the movement of the turtle. Click on the tab labeled "Grammar Turtle." In this window we can type in the start symbol (axiom) and a list of productions. Type in "X" for the axiom and then type in the two productions as shown, using "=" to designate the left and right sides of the rule. Type in "1" in the Rewrite Level box and hit the Rewrite button. The computer will rewrite the axiom, using the two production (replacement) rules.

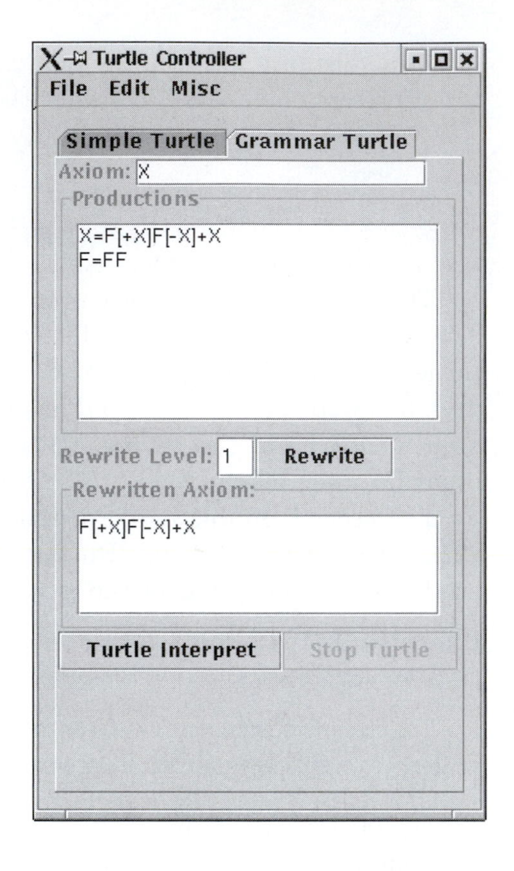

Now, click on the "Turtle Interpret" button. The turtle will carry out the commands in the rewritten word.

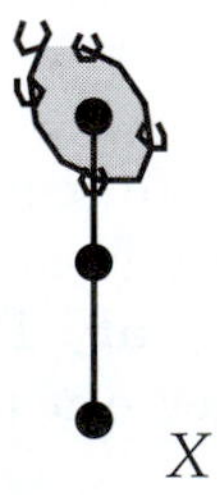

We do not see any actual branches yet, as the three new X's in the rewritten word are *potential* branches. Undo the turtle back to its start position by typing "Ctrl-U" repeatedly. Then, in the grammar window type in "2" for the rewrite level and hit "Rewrite." Click on "Turtle Interpret" again to see the plant starting to take shape.

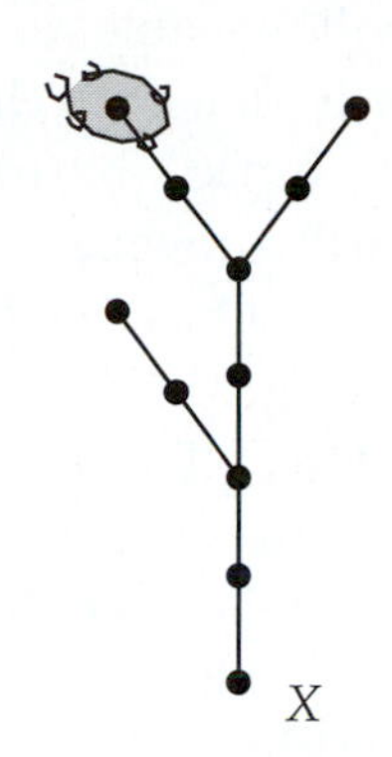

The figure drawn by the turtle is clearly a branched structure, but is not really much like a plant. To more fully develop the branching pattern, we need to rewrite the axiom to a higher level.

Undo the turtle back to its start position and move E close to D, so that the turtle moves only a short distance each time it changes position. Change the rewrite level to 4 in the Turtle Controller and hit Return. Note how the new sentence has expanded. Click on "Turtle Interpret" and watch the turtle interpret the rewritten sentence. This may take awhile—anywhere from a few seconds to several minutes, depending on the speed of your computer. There are a tremendous number of actions that the turtle needs to carry out as it interprets the sentence.

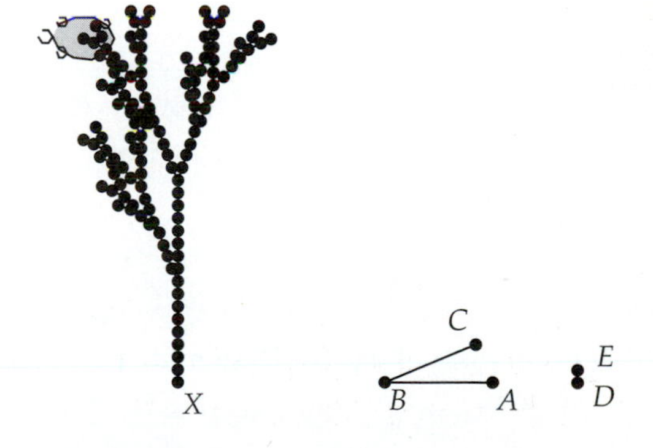

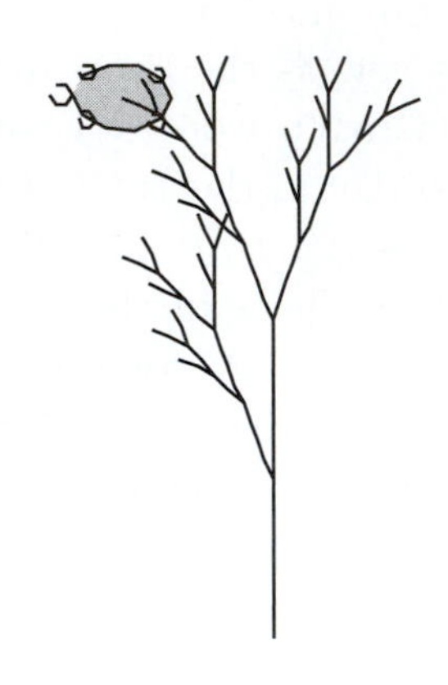

You will be able to tell when the turtle is done by the state of the "Stop Turtle" button. If the turtle is still drawing, the button will be active. Once the turtle completes drawing, the button will become inactive. The image is very "blotchy" with all of the points visible that were drawn by the turtle. Let's hide all of these points by choosing **Hide All** (**View** menu) and then choosing the **Points** submenu. Now the figure looks like the bushy branch structure of a plant (minus the leaves).

Exercise 9.10.2. Find a set of two productions that will generate the branching pattern shown here, from an initial start symbol of X.

Exercise 9.10.3. Design an L-system that will have a thick, bushy branching system.

The analysis of complex branching patterns is quite an interesting subject, one for which we have only scratched the surface. For more information on this subject, consult the Prusinkiewicz and Lindenmayer text [29].

It is interesting to note how the use of grammar rewriting systems mirrors our discussion of axiomatic systems from Chapter 1. The start symbol and production rules can be thought of as abstract axioms or postulates. The words of the language produced by the grammar are akin to *theorems* in the axiomatic system, as they are generated using the axioms of the system. It seems that our investigation of geometry has returned full circle to where it began with the Greeks and their axiomatic system of reasoning.

Appendix A

Book I of Euclid's *Elements*

A.1 Definitions

1. A *point* is that which has no part.
2. A *line* is breadthless length.
3. The extremities of a line are points.
4. A *straight line* is a line that lies evenly with the points on itself.
5. A *surface* is that which has length and breadth only.
6. The extremities of a surface are lines.
7. A *plane surface* is a surface that lies evenly with the straight lines on itself.
8. A *plane angle* is the inclination to one another of two lines in a plane that meet one another and do not lie in a straight line;
9. And when the lines containing the angle are straight, the angle is called *rectilinear*.
10. When a straight line set up on a straight line makes the adjacent angles equal to one another, each of the equal angles is *right*, and the straight standing on the other is called a *perpendicular* to that on which it stands.
11. An *obtuse angle* is an angle greater than a right angle.
12. An *acute angle* is an angle less than a right angle.

13. A *boundary* is that which is an extremity of anything.

14. A *figure* is that which is contained by any boundary or boundaries.

15. A *circle* is a plane figure contained by one line such that all the straight lines falling upon it from one point among those lying within the figure are equal to one another;

16. And the point is called the *center* of the circle.

17. A *diameter* of the circle is any straight line drawn through the center and terminated in both directions by the circumference of the circle, and such a straight line also bisects the circle.

18. A *semicircle* is the figure contained by the diameter and the circumference cut off by it. And the center of the semicircle is the same as that of the circle.

19. *Rectilinear figures* are those that are contained by straight lines, *trilateral* figures being those contained by three, *quadrilateral* those contained by four, and *multilateral* those contained by more than four straight lines.

20. Of trilateral figures, an *equilateral triangle* is that which has its three sides equal, an *isosceles triangle* that which has two of its sides equal, and a *scalene triangle* that which has its three sides unequal.

21. Further, of trilateral figures, a *right-angled triangle* is that which has a right angle, an *obtuse-angled triangle* that which has an obtuse angle, and an *acute-angled triangle* that which has its three angles acute.

22. Of quadrilateral figures, a *square* is that which is both equilateral and right-angled; an *oblong* that which is right-angled but not equilateral; a *rhombus* that which is equilateral but not right-angled; and a *rhomboid* that which has its opposite sides and angles equal to one another but is neither equilateral nor right-angled. And let quadrilaterals other than these be called *trapezia.*

23. *Parallel* straight lines are straight lines that, being in the same plane and being produced indefinitely in both directions, do not meet one another in either direction.

A.2 The Postulates (Axioms)

1. To draw a straight line from any point to any point.

2. To produce a finite straight line continuously in a straight line.

3. To describe a circle with any center and distance.

4. That all right angles are equal to one another.

5. That, if a straight line falling on two straight lines make the interior angles on the same side less than two right angles, the two straight lines, if produced indefinitely, meet on that side on which are the angles less than two right angles.

A.3 Common Notions

1. Things that are equal to the same thing are also equal to one another.

2. If equals be added to equals, the wholes are equal.

3. If equals be subtracted from equals, the remainders are equal.

4. Things that coincide with one another are equal to one another.

5. The whole is greater than the part.

A.4 Propositions (Theorems)

I-1 On a given straight line to construct an equilateral triangle.

I-2 To place at a given point (as an extremity) a straight line equal to a given straight line. [Given a length and a point, we can construct a segment of that length from the point.]

I-3 Given two unequal straight lines, to cut off from the greater a straight line equal to the less. [Given two segments of different lengths, we can cut off from the larger a segment equal to the smaller.]

I-4 If two triangles have the two sides equal to two sides, respectively, and have the angles contained by the equal straight lines equal, they will also have the base equal to the base, the triangle will be equal to the triangle, and the remaining angles will be equal to the remaining angles, respectively, namely those that the equal sides subtend. [SAS Congruence]

I-5 In isosceles triangles the angles at the base are equal to one another, and if the equal straight lines be produced further, the angles under the base will be equal to one another.

I-6 If in a triangle two angles be equal to one another, the sides that subtend the equal angles will also be equal to one another.

I-7 Given two straight lines constructed on a straight line (from its extremities) and meeting in a point, there cannot be constructed on the same straight line (from its extremities) and on the side of it, two other straight lines meeting in another point and equal to the former two, respectively, namely each to that which has the extremity with it. [Given a base length and two other lengths, there is only one triangle possible on a particular side of the base.]

I-8 If two triangles have the sides equal to two sides, respectively, and have also the base equal to the base, they will also have the angles equal that are contained by the equal straight lines. [SSS Congruence]

I-9 To bisect a given rectilinear angle.

I-10 To bisect a given finite straight line.

I-11 To draw a straight line at right angles to a given straight line from a given point on it. [Given a line and a point on the line, we can construct a perpendicular to the line at the point.]

I-12 To a given infinite straight line, from a given point that is not on it, to draw a perpendicular straight line. [Given a line and a point not on the line, we can construct a perpendicular to the line through the point.]

I-13 If a straight line set up on a straight line make angles, it will make either two right angles or angles equal to two right angles. [Supplementary angles add to 180.]

I-14 If with any straight line, and at a point on it, two straight lines not lying on the same side make the adjacent angles equal to two right angles, the two straight lines will be in a straight line with one another. [Given two angles that share a line as a common side, if the angles add to 180, then the non-shared sides of the two angles must be coincident on a line.]

I-15 If two straight lines cut one another, they make the vertical angles equal to one another. [Vertical Angle Theorem]

I-16 In any triangle, if one of the sides be produced, the exterior angle is greater than either of the interior and opposite angles. [Exterior Angle Theorem]

I-17 In any triangle two angles taken together in any manner are less than two right angles.

I-18 In any triangle the greater side subtends the greater angle. [In a triangle the larger side is opposite the larger angle.]

I-19 In any triangle the greater angle is subtended by the greater side. [In a triangle the larger angle is opposite the larger side.]

I-20 In any triangle two sides taken together in any manner are greater than the remaining one. [Triangle Inequality]

I-21 If on one of the sides of a triangle, from its extremities, there be constructed two straight lines meeting within the triangles, the straight lines so constructed will be less than the remaining two sides of the triangle, but will contain a greater angle. [Given triangle ABC, if we construct triangle DBC with D inside ABC, then $DB < AB$, $DC < AC$, and the angle at D will be greater than the angle at A.]

I-22 Out of three straight lines, which are equal to three given straight lines, to construct a triangle: thus it is necessary that two of the straight lines taken together in any manner should be greater than the remaining one. [To construct a triangle from three lengths, it is necessary that when you add any pair of lengths, the sum is greater than the other length.]

I-23 On a given straight line and at a point on it, to construct a rectilinear angle equal to a given rectilinear angle. [Angles can be copied.]

I-24 If two triangles have the two sides equal to two sides, respectively, but have one of the angles contained by the equal sides greater than the other, they will also have the base greater than the base.

I-25 If two triangles have the two sides equal to two sides, respectively, but have the base greater than the base, they will also have one of the angles contained by the equal straight lines greater than the other.

I-26 If two triangles have the two angles equal to two angles, respectively, and one side equal to one side, namely, either the side adjoining the equal angles, or that subtending one of the equal angles, they will also have the remaining sides equal to the remaining sides and the remaining angle to the remaining angle. [AAS and ASA Congruence]

I-27 If a straight line falling on two straight lines make the alternate angles equal to one another, the straight lines will be parallel to one another. [Given two lines cut by a third, if the alternate interior angles are congruent then the lines are parallel.]

I-28 If a straight line falling on two straight lines make the exterior angle equal to the interior and opposite angle on the same side, or the interior angles on the same side equal to two right angles, the straight lines will be parallel to one another. [Given two lines cut by a third, if the exterior and opposite interior angles on the same side of the cutting line are congruent, the lines are parallel. Or, if the interior angles on the same side add to 180, the lines are parallel.]

The first 28 propositions listed are independent of Euclid's fifth postulate. They are called *neutral* propositions. Proposition 29 is the first proposition where Euclid explicitly requires the fifth postulate to carry out the proof.

I-29 A straight line falling on parallel lines makes the alternate angles equal to one another, the exterior angle equal to the interior and opposite angle, and the interior angles on the same side equal to two right angles.

I-30 Straight lines parallel to the same straight line are parallel to one another.

I-31 Through a given point, to draw a straight line parallel to a given straight line. [Given a line and a point not on the line, we can construct a parallel to the line through the point.]

I-32 In any triangle, if one of the sides be produced, the exterior angle is equal to the two opposite and interior angles, and the three interior angles of the triangle are equal to two right angles. [The sum of the angles of a triangle is 180 degrees. The exterior angle is equal to the sum of the opposite interior angles.]

I-33 The straight lines joining equal and parallel straight lines (at the extremities that are) in the same directions (respectively) are themselves equal and parallel. [Given a quadrilateral $ABCD$ with $\overline{AB}$ congruent and parallel to $\overline{CD}$, then $\overline{AD}$ must be congruent and parallel to $\overline{BC}$.]

I-34 In parallelogrammic areas the opposite sides and angles are equal to one another, and the diameter bisects the areas. [Given parallelogram $ABCD$, both pairs of opposite sides are congruent, and both pairs of opposite angles are congruent. Also, the diagonals split the parallelogram into two equal parts.]

I-35 Parallelograms that are on the same base and in the same parallels are equal to one another. [Given two parallelograms $ABCD$ and $ABEF$ with C, D, E, F collinear, then the parallelograms have the same area.]

I-36 Parallelograms that are on equal bases and in the same parallels are equal to one another. [Given two parallelograms $ABCD$ and $EFGH$, with $\overline{AB}$ congruent to $\overline{EF}$, A, B, E, F collinear, and C, D, G, H collinear, then the parallelograms have the same area.]

I-37 Triangles that are on the same base and in the same parallels are equal to one another. [Given two triangles ABC and ABD, with $\overline{CD}$ parallel to $\overline{AB}$, then the triangles have the same area.]

I-38 Triangles that are on equal bases and in the same parallels are equal to one another. [Given two triangles ABC and DEF, with $\overline{AB}$ congruent to $\overline{DE}$, A, B, D, E collinear, and $\overline{CF}$ parallel to $\overline{AB}$, then the triangles have the same area.]

I-39 Equal triangles that are on the same base and on the same side are also in the same parallels. [Given two triangles ABC and ABD having the same area and on the same side of $\overline{AB}$, then $\overline{CD}$ must be parallel to $\overline{AB}$.]

I-40 Equal triangles that are on equal bases and on the same side are also in the same parallels. [Given two triangles ABC and DEF having the same area, with $\overline{AB}$ congruent to $\overline{DE}$, A, B, D, E collinear, and the two triangles being on the same side of $\overline{AB}$, then $\overline{CF}$ must be parallel to $\overline{AB}$.]

I-41 If a parallelogram have the same base with a triangle and be in the same parallels, the parallelogram is double that of the triangle. [The area of a triangle is half that of a parallelogram with the same base and height.]

I-42 To construct, in a given rectilinear angle, a parallelogram equal to a given triangle. [It is possible to construct a parallelogram with a given angle having the same area as a given triangle.]

I-43 In any parallelogram the complements of the parallelograms about the diameter are equal to one another.

[Given parallelogram $ABCD$ and its diameter $\overline{AC}$, let K be a point on the diameter. Draw parallels $\overline{EF}$ to $\overline{BC}$ and $\overline{GH}$ to $\overline{AB}$, both through K. Then, complements $EKBG$ and $HKFD$ have the same area.]

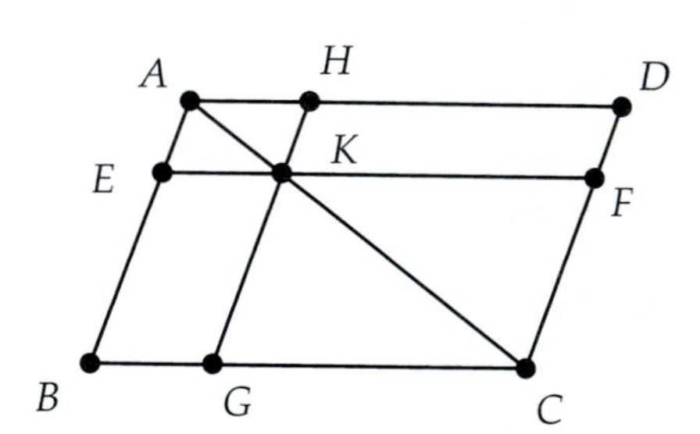

I-44 To a given straight line, to apply in a given rectilinear angle a parallelogram equal to a given triangle. [It is possible to construct a parallelogram on a given segment with a given angle having the same area as a given triangle.]

I-45 To construct in a given rectilinear angle a parallelogram equal to a given rectilinear figure. [It is possible to construct a parallelogram with a given angle having the same area as a given polygon.]

I-46 On a given straight line to describe a square. [It is possible to construct a square on a given segment.]

I-47 In right-angled triangles the square on the side subtending the right angle is equal to the squares on the sides containing the right angle. [Pythagorean Theorem]

I-48 If in a triangle the square on one of the sides be equal to the squares on the remaining two sides of the triangle, the angle contained by the remaining two sides of the triangle is right. [Converse to the Pythagorean Theorem]

Appendix B

Brief Guide to *Geometry Explorer*

Geometry Explorer is designed as a geometry laboratory where one can create geometric objects (like points, circles, polygons, areas, and the like), carry out transformations on these objects (dilations, reflections, rotations, and translations), and measure aspects of these objects (like length, area, radius, and so on). As such, it is much like doing geometry on paper (or sand) with a ruler and compass. However, on paper such constructions are static—points placed on the paper can never be moved again. In *Geometry Explorer* all constructions are *dynamic*. One can draw a segment and then grab one of the endpoints and move it around the canvas with the segment moving accordingly. Thus, one can create a construction and test out hypotheses about the construction with numerous variations of the original construction. *Geometry Explorer* is just what the name implies—an environment to explore geometry.

Non-Euclidean geometry can easily be explored using *Geometry Explorer*. Constructions can be carried out in a Euclidean or non-Euclidean (Poincaré model) environment using the same user interface. Almost all actions that apply in the Euclidean environment can be carried out in the non-Euclidean environment (with a few important exceptions that depend on the parallel postulate).

Fractal geometry can be explored using turtle graphics and grammatical descriptions of fractals. In turtle graphics, one controls a "turtle" on the screen by telling it to move, draw, rotate, change color, and so forth.

This appendix provides a brief introduction to the capabilities of *Geometry Explorer*. As such this guide is necessarily incomplete. Only the basic functionality of the program is covered. A complete user's manual for the program is available from the author.

B.1 The Main *Geometry Explorer* Window

Upon starting *Geometry Explorer* you will see the main *Geometry Explorer* Euclidean window appear on the screen (Fig. B.1).

There are four important areas within this window.

1. The *Canvas* is where geometry is created and transformed. This is the large white area on the right side of the main window.

2. The *Tool panel* is where geometric tools are located. The Tool panel is directly to the left of the Canvas. It consists of a set of iconic buttons that represent tools used to create and modify geometric figures. The icons (pictures) on the buttons depict the function that the particular button serves. Sometimes this function is quite clear, other times it is less intuitive, but the pictures serve as reminders as to what the buttons can do. The Tool panel is split into four sub-panels: Create, Construct, Transform, and Color Palette. Note that the cursor is over the *Info* tool (the one with the question mark). A small box with the words *Get Info on Object* appears below the button. This box is called a *Tool Tip*. Tool tips appear as the cursor sits over a button for a second or two. Tool tips are designed to give quick information on a button's purpose.

3. The *Menu Bar* includes nine menus: **File**, **Edit**, **View**, **Measure**, **Graph**, **Misc**, **Turtle**, **Windows**, and **Help**. Each of these menus controls specific user actions.

4. The *Message box* is where detailed information will be shown concerning various tools that one may wish to use. In Fig. B.1 the mouse cursor is over the Info tool. In the Message box we see information concerning how this tool should be used. (Other system information may also appear in the Message box.) The Message box is located below the Canvas.

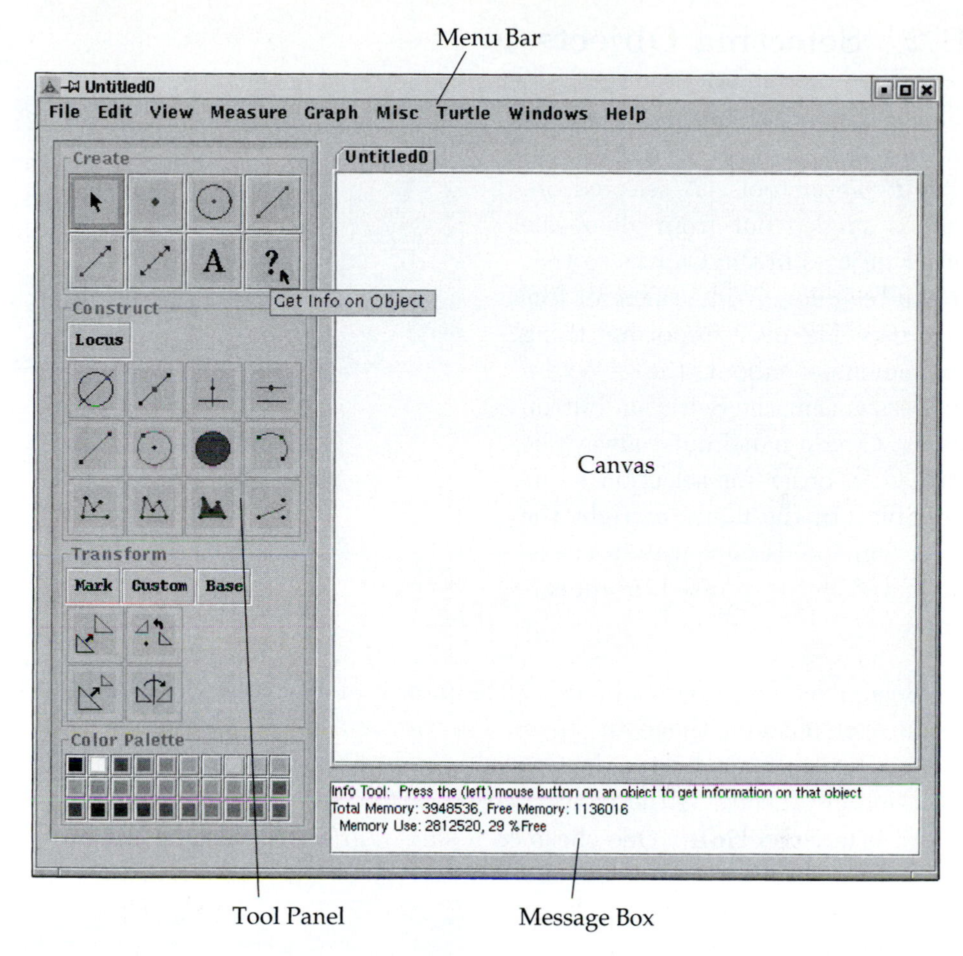

Fig. B.1 The *Geometry Explorer* Main (Euclidean) Window

B.2 Selecting Objects

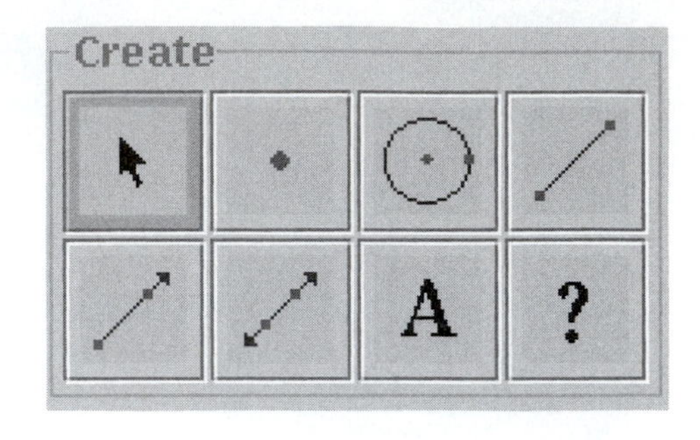

The *Selection* tool is perhaps the most widely used of all the *Geometry Explorer* tools. A selected object is singled out from all of the other objects in the Canvas so that it can be uniquely identified for further use. The most important thing to remember about the selection process is that the Selection button in the Create panel must always be clicked in order for selection to be possible. In the figure at right the Selection tool is currently in use as indicated by its pressed-in appearance.

Selections are carried out using the mouse or by using the **Select All** menu option under the **Edit** menu. All mouse actions use the left mouse button, or a single button for those mice having just one button. The selection of objects via the mouse can be carried out in three ways.

Single Selection One clicks on a single object to select it.

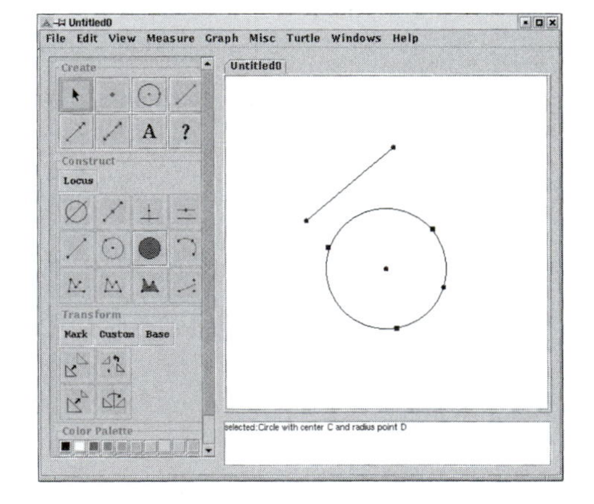

In the figure at right we have a circle and a segment. Suppose we wish to select the circle. We would first click on the Selection button (in the Create panel) to make the Selection tool active. Then we would move the mouse to the Canvas and click somewhere along the circle. In the figure, we see that the circle is selected. Note the rectangular boxes that appear on the circle. These boxes are used to visually signify that the circle is currently selected. Also note that a message appears in the Message box telling the user what object is being selected.

Multiple Selection To do a multiple selection we hold down the Shift key as we click on objects.

In the figure at right we have created a line segment and a point. Suppose we want to construct a parallel line to the segment through the point. Since a parallel is constructed from a line *and* a point, we need to select these two items. Make the Selection tool active (i.e., click on it) and click on a white area in the Canvas to unselect all objects. To do a multiple selection, hold down the Shift key and click on the segment and then the point. The Parallel and Perpendicular tools should now be active in the Construct panel. A button will be *active* when a beveled white line appears around it. You should see something like the figure at right.

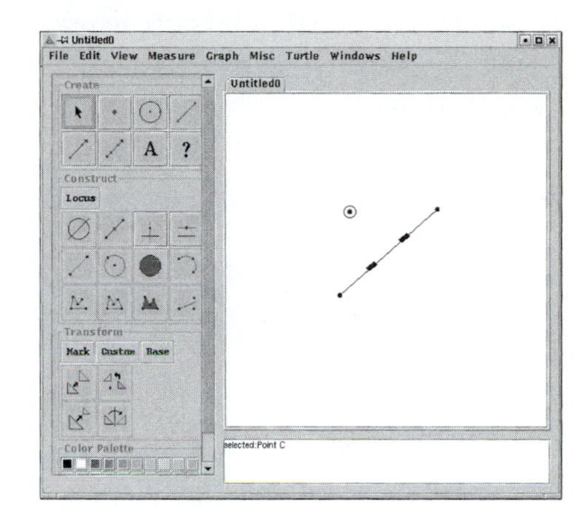

At this point you can click on the Parallel tool (the fourth button in the top row in the Construct panel). *Geometry Explorer* will then carry out the construction. Note how the newly constructed parallel line is created in a selected state.

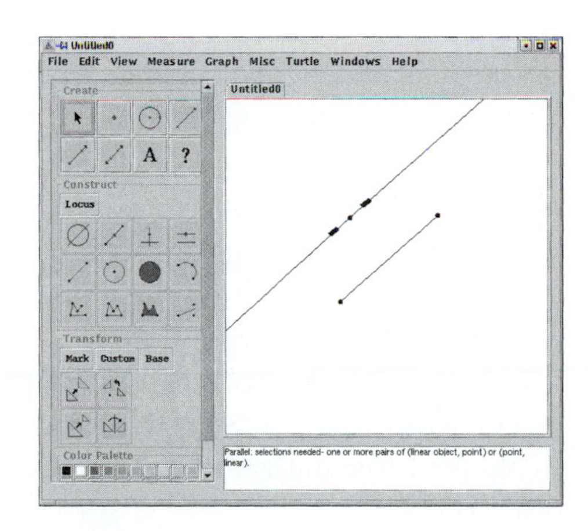

Box Selection One can draw a Selection box about a set of objects to select all of the objects enclosed in the box.

In the figure at right, a series of points have been created. Suppose that we wanted to draw a polygon through this set of points. It would be tedious to do a multiple selection of each point.

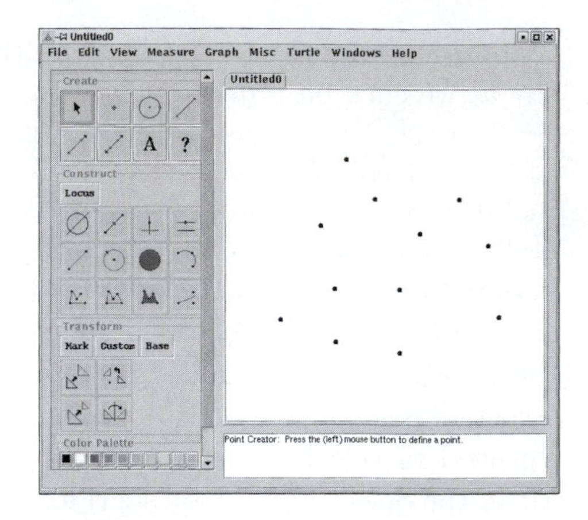

We can select all the points using the Selection box. First make sure that the Selection tool is active. Then click in the upper left-hand corner of the Canvas and drag to create a Selection box surrounding all of the points. The Selection box will be visually identified by its red appearance.

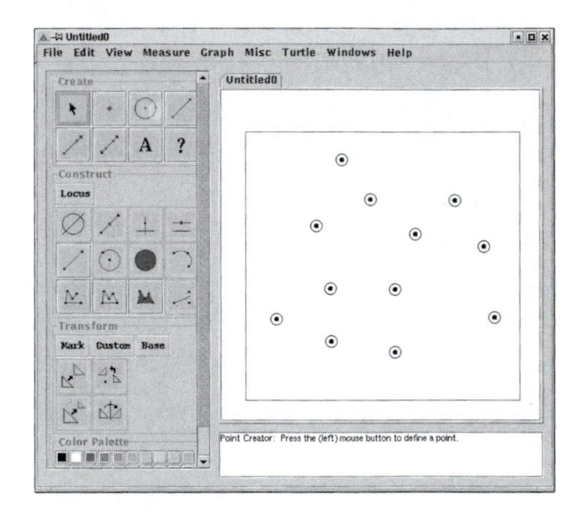

At this point the Closed Polygon tool (second from left in bottom row of the Construct panel) will be active. Click on it to draw the desired polygon. You should see a figure similar to that of the one at right. Note that the Selection box remains visible until we select some other object.

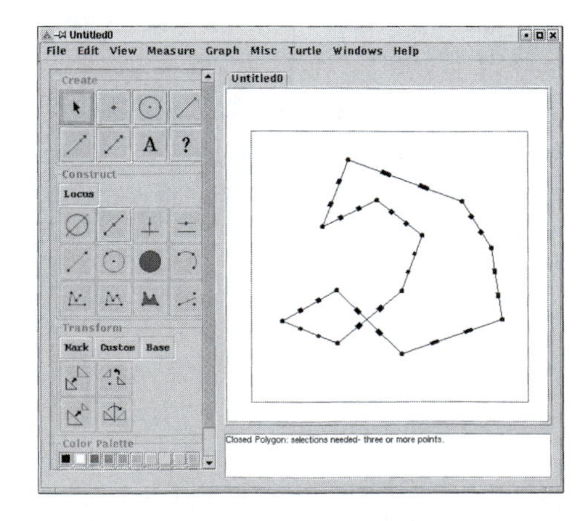

B.3 Active vs. Inactive Tools

We have talked a lot about how to make tools *active* so that they can be utilized. Some tools are always active. Others can change from active to inactive and vice versa depending on user actions. For example, tools in the Create panel are always available for use—they are always in an active state. Most other tools will start out in an inactive state. An inactive tool can be visually identified by its grayed-out appearance. When a tool is in an inactive state, clicking on that tool will have no effect. To activate an inactive tool one needs to select the kinds of objects that the tool needs in order to function. For example, to activate the Midpoint tool (second one in first row of the Construct panel), one needs to select a segment first and then the Midpoint tool will become active.

B.4 Labels

All objects are created with labels, but the label is invisible at first. To make a label visible, use the Text/Label tool in the Create panel (the one with the "A" on it).

In the figure at right we have created several objects. The labels on these objects are made visible by first clicking on the Text/Label tool in the Create panel and then clicking on the object to make its label visible. Labels are created in alphabetical order based on order of object creation. Points A and B were created first, then circle a, then line b. Note that points are created with capital letters, whereas lines, circles, and arcs have lowercase labels.

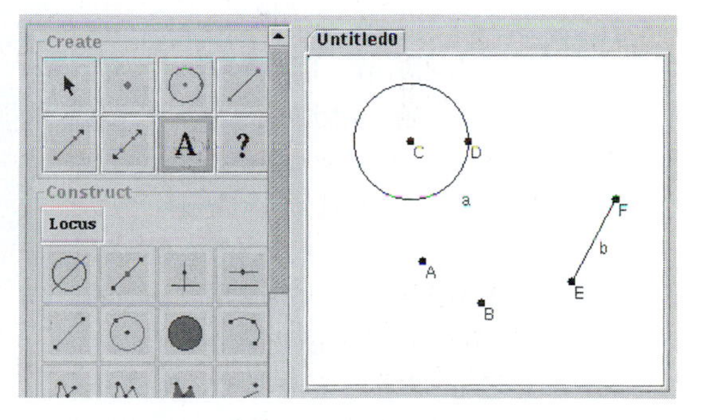

If we want to edit a label, we double-click on an object using the Text/Label tool. If we double-click on point B, we get a label edit dialog window as shown.

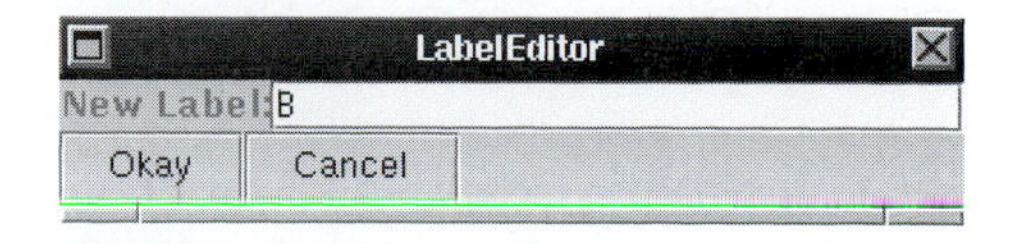

We can change the label to another name, say "cool point," and hit "Okay" to change the label in the Canvas.

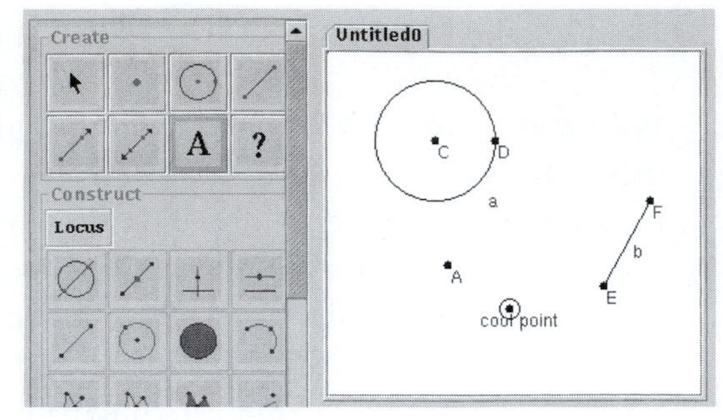

Sometimes a label can get partially obscured by other objects. In the figure at right the label "a" for the line segment $\overline{AB}$ is partially obscured by the label for point A.

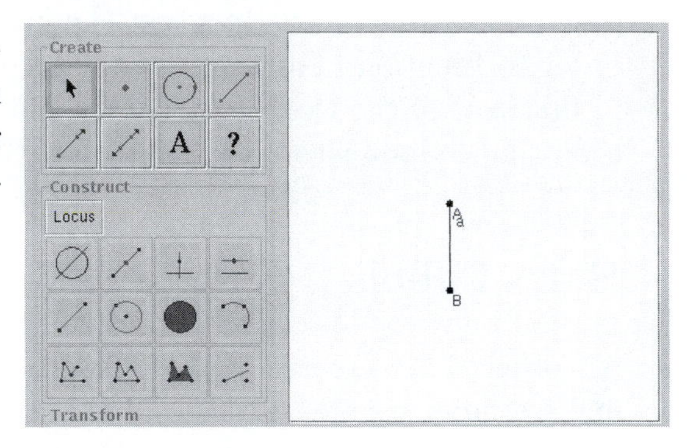

To move a label, we use the Text/Label tool in the Create panel. Click on this button and click near the middle letter of the label on the Canvas. Then drag the label to the desired position. Note that the label cannot be placed anywhere. Labels can only be moved within a limited area around the object to which they are attached. In the figure at right we have placed the labels in a better position.

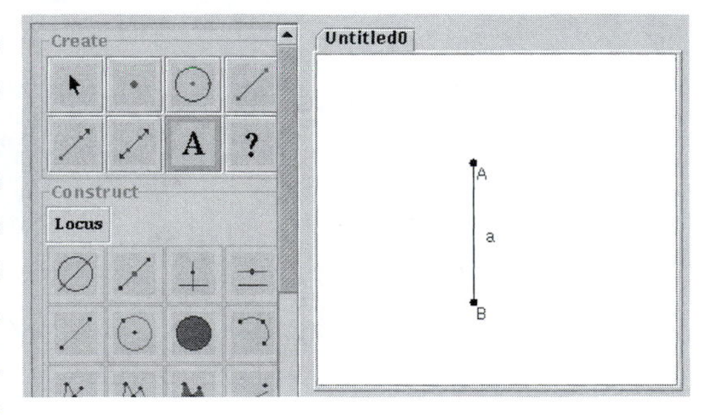

B.5 Object Coloring

We can change the color of an object that exists in the Canvas by using the Color Palette in the Tool panel. The Color Palette consists of a set of color squares on the bottom left of the main window. To change an object's color we first select the object and then click on a color square to immediately change that object's color. If we select a group of objects (using multiple selection), then all objects in that group will have their color changed to the desired color.

The color of the label of an object can be changed by first clicking on the object with the Text/Label tool and then clicking on a color in the Color Palette.

B.6 Online Help

There is an extensive online help system that can be accessed via the **Help** menu item in the Menu bar at the top of the main *Geometry Explorer* window. Click on this menu item and then on the **Help** sub-menu to start the help system. The help system is designed as a series of Web pages that are viewed by an Internet browser that is built into *Geometry Explorer*. No additional software is needed to view these Web pages. The help system is organized into categories that roughly correspond to the visual areas in the *Geometry Explorer* window—panels, menus, and so on. There are many examples available in the help system from an introductory to advanced level.

B.7 Undo/Redo of Actions

Geometry Explorer provides the user with the ability to *undo* almost any action that arises from some geometric construction.

For example, in the figure at right we have created (in order of creation) a circle, a segment, and the midpoint of the segment.

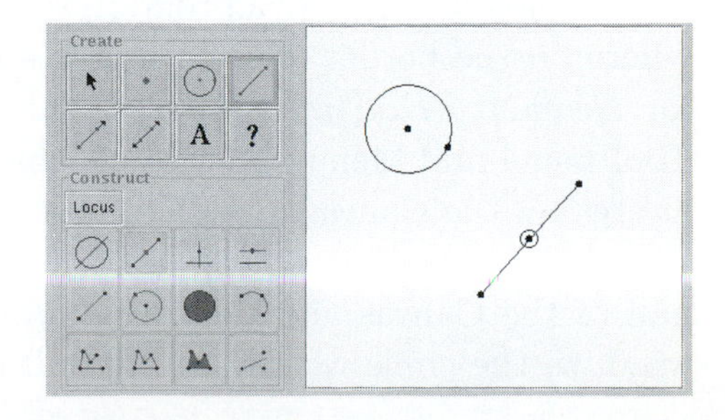

Suppose we decide the midpoint construction is not needed. We can undo the midpoint action by choosing **Undo midpoint** (**Edit** menu). The midpoint construction will be undone, leaving just the circle and segment.

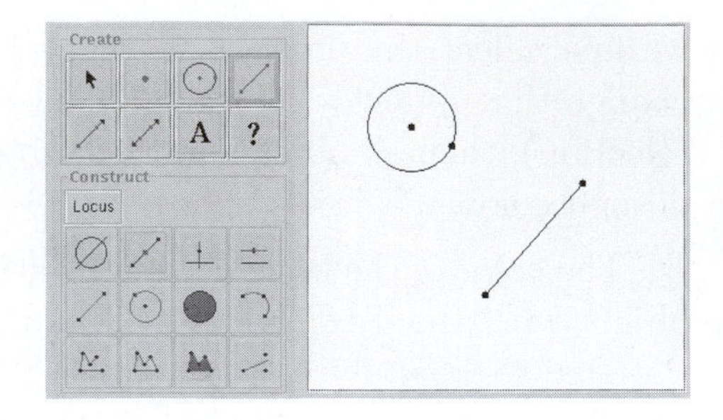

At this point if we decided that the segment was also a mistake, we could undo again to get rid of the segment. Undoing yet another time would erase the circle and leave a totally blank Canvas.

Now suppose we decided that we really did like the circle, segment, and midpoint that we had initially constructed. Then we could *redo* all of the steps that we just undid. This is done by choosing **Redo** (**Edit** menu).

Geometry Explorer provides the user with an unlimited ability to undo and redo steps. This capability is very useful for showing someone the sequence of steps that produced a geometric figure.

Note that objects can only be undone/redone in the order in which they were created. If you want to simply hide an object, select the object and choose **Hide Object** (**View** menu).

B.8 Clearing and Resizing the Canvas

To clear the Canvas of all objects currently constructed, choose **Clear** (**Edit** menu). This action will clear all currently defined objects. Note that this is different than "undoing" the constructions. When we clear the Canvas all objects are immediately removed. However, clearing the screen is itself an action that can be undone. Thus, if we clear the screen and then change our mind we can always undo or redo this action.

On most computers a program's window can be resized by clicking somewhere on the border of the window and dragging. If the boundary window for *Geometry Explorer* is resized, the Canvas will also change size, but the Tool panel and Menu bar will not change size. As the Canvas changes size, figures on the Canvas also change so that the size of objects relative to the size of the window stays the same. For example, if we had a circle that filled half of the Canvas and then we doubled the length and width of the main window, the circle would still fill half of the new expanded Canvas.

The reason for this is that all of the mathematical calculations for the program are done on a "virtual" Canvas that has the dimensions of a square. The virtual coordinates of this square Canvas are transformed to screen pixel coordinates and then displayed on the screen. The virtual Canvas is always fixed in size, but as the screen area changes, the transformation from the virtual Canvas to the screen Canvas preserves relative distances.

Expanding the size of the main window will have the effect of increasing the resolution of your figure. If objects are too close, then expanding the window size will be like putting the figure under a microscope.

If we expand the window in such a way that the Canvas can no longer be displayed inside of a perfect square, then the Canvas will be placed inside of a scrolling window.

Sometimes a construction will be so large that it leaves the boundaries of the Canvas. Choose **Rescale Geometry in Canvas...** (**View** menu) to rescale the figures in the Canvas so that the image will shrink or grow.

B.9 Saving Files as Images

It is often desirable to save the contents of the Canvas to an image file. This is useful, for example, if one wants to add a picture of the Canvas to a Web page or insert a picture of the Canvas into a word processing document.

As an example, suppose that we have constructed the equilateral triangle shown at the right and wish to save it as a GIF file.

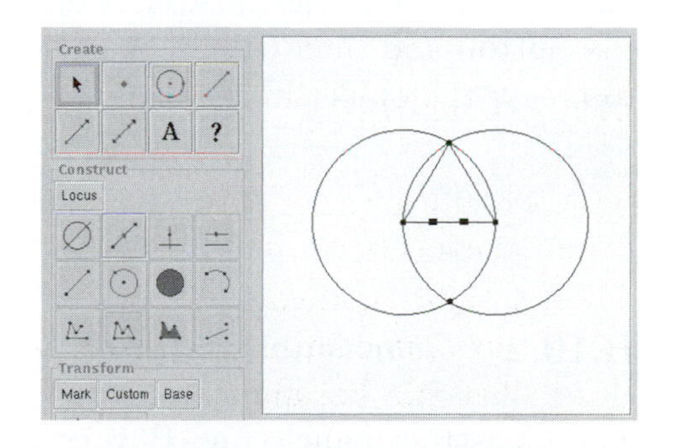

To save the Canvas as an image choose **Save as Image...** (**File** menu). A dialog box will pop up as shown at the right.

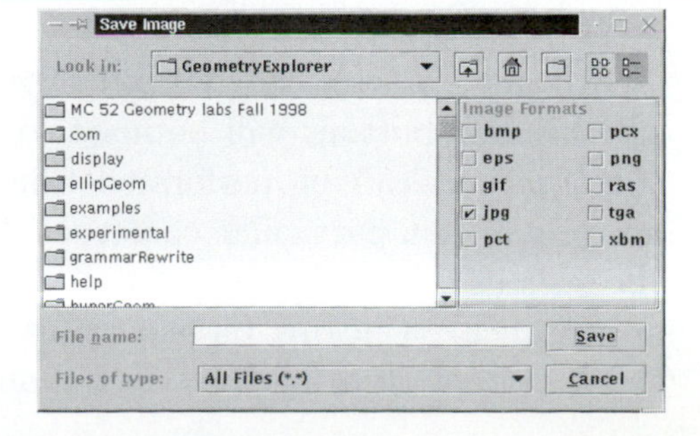

Note the rows of buttons to the right of the directory window. These allow one to specify the image format to which the Canvas will be saved. Supported image types include most of the commonly used formats: bmp (Windows bitmap), eps (Encapsulated Postscript), gif (Graphics Interchange Format), jpg (JPEG format), pcx (PC Paintbrush), png (Portable Network Graphics), ras (Sun Raster), tga (Targa), and xbm (X Windows Bitmap). The default image format is the JPEG format.

One important note about EPS files is that a preview image is stored with an EPS file so that the image can be inserted into another program, such as a word processor. However, the image quality will typically be much lower than the real Postscript image. The image will have the original Postscript quality when printed with a Postscript-compatible printer.

B.10 Main Window Button Panels

B.10.1 Create Panel

The tools in the Create panel are used to make points, circles, segments, lines, and rays. To carry out any of these activities, just click on the button and then use the mouse to create the object in the Canvas.

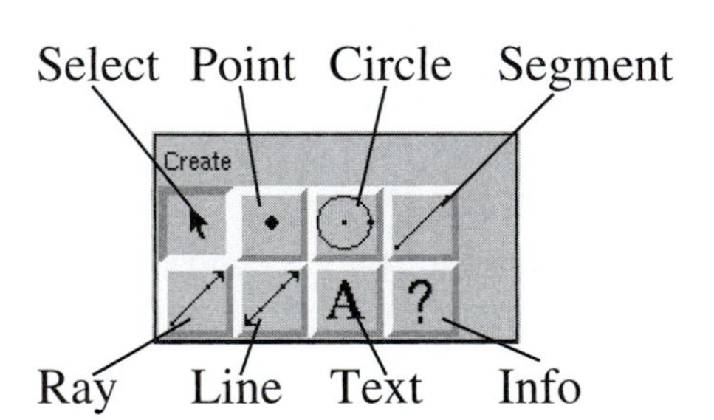

B.10.2 Construct Panel

The Construct panel (Fig. B.2) controls the construction of geometric objects that depend upon already existing objects.

Initially, when there are no objects defined on the Canvas, the Construct panel buttons will be inactive (i.e., grayed-out) because none of the constructions can be performed from scratch. Once the objects that are necessary for a particular construction have been built, and are selected in the correct order, that particular button will become active (i.e., darker in appearance). Clicking on the activated button will automatically perform the construction using the selected objects.

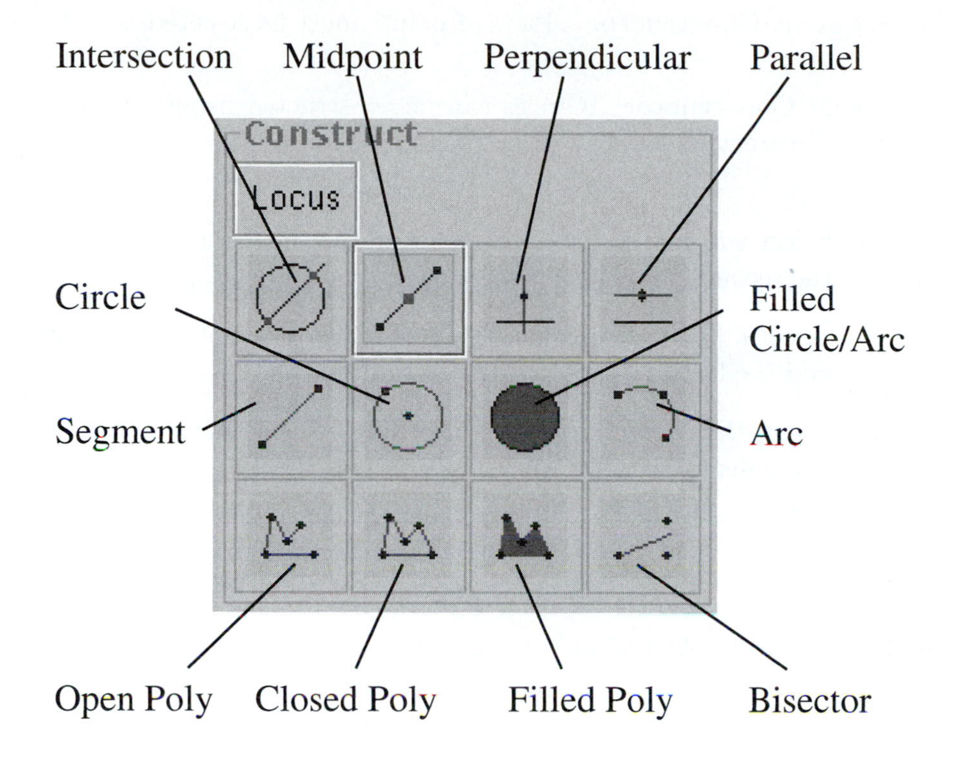

Fig. B.2 The Construct Panel

To get a quick idea of what needs to be selected to activate a tool, pass the mouse cursor over that button and information will appear in the Message box.

We now look at each construction tool in detail. For each tool we list the objects from which it is built.

The Intersection Constructor Two objects must be selected. The objects that can be used for intersections include lines, rays, segments, circles, and arcs.

The Midpoint Constructor One or more segments must be selected.

The Perpendicular Line Constructor Pairs of objects must be selected. Each pair must consist of a linear object (line, segment, or ray) and a point.

The Parallel Line Constructor Pairs of objects must be selected. Each pair must consist of a linear object (line, segment, or ray) and a point.

The Segment Constructor Pairs of points must be selected.

The Circle Constructor Circles can be constructed using this tool in three different ways.

(a) Select two points. The first point is the center of the circle and the second is a point on the circle.

(b) Select a point and segment. The point is the center of the circle and the length of the segment is the radius.

(c) Select three points. The points will be located on the circle's circumference.

(Note: In hyperbolic geometry, only the first two options apply.)

The Filled Circle/Arc Constructor Circles or arcs must be selected. Clicking on this button will fill in the interior of the circle or arc. Note that arcs can be filled in two ways, by filling in the *chord* of the circle or by filling in the entire *sector* of the circle defined by the arc.

Here is an arc example where the chord of an arc has been filled in.

Here is the same arc where the sector defined by the arc has been filled in.

After clicking on the Filled Circle/Arc tool and then selecting an arc, a dialog box will pop up asking which type of filled arc is desired.

The Arc Constructor Arcs can be constructed in three different ways.

(a) Select two points. The first point is the center of the arc and the second is a point on the arc.

(b) Select two points *attached* to an existing circle. These will define an arc on that circle.

(c) Select three points, all of which will be located on the arc's circumference. The first and last points will become endpoints of the arc.

Clicking on the arc button will create an arc. However, in the first option a dialog box will pop up asking for the initial and terminal angles of the arc (in degrees). (Note: In hyperbolic geometry, only the first two options apply.)

The Open Polygon Constructor Select a set of points for an open polygon (three or more points are necessary). Clicking on this button will cause a series of segments to be drawn: from the first point selected to the second, from the second to the third, and so on. If the points are selected using the box selection method, the order of connection will be the order in which the points were created.

The Closed Polygon Constructor Selections are the same as for the open polygon. The only difference between the closed polygon and the open polygon constructors is that the closed polygon constructor draws a segment between the last point selected and the first point selected, thereby closing the figure.

The Filled Polygon Constructor Selections are the same as for the open polygon. Clicking on this button will result in the computer filling the interior of the polygon.

The Angle Bisector Constructor Select three points. The first point will be on the initial ray of the angle. The second point will be the vertex of the angle, and the third point will be on the terminal ray of the angle. The angle bisector construction is oriented, which means that if the points are selected in reverse order, a ray will be drawn in the opposite direction.

The Locus Constructor This button hides a pop-down menu for the construction of geometric loci. A locus is a geometric construction that is defined by two objects: a point that is *attached* to a one-dimensional object (line, ray, segment, circle, or arc) and any other geometric object (called the "primitive"). The locus of the primitive will be a set of copies of that primitive produced as the attached point moves along its one-dimensional path.

B.10.3 Transform Panel

The Transform panel controls four different transformations on geometric objects. These include translations, rotations, dilations, and reflections. Transformations are carried out in a two-stage process. First, the geometric information that defines a transformation must be specified. Then, the objects to be transformed must be selected, and then the appropriate transform button must be clicked. There are three pop-down menus used to define necessary geometric information for transformations. These are hidden under the **Mark**, **Custom**, and **Base** buttons (Fig. B.3).

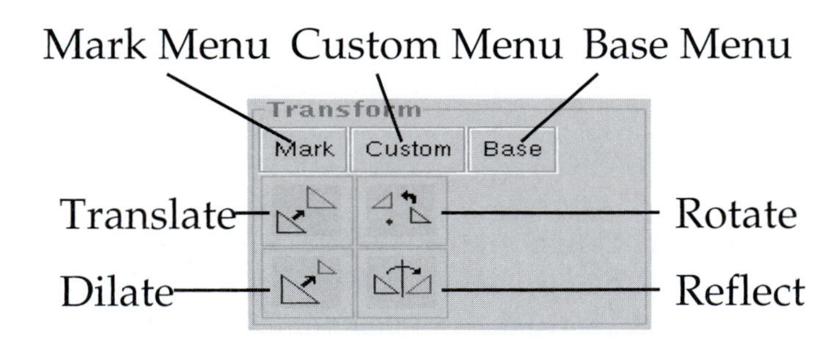

Fig. B.3 The Transform Panel

Here is a quick description of the functionality of the four transformation buttons:

1. **Translation** To translate (that is, move) an object in the Canvas, define a *vector* as described in the next section, or define a *custom*

translation using the **Custom** pull-down menu. Once the definition is complete, select the object(s) to translate and click on the Translate button in the Transform panel.

2. **Rotation** To rotate an object in the Canvas, define an angle and a center of rotation as described in the next section, or define a *custom* rotation using the **Custom** pull-down menu. Once the definition is complete, select the object(s) to rotate and click on the Rotate button in the Transform panel.

3. **Dilation** To dilate an object in the Canvas, define a ratio and a center of dilation as described in the next section, or define a *custom* dilation using the **Custom** pull-down menu. Once the definition is complete, select the object(s) to dilate and click on the Dilate button in the Transform panel.

4. **Reflection** To reflect an object in the Canvas about a line, define a mirror as described in the next section. Once the definition is complete, select the object(s) to reflect and click on the Reflect button in the Transform panel. One cannot define a custom reflection.

Note that all the transformations require predefined geometric information, for example, vectors or mirrors of reflection, and the like. Here is a quick guide to defining these items.

Setting Geometric Transformation Data

The **Mark** pull-down menu is used to define *geometric* data that are needed to specify a transformation.

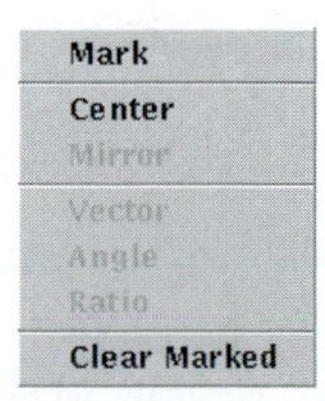

Each item in this menu is defined by selecting a set of geometric objects in the Canvas as follows:

1. **Center** To define either a center of rotation or a center of dilation, select a point and then choose **Center**.

2. **Mirror** To define a mirror of reflection, select a linear object (line, ray, or segment) and then choose **Mirror**.

3. **Vector** To define a vector of translation select two points and then choose **Vector**. A dialog box will pop up asking whether to interpret this vector as a simple vector in rectangular coordinates, or if it should be interpreted as a magnitude to be used in polar coordinates. If one chooses the polar coordinates option, then an angle must also be defined.

4. **Angle** To define an angle, select three points (the initial, vertex, and terminal points of the angle). Then choose **Angle**.

5. **Ratio** To define a ratio, select two segments; the length of the first will be the numerator in the ratio and the length of the second will be the denominator. Then choose **Ratio**.

6. **Clear Marked** Use this to clear all defined geometric data.

Custom Transformations

The **Custom** pull-down menu is used to define special types of transformations.

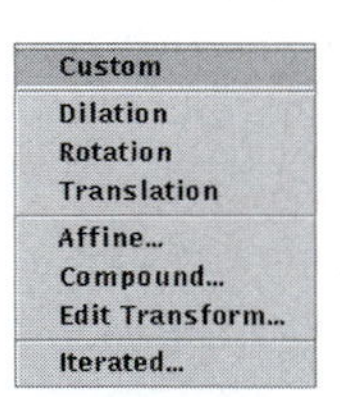

There are seven possible menu options under the **Custom** menu. When choosing any of the first three options, a dialog box will pop up asking for the appropriate numerical data needed to define that transformation. The fourth option is used to define a general type of transformation known as an *affine transformation.* The fifth option is used for defining transformations built from existing transformations. The sixth option is used to edit previously defined transformations. Finally, the seventh menu option is used to carry out multiple iterations of a transformation.

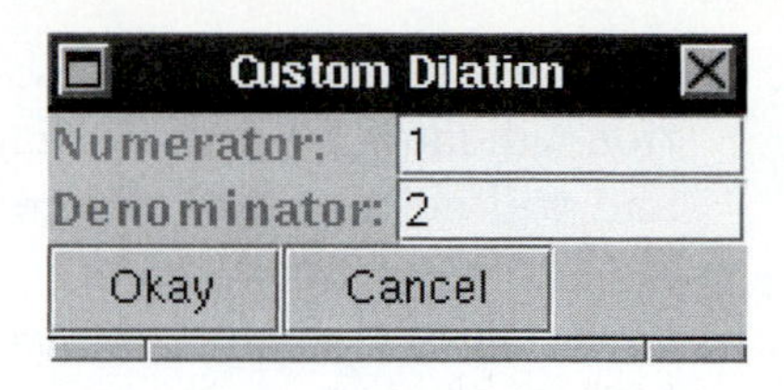

As an example suppose we want to define a *dilation* that scales objects by a numerical ratio of $\frac{1}{2}$. We first define a center of dilation using the **Mark** menu. Then we choose the **Dilation** option under the **Custom** menu. A dialog box like the one at right will pop up. Here we have defined a dilation with a scale ratio of $\frac{1}{2}$.

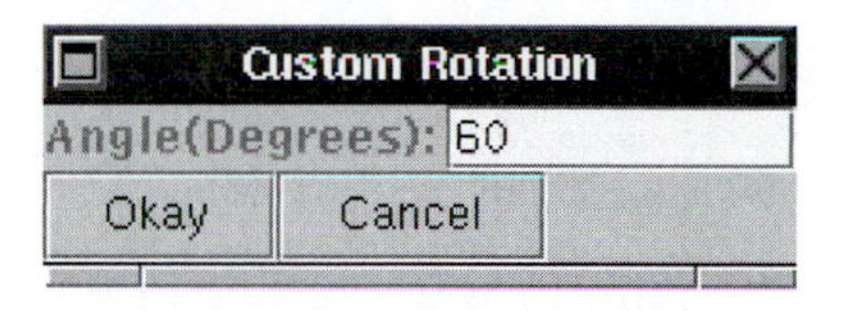

For a custom *rotation* we need to specify an angle of rotation (in degrees). In the figure at the right we have defined a rotation of 60 degrees. Note that a center of rotation must still be defined before the rotation can be applied. Do this by using the **Mark** menu.

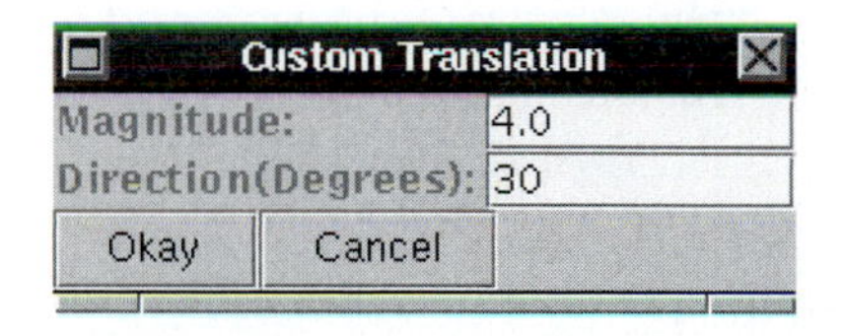

For a custom *translation* we need to specify a vector by defining the magnitude and direction (in degrees) of the vector. Here we have defined a translation that will translate objects 4.0 units in the direction that is 30 degrees up from the horizontal.

Transformations Based on Measurements

It is often useful to define transformations in terms of measurements such as distance, slope, area, and the like. One can use measurements as the basis for rotations, translations, and dilations.

Transformations can be defined in terms of measurements in three ways:

1. **Translate** Translations can be defined in terms of a single measurement or in terms of two measurements. If a single measurement is

selected and **Vector** is chosen from the **Mark** menu, then the translation will move a figure by that measurement amount in the x-direction, and will not move the figure in the y-direction.

If two measurements are selected and **Vector** is chosen, then the translation will move a figure in one of two ways. The two measurements will either determine the x and y directions through which a figure is moved or will determine the angle of movement (from the horizontal) and the distance traveled in the direction of that angle. Upon choosing **Vector** a dialog box will pop up asking which of these two translation types is desired.

2. **Rotate** Rotations can be defined by first selecting a point to act as the center of rotation and then choosing **Center** from the **Mark** menu in the Transform panel. Then select a measurement to serve as the angle of rotation and choose **Angle** from the **Mark** menu. (Note: The angle will be interpreted as measured in degrees.)

3. **Dilate** Dilations can be defined by first selecting a point to act as the center of dilation and then choosing **Center** from the **Mark** menu in the Transform panel. Then select a measurement to serve as the ratio of dilation and choose **Ratio** from the **Mark** menu.

(Note: Transformations defined by measured values are available only in the Euclidean Canvas.)

B.11 Measurement in *Geometry Explorer*

In *Geometry Explorer* measurements are handled by the use of the **Measure** menu in the Menu bar (Fig. B.4). The items in this menu are split into three groups. The top group includes the basic measurements that one can perform on geometric objects. The middle group consists of a single "user-input" measurement. The bottom group controls the creation and modification of tables of measurements.

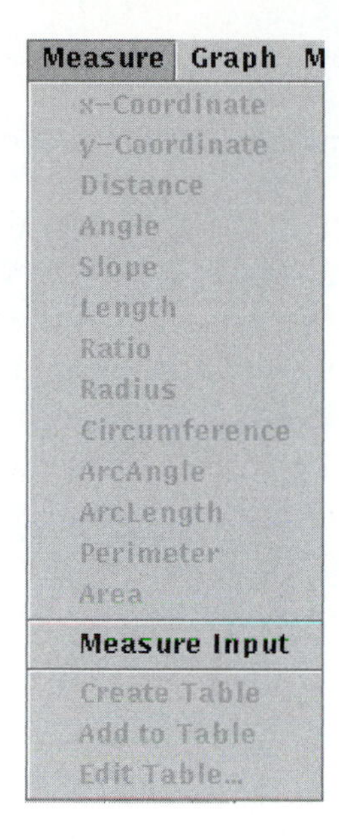

Fig. B.4 The Measure Menu

We will now review the measurement types in detail. They fall into three groups: those measurements that are applicable to either Euclidean or hyperbolic geometry, measurements applicable only to Euclidean geometry, and measurements applicable only to hyperbolic geometry.

B.11.1 Neutral Measurements

Neutral measurements are applicable to either Euclidean or hyperbolic geometry. They can be sub-classified into the following groups:

Point Measurements

- **Distance** Distance can be measured in two ways—between two points or from a point to a line. Either select two points or a point and a line and then choose **Distance** (**Measure** menu).

- **Angle** Select three points: a point on the initial ray, a point at the vertex, and a point on the terminal ray. Then choose **Angle** (**Measure** menu). Note that angles are always measured as *oriented* angles. That is, it matters what order is specified for the initial and terminal points.

Segment Measurements

- **Length** Select a segment and choose **Length** (**Measure** menu).

- **Ratio** A ratio is a proportion of two distances or lengths. Ratios can be defined in three ways: (1) Select two segments. The ratio measurement will calculate the ratio of the first segment's length to the second. (2) Select three points A, B, and C. The ratio measurement will calculate the ratio of the distance from A to B to the distance from B to C. (3) Select four points A, B, C, and D. The ratio measurement will calculate the ratio of the distance from A to B to the distance from C to D. In all cases, once the appropriate data is selected, choose **Ratio** (**Measure** menu).

Circle Measurements

- **Radius** Select a circle or an arc and choose **Radius** (**Measure** menu).

- **Circumference** Select a circle and choose **Circumference** (**Measure** menu).

Arc Measurements

- **ArcAngle** Select an arc and choose **ArcAngle** (**Measure** menu). The angle will be measured in degrees.

- **ArcLength** Select an arc and choose **ArcLength** (**Measure** menu).

Filled Object Measurements

- **Perimeter** Select a filled polygon and choose **Perimeter** (**Measure** menu).

- **Area** Select a filled polygon, filled circle, or filled arc and choose **Area** (**Measure** menu).

B.11.2 Euclidean-only Measurements

Point Measurements

- **x-Coordinate** Select a point and choose **x-Coordinate** (**Measure** menu).

- *y*-**Coordinate** Select a point and choose *y*-**Coordinate** (**Measure** menu).

Linear Object Measurements

A *linear object* is a line or a part of a line, namely, a ray or segment.

- **Slope** Select a linear object and choose **Slope** (**Measure** menu).

B.11.3 Hyperbolic-only Measurements

The only measurement that is applicable solely to hyperbolic geometry is the **Defect** measurement. The defect measurement is defined on a set of three points in the hyperbolic plane. If one considers these three points as being the vertices of a hyperbolic triangle, then the defect measures the difference between 180 degrees and the angle sum of a triangle in hyperbolic geometry. To measure the defect, select three points and choose **Defect** (**Measure** menu).

B.11.4 User Input Measurements

Under the **Measure** menu you will find an option labeled **Measure Input**. This option can be used to create an input box in the canvas. Once this input box is created, numerical values can be typed in and used as any other measurement can be used.

B.12 Using Tables

Tables of measurements are useful for analyzing relationships between measurements. For example, if we consider the interior angles of a triangle, then there is a relationship for these angles, namely that their sum is always 180 degrees.

To create tables, use the bottom group of three menu items located under the **Measure** menu in the main window.

- **Create Table** To create a table of measurements, first select all of the measurements that are to be tabulated. Then choose **Create Table** (**Measure** menu).

- **Add to Table** To add another column of data values to an existing table, first select the table and then choose **Add to Table** (**Measure** menu).

- **Edit Table...** To edit an existing table, first select the table and then choose **Edit Table...** (**Measure** menu).

B.13 Using the Calculator

The Calculator (Fig. B.5) allows one to create complex mathematical expressions using measurements from the Canvas, numerical quantities, and built-in mathematical functions.

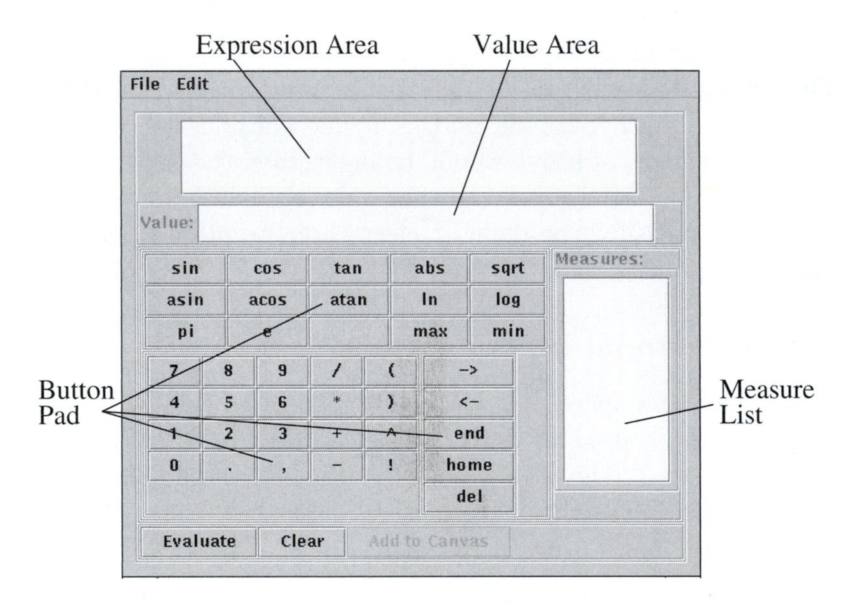

Fig. B.5 The *Geometry Explorer* Calculator Window

The Calculator window is organized into four main sections: The Expression Area, Value Area, Button Pad, and Measure List. Additionally, there are two menus, the **File** and **Edit** menus, and three buttons on the bottom of the window: Evaluate, Clear, and Add to Canvas.

The calculator interface is designed similar to that of a modern scientific calculator. The large Expression area at the top of the Calculator is where mathematical expressions are visually displayed. The Button pad consists of a series of buttons that represent numerical values, mathematical functions, mathematical operators (+, − , and so on), and editing buttons.

One difference between this Calculator and a handheld calculator is the Measure List area. As measurements are made in the Canvas, they will appear in this list. Then one can select them from the list and add them to a current expression in the Expression area. This way, compound measurements can be created.

Another difference is the Value area. The paradigm for the Calculator is that expressions are built up in the Expression area as *symbolic* expressions. Once the Evaluate key is pressed, the expression is numerically evaluated and the numerical result is displayed in the Value area section.

For more detailed help on using the Calculator, see the online help page, accessed from the **Help** menu of *Geometry Explorer*.

B.14 Hyperbolic Geometry

In Euclidean geometry given a line and a point not on the line, there is only one line parallel to the given line through the point. In hyperbolic geometry there are many lines parallel. *Geometry Explorer* provides a non-Euclidean Canvas with which to explore hyperbolic geometry. The model of hyperbolic geometry it uses is the Poincaré model.

To open a hyperbolic geometry window, choose **New** (**File** menu). A dialog box will pop up asking for the type of geometry in which you wish to work. Choose the hyperbolic model. A window will appear like that shown in Fig. B.6.

This window looks almost identical to a Euclidean window. Working in hyperbolic geometry with *Geometry Explorer* is essentially no different than working in Euclidean geometry. Almost all of the tools work in both environments, with a few notable exceptions:

1. In the Euclidean Canvas circles and arcs can be defined using three points. This construction depends on the Euclidean parallel postulate (i.e., the uniqueness of parallels) and thus is not available in the hyperbolic Canvas.

2. There is no Graph menu in the hyperbolic window.

3. Some measurements are different. There is no x- or y-coordinate measure and no slope measure. These depend on a coordinate system. However, there is a new measure: the *defect* measure. The defect is the difference between 180 degrees and the angle sum of a triangle in hyperbolic geometry (more on this below).

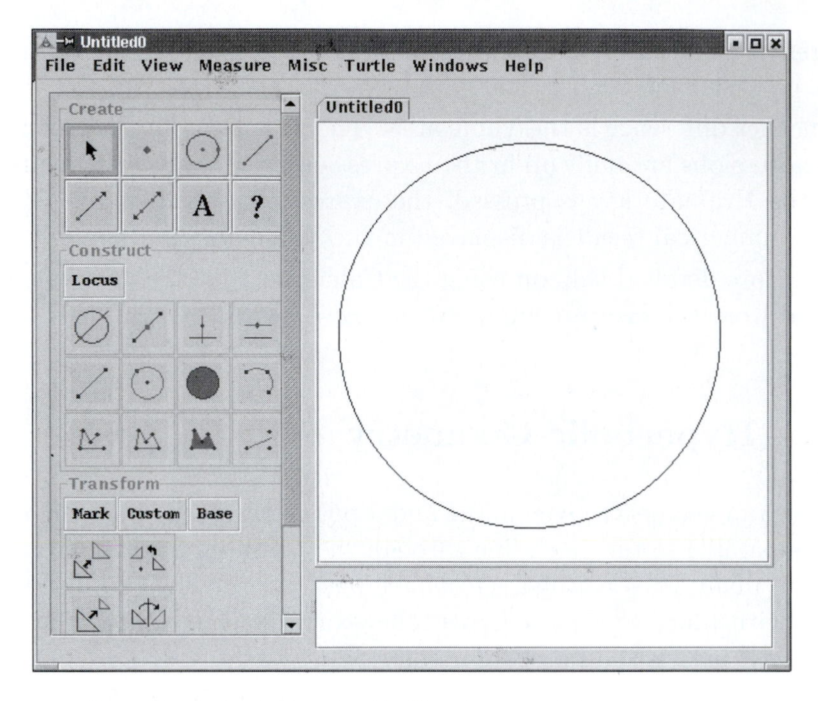

Fig. B.6 The Hyperbolic Workspace Main Window

4. In the Euclidean Canvas the Parallel button in the Tool panel is used to construct the *unique* parallel for a line and a point not on the line. In hyperbolic geometry there are no unique parallels. In the hyperbolic environment using the Parallel tool (with the same selection of a linear object and a point) will result in the creation of two parallels called *limiting parallels*. In Fig. B.7 we see the two (unique) limiting parallels to line a through point A (the parallels are the two lines passing through A). These are parallels since they are lines through A that do not intersect line a. (Although they do intersect at the boundary, they are still parallel as the boundary is not considered part of the hyperbolic plane.)

This similarity of user environments for the two geometries was deliberately designed to give the user the maximum opportunity to explore and contrast these two different geometric "universes" using similar basic geometric ideas, such as points, lines, perpendiculars, rotations, measurements, and so on. The goal in working in these geometries is to develop an intuition for how it "feels" to live in one geometry versus another.

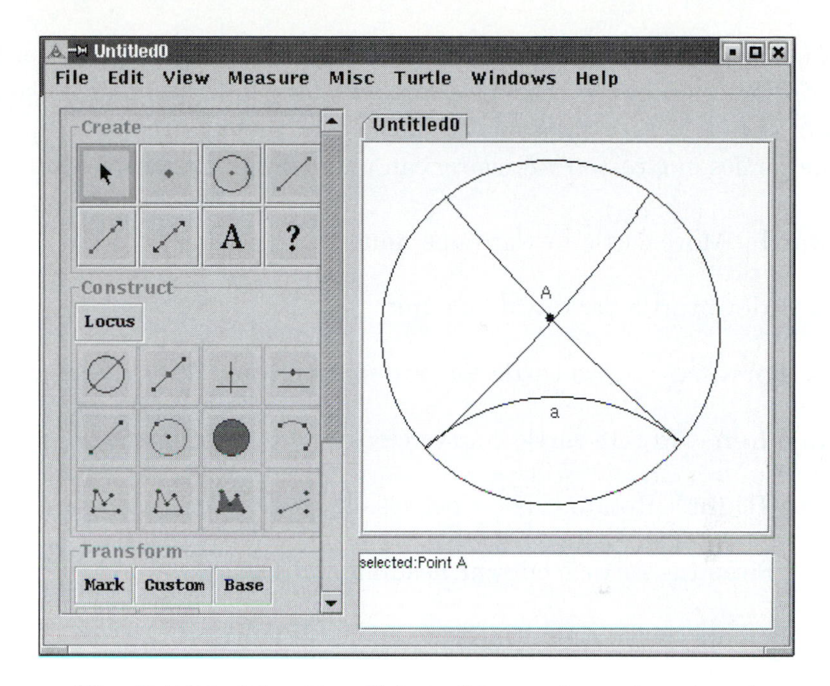

Fig. B.7 Limiting Parallels to Line a through point A

B.15 Analytic Geometry

Using *Geometry Explorer* one can graph the relationship between two measured (Euclidean) quantities. A *graph* consists of two coordinate axes (x and y) and points plotted in relation to these axes. The *coordinate system* is the system by which a point is located on the graph.

The **Graph** menu controls the user interface to the graphing capability of *Geometry Explorer*. There are eight options under this menu that control graphing: **(Show/Hide) Axes**, **Grid (On/Off)**, **Add Function to Graph...**, **Add As (x,y) Point from Measures**, **Add Point on Function from x-Point**, **Iterate Function from Point...**, **Derivative of Function**, and **Input Box for Function**. For more detailed examples of the analytic geometry capabilities of *Geometry Explorer*, see the online help page on analytic geometry.

B.16 Turtle Geometry

Turtle geometry was created as part of the development of the LOGO programming language. LOGO was designed in part to give children a relatively easy way to program a computer. In turtle geometry one imagines a small turtle on the computer screen. This turtle understands commands like move forward, turn left, turn right, and change color, among others.

The turtles in *Geometry Explorer* can understand these basic commands:

Forward Move turtle forward one unit.

Back Move turtle backward one unit.

Draw Forward Move turtle forward one unit and draw a segment.

Rotate Left Rotate turtle counterclockwise by a set angle.

Rotate Right Rotate turtle clockwise by a set angle.

Push Store the turtle's current heading and length.

Pop Restore the turtle's stored heading and length.

The turtle starts out with a specified heading and length. The *heading* is the direction in which the turtle will move. The *length* is how far the turtle should move when told to go forward or backward. The heading and length are given by a vector. The vector's length is just the distance between the points, and the vector's heading is given by an arrow from the first point toward the second.

A turtle must also know the angle by which to turn. This is specified by a set of three points—the initial, vertex, and terminal points of an angle.

The items under the **Turtle** menu in the main window are used to define a turtle on the Canvas and also to control turtle movements. There are five items under the **Turtle** menu:

Turtle Heading Vector This menu item will be activated once two points are selected in the Canvas. The direction determined by the vector between these points will determine the direction the turtle moves forward, and the distance between these points will determine how far the turtle moves in that direction. After selecting two points and clicking on this menu item, the vector defined by the two points will be stored for use in creating a turtle.

Turtle Turn Angle This menu item will be activated once three points are selected in the Canvas. These points will be the initial, vertex, and terminal points of an angle. This angle will determine how the turtle turns when directed to do so. After selecting three points and clicking on this menu item, the angle defined by the three points will be stored for use in creating a turtle.

Create Turtle At Point This menu item will be activated once a vector and angle have been defined (see the previous two items) and a point on the Canvas has been selected. This point will be the point at which the turtle will be located. After clicking on this menu item, a turtle will be created in the Canvas at the position given by the point, and a Turtle Controller panel will pop up.

Control Panel... This menu item will be activated once a turtle has been created. After clicking on this menu item a Turtle Controller panel will pop up. This panel contains tools for controlling the movement of the turtle.

Create Simple Turtle This menu item is always active. After clicking on this menu item, a turtle will be created at the center of the Canvas. It will initially be oriented in the vertical direction, make moves of 1 unit of length, and have a turn angle of 90 degrees.

In the figure at right, $\angle ABC$ and vector DE determine the required angle and vector for a turtle. We select A, B, and C and choose **Turtle Turn Angle** (**Turtle** menu). Then we select D and E and choose **Turtle Heading Vector** (**Turtle** menu).

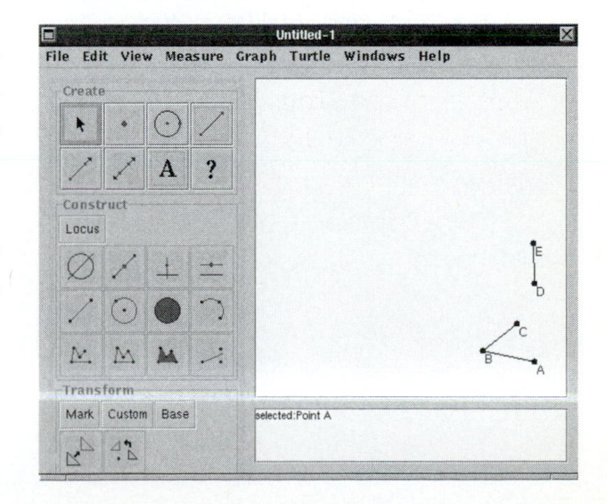

Next, we create a point that will be the starting location of our turtle. Here point F will be this starting point. Select F and choose **Create Turtle At Point** (**Turtle** menu). A turtle will be created at F as shown.

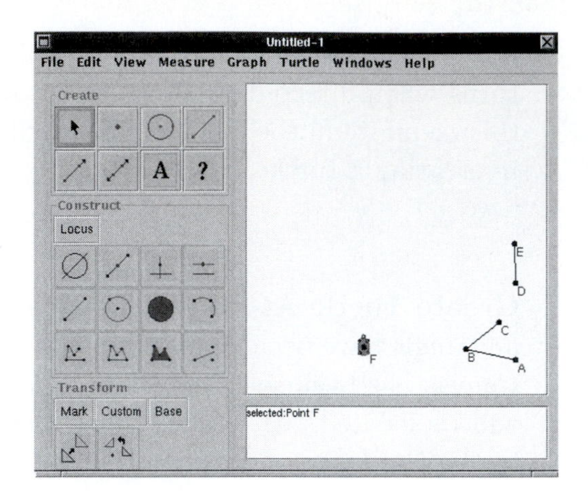

The turtle is graphically displayed as a little green turtle. It looks a bit small in the previous figure. Here is a bigger version.

Once the turtle is created, another window automatically pops up. This is the Turtle Controller panel. This window controls the movement of the turtle. The Controller has two tabbed panels that are labeled "Simple Turtle" and "Grammar Turtle."

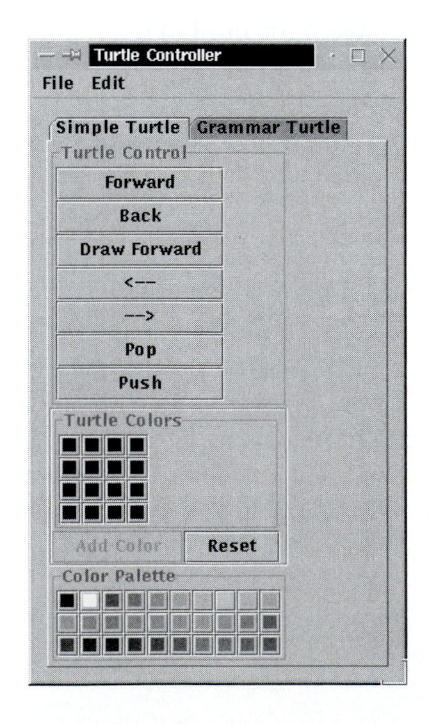

The Simple Turtle panel consists of three areas labeled "Turtle Control," "Turtle Colors," and "Turtle Palette." In the Turtle Control area there are seven buttons: Forward, Back, Draw Forward, <—, —>, Pop, and Push.

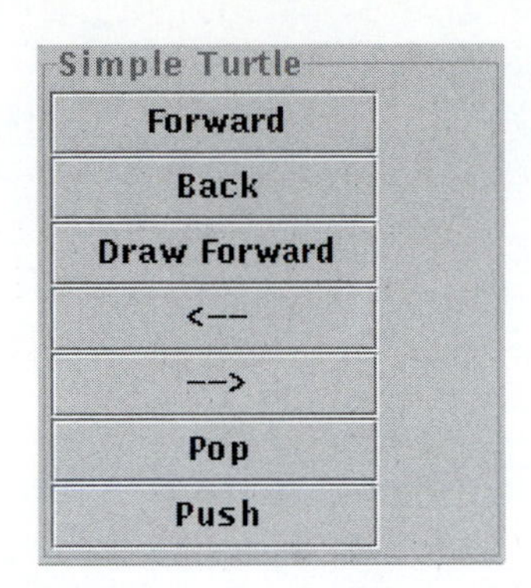

Here the Forward button and then the Draw Forward button have been pushed. The turtle carries out these commands on the Canvas.

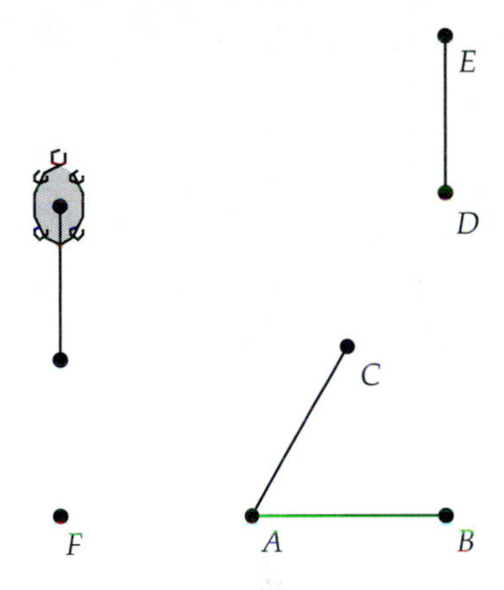

Appendix C

Birkhoff's Axioms for Euclidean Geometry

Undefined terms are *point* and *line*, as well as two real-valued functions—a *distance* function, $d(A, B)$, which takes two points and returns a non-negative real number, and an *angle* function, $m(\angle A, O, B)$, which takes an ordered triple of points ($\{A, O, B\}$ with $A \neq O$ and $B \neq O$) and returns a number between 0 and 2π. The point O is called the *vertex* of the angle.

The Ruler Postulate The points of any line can be put into one-to-one correspondence with the real numbers x so that if x_A corresponds to A and x_B corresponds to B, then $|x_A - x_B| = d(A, B)$ for all points A, B on the line.

The Euclidean Postulate One and only one line contains any two distinct points.

The Protractor Postulate Given any point O, the rays emanating from O can be put into one-to-one correspondence with the set of real numbers (mod 2π) so that if a_m corresponds to ray m and a_n corresponds to ray n, and if A, B are points (other than O) on m, n, respectively, then $m(\angle NOM) = a_m - a_n$ (mod 2π). Furthermore, if the point B varies continuously along a line not containing O, then a_n varies continuously also.

The SAS Similarity Postulate If in two triangles ABC and $A'B'C'$ and for some real number $k > 0$, we have $d(A', B') = k\, d(A, B)$, $d(A', C') = k\, d(A, C)$, and $m(\angle BAC) = m(\angle B'A'C')$, then the remaining angles are pair-wise equal and $d(B', C') = k\, d(B, C)$.

Other definitions:

A point B is said to be *between* points A and C $(A \neq C)$ if $d(A, B) + d(B, C) = d(A, C)$. A *segment* $\overline{AB}$ consists of the points A and B along with all points between A and B. The *ray* with endpoint O, defined by two points O and A in line l, is the set of all points B on l such that O is not between A and B. Two distinct lines having no point in common are called *parallel.*

Two rays m, n through O form a *straight angle* if $m(\angle MON) = \pi$, where M and N are points on m, n, respectively. The rays form a *right angle* if $m(\angle MON) = \pm\frac{\pi}{2}$. If the rays form a right angle, we say the lines defining the rays are *perpendicular.*

Appendix D

Hilbert's Axioms for Euclidean Geometry

Undefined terms are *point*, *line*, *between*, *on* (or *incident*), and *congruent*.

Incidence Axioms

I-1 Through any two distinct points A and B there is always a line m.

I-2 Through any two distinct points A and B, there is not more than one line m.

I-3 On every line there exist at least two distinct points. There exist at least three points not all on the same line.

I-4 Through any three points not on the same line, there is one and only one plane.

Betweeness Axioms

II-1 If B is a point between A and C (denoted $A * B * C$) then A, B, and C are distinct points on the same line and $C * B * A$.

II-2 For any distinct points A and C, there is at least one point B on the line through A and C such that $A * C * B$.

II-3 If A, B, and C are three points on the same line, then exactly one is between the other two.

II-4 (Pasch's Axiom) Let A, B, and C be three non-collinear points and let m be a line in the plane that does not contain any of these points. If m contains a point of segment $\overline{AB}$, then it must also contain a point of either $\overline{AC}$ or $\overline{BC}$.

Congruence Axioms

III-1 If A and B are distinct points and A' is any other point, then for each ray r from A' there is a unique point B' on r such that $B' \neq A'$ and $\overline{AB} \cong \overline{A'B'}$.

III-2 If $\overline{AB} \cong \overline{CD}$ and $\overline{AB} \cong \overline{EF}$ then $\overline{CD} \cong \overline{EF}$. Also, every segment is congruent to itself.

III-3 If $A * B * C$, $A' * B' * C'$, $\overline{AB} \cong \overline{A'B'}$, and $\overline{BC} \cong \overline{B'C'}$, then $\overline{AC} \cong \overline{A'C'}$.

III-4 Given $\angle ABC$ and given any ray $\overrightarrow{A'B'}$, there is a unique ray $\overrightarrow{A'C'}$ on a given side of $\overleftrightarrow{A'B'}$ such that $\angle ABC \cong \angle A'B'C'$.

III-5 If $\angle ABC \cong \angle A'B'C'$ and $\angle ABC \cong \angle A''B''C''$ then $\angle A'B'C' \cong \angle A''B''C''$. Also, every angle is congruent to itself.

III-6 Given two triangles ABC and $A'B'C'$, if $\overline{AB} \cong \overline{A'B'}$, $\overline{AC} \cong \overline{A'C'}$, and $\angle BAC \cong \angle B'A'C'$, then the two triangles are congruent.

Continuity Axiom

IV-1 (Dedekind's Axiom) If the points on a line l are partitioned into two nonempty subsets Σ_1 and Σ_2 (i.e., $l = \Sigma_1 \cup \Sigma_2$) such that no point of Σ_1 is between two points of Σ_2 and vice-versa, then there is a unique point O lying on l such that $P_1 * O * P_2$ if and only if one of P_1 or P_2 is in Σ_1, the other is in Σ_2, and $O \neq P_1$ or P_2.

Parallelism Axiom

V-1 Given a line l and a point P not on l, it is possible to construct one and only one line through P parallel to l.

Appendix E

The 17 Wallpaper Groups

Here is a listing of the wallpaper patterns for the Euclidean plane. The seventeen groups have traditionally been listed with a special notation consisting of the symbols p, c, m, g, and the integers 1, 2, 3, 4, 6. This is the crystallographic notation adopted by the International Union of Crystallography (IUC) in 1952.

In the IUC system the letter "p" stands for *primitive*. A lattice is generated from a cell that is translated to form the complete lattice. In the case of oblique, rectangular, square, and hexagonal lattices, the cell is precisely the original parallelogram formed by the lattice vectors v and w and, thus, is primitive.

In the case of the centered-rectangle lattice, the cell is a rectangle, together with an *interior* point that is on the lattice. The rectangular cell is larger than the original parallelogram and not primitive. Thus, lattice types can be divided into two classes: primitive ones designated by the letter "p," and non-primitive ones designated by the letter "c."

A reflection is symbolized by the letter "m" and a glide reflection by the letter "g."

The numbers 1, 2, 3, 4, and 6 are used to represent rotations of those orders. For example, 1 would represent a rotation of 0 degrees, while 3 would represent a rotation of 120 degrees.

The symmetries of the wallpaper group are illustrated in the second and third columns by lines and polygons. Rotations are symbolized by diamonds (◇) for 180-degree rotations, triangles (△) for 120-degree rotations, squares (□) for 90-degree rotations, and hexagons (⬡) for 60-degree rotations. Double lines show lines of reflection and dashed lines show the lines for glide reflections.

The symbolization used here is taken from a Web page on the wallpaper groups developed by Xah Lee [27].

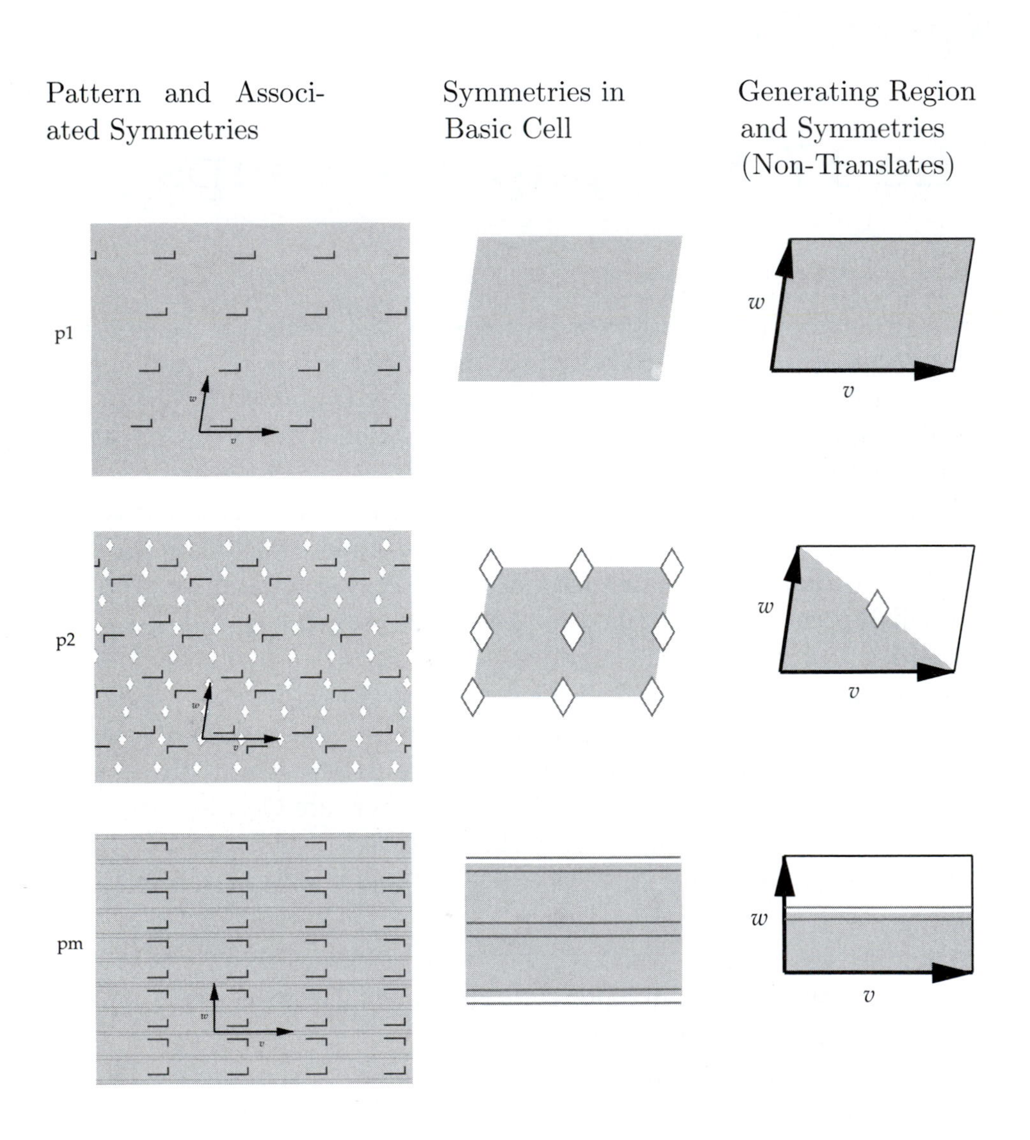

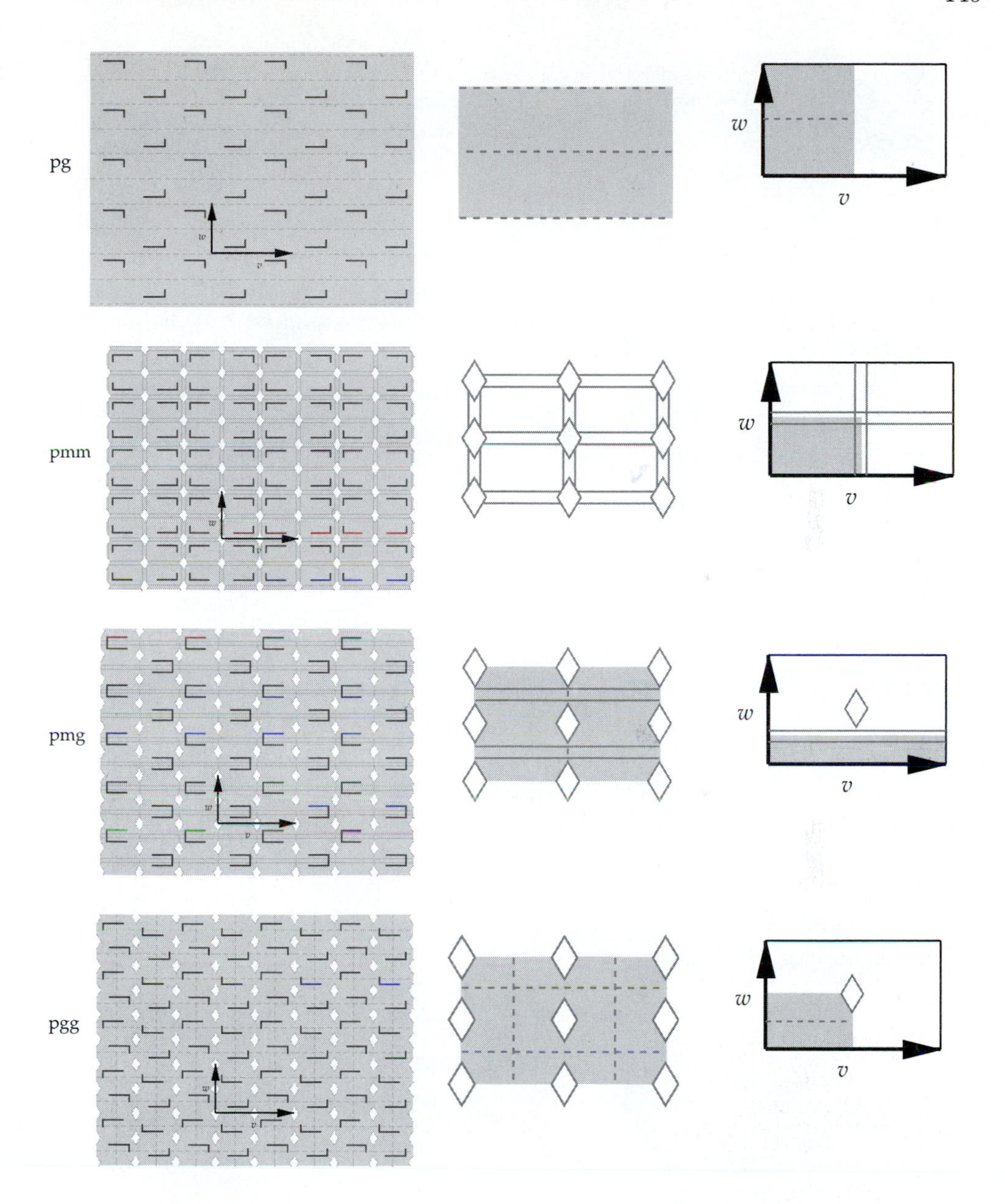
pg
w
v
pmm
w
v
pmg
w
v
pgg
w
v

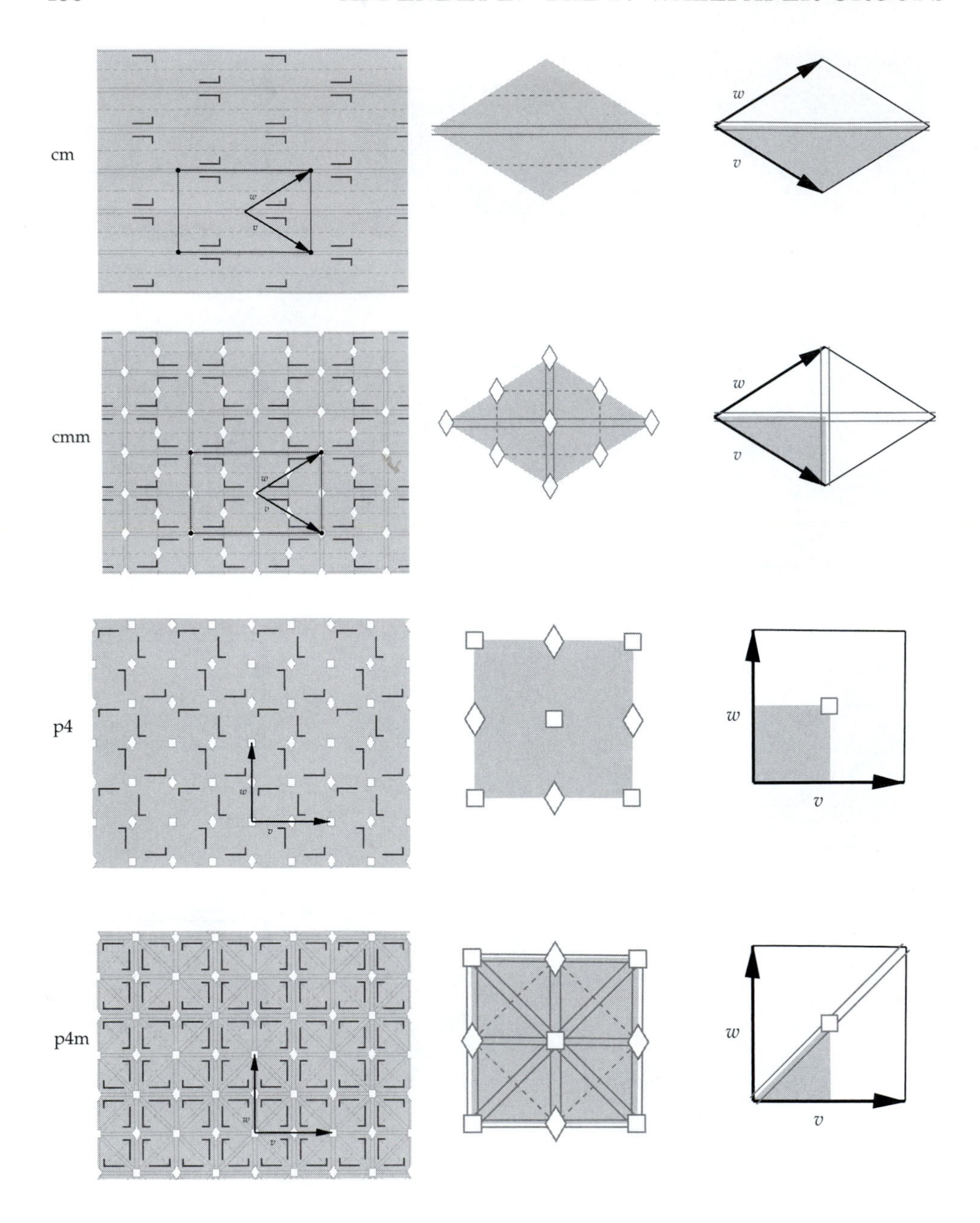
cm
w
v
cmm
p4
p4m

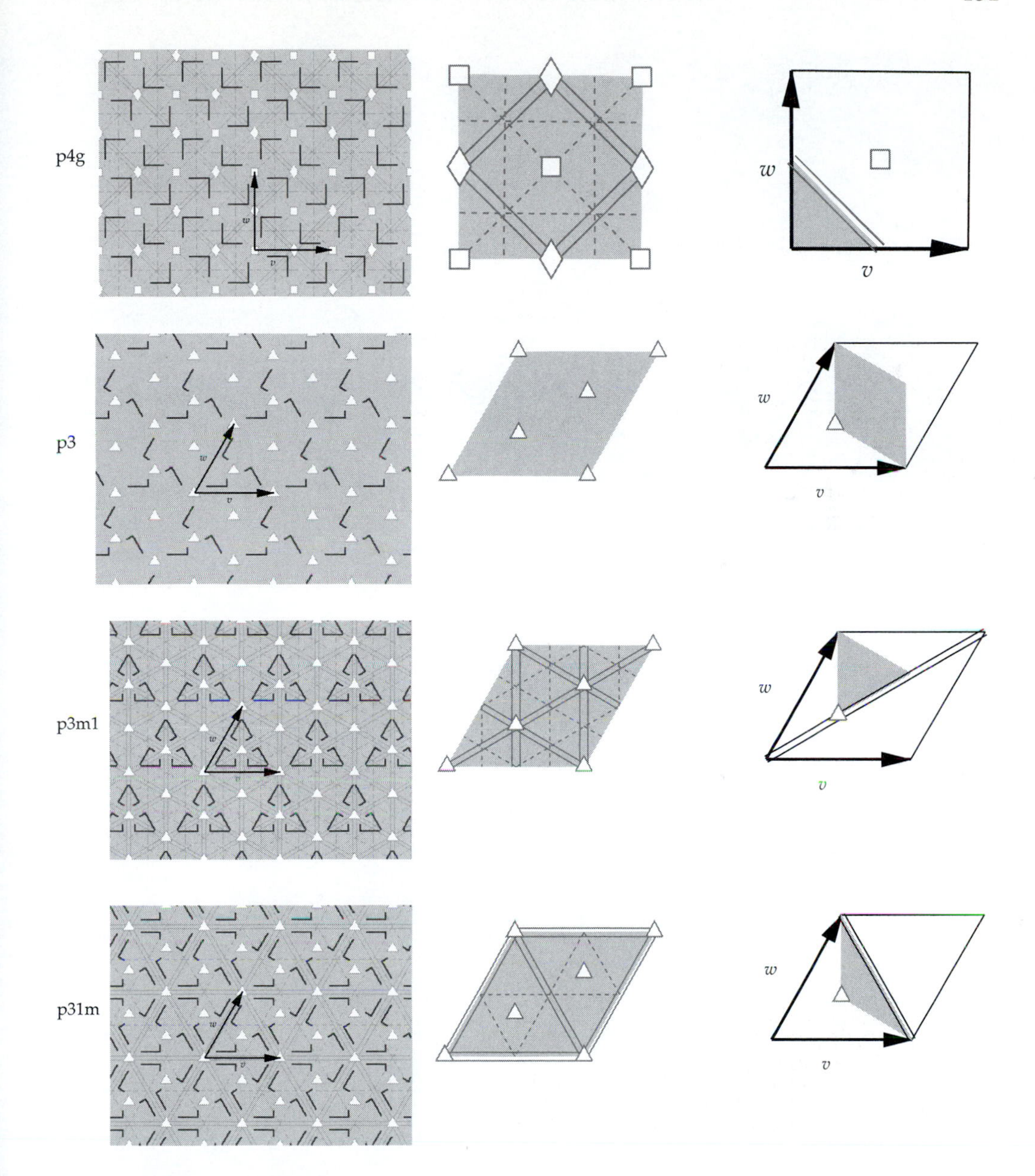
p4g
w
v
w
v
p3
w
v
w
v
p3m1
w
v
w
v
p31m
w
v
w
v

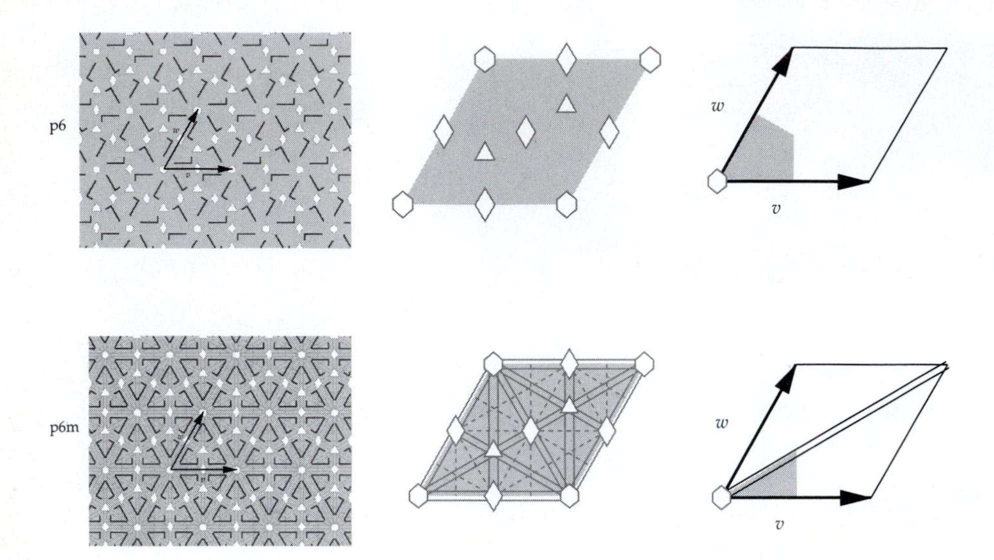
p6
w
v
p6m
w
v

Bibliography

[1] Lars V. Ahlfors. *Complex Analysis.* McGraw-Hill, New York, 1979.

[2] Roger C. Alperin. A mathematical theory of origami constructions and numbers. *The New York Journal of Mathematics*, 6:119–133, 2000.

[3] M. A. Armstrong. *Groups and Symmetry.* Springer-Verlag, New York, 1988.

[4] Michael Barnsley. *Fractals Everywhere.* Academic Press, San Diego, 1988.

[5] Tom Bassarear. *Mathematics for Elementary School Teachers: Explorations.* Houghton Mifflin Company, Boston, second edition, 2001.

[6] G. D. Birkhoff. A set of postulates for plane geometry, based on scale and protractor. *The Annals of Mathematics*, 33:329–345, April 1932.

[7] Carl B. Boyer and Uta C. Merzbach. *A History of Mathematics.* John Wiley and Sons, New York, second edition, 1968.

[8] John B. Conway. *Functions of One Complex Variable.* Springer-Verlag, New York, second edition, 1973.

[9] H. S. M. Coxeter. *Introduction to Geometry.* John Wiley and Sons, New York, second edition, 1961.

[10] H. S. M. Coxeter. The Non-Euclidean Symmetry of Escher's Picture Circle Limit III. *Leonardo*, 12:19–25, 1979.

[11] H. S. M. Coxeter and S. L. Greitzer. *Geometry Revisited.* The Mathematical Association of America, Washington, D. C., 1967.

[12] Robert Dixon. *Mathographics.* Dover, New York, 1987.

[13] William Dunham. *Journey Through Genius.* Penguin Books, New York, 1991.

[14] Howard Eves. *Great Moments in Mathematics (Before 1650).* Mathematical Association of America, Washington, D. C., 1980.

[15] Richard L. Faber. *Foundations of Euclidean and Non-Euclidean Geometry.* Marcel-Dekker, Inc., New York, 1983.

[16] J. D. Foley, A. Van Dam, S. K. Feiner, and J. F. Hughes. *Computer Graphics, Principles and Practice.* Addison-Wesley, Reading, Massachusetts, 1990.

[17] Marvin Jay Greenberg. *Euclidean and Non-Euclidean Geometries.* W. H. Freeman and Co., New York, second edition, 1980.

[18] Ernst Haeckel. *Art Forms in Nature.* Prestel-Verlag, Munich, Germany, 1998.

[19] Robin Hartshorne. *Geometry: Euclid and Beyond.* Springer-Verlag, New York, 2000.

[20] J. L. Heilbron. *Geometry Civilized.* Clarendon Press, Oxford, UK, 1998.

[21] David Hilbert. *Foundations of Geometry.* Open Court Press, LaSalle, Illinois, 1971.

[22] David Hilbert and S. Cohn-Vossen. *Geometry and the Imagination.* Chelsea Publishing Co., New York, 1952.

[23] F. S. Hill. *Computer Graphics Using OpenGL.* Prentice-Hall, Upper Saddle River, New Jersey, second edition, 1990.

[24] Einar Hille. *Analytic Function Theory, Volume I.* Blaisedell Publishing, New York, 1959.

[25] H. E. Huntley. *The Divine Proportion: A Study in Mathematical Beauty.* Dover Books, New York, 1970.

[26] Humiaki Huzita. Understanding Geometry through Origami Axioms. In J. Smith, editor, *Proceedings of the First International Conference on Origami in Education and Therapy (COET91)*, pages 37–70. British Origami Society, 1992.

[27] Xah Lee. The discontinuous groups of rotation and translation in the plane. Web page, 1997. `http://www.xahlee.org/Wallpaper_dir/c5_17WallpaperGroups.html`.

[28] Norman Levinson and Raymond M. Redheffer. *Complex Variables.* Holden-Day, San Francisco, 1970.

[29] Astrid Lindenmayer and Przemyslaw Prusinkiewicz. *The Algorithmic Beauty of Plants.* Springer-Verlag, New York, 1990.

[30] Benoit Mandelbrot. *The Fractal Geometry of Nature.* W. H. Freeman and Co., New York, 1977.

[31] George E. Martin. *The Foundations of Geometry and the Non-Euclidean Plane.* Springer-Verlag, New York, 1975.

[32] Edwin E. Moise. *Elementary Geometry from an Advanced Standpoint.* Addison-Wesley, Reading, Massachusetts, second edition, 1974.

[33] Jackie Nieder, Tom Davis, and Mason Woo. *OpenGL Programming Guide.* Addison-Wesley, Reading, Massachusetts, 1993.

[34] Seymour Papert. *Mindstorms: Children, Computers, and Powerful Ideas.* Basic Books Inc., New York, 1980.

[35] G. Polya. *How to Solve It: A New Aspect of Mathematical Method.* Princeton University Press, Princeton, New Jersey, second edition, 1957.

[36] Helen Whitson Rose. *Quick-and-Easy Strip Quilting.* Dover Publications, Inc., New York, 1989.

[37] Doris Schattschneider. *M. C. Escher, Visions of Symmetry.* W. H. Freeman and Co., New York, 1990.

[38] Thomas Q. Sibley. *The Geometric Viewpoint.* Addison-Wesley, Reading, Massachusetts, 1998.

[39] University of St. Andrews School of Mathematics. The MacTutor History of Mathematics Archive. Web page, 2002. `http://www-gap.dcs.st-and.ac.uk/~history/`.

[40] Hermann Weyl. *Symmetry.* Princeton University Press, Princeton, New Jersey, 1952.

[41] Harold E. Wolfe. *Non-Euclidean Geometry*. Henry Holt and Co., New York, 1945.

Index